PARTICLES AND FIELDS

Previous Proceedings in the Series of Particles and Fields Schools

	Year	Held in	Publisher	ISBN
VIII	1998	Oaxaca, México	AIP Conf. Proceedings vol. 490	1-56396-895-9
VII	1996	Mérida, Yucatán	AIP Conf. Proceedings vol. 400	1-56396-686-7
VI	1994	Tabasco, Villahermosa	World Scientific	981-02-2121-5
V	1992	Guanajuato, México	AIP Conf. Proceedings vol. 317	1-56396-378-7

Previous Proceedings in the Series of Particles and Fields Workshops

	Year	Held in	Publisher	ISBN
7th	1999	Mérida, Yucatán	AIP Conf. Proceedings vol. 531	1-56396-954-8
6th	1997	Morelia, Michoacán	AIP Conf. Proceedings vol. 445	1-56396-791-X
5th	1995	Puebla, Puebla	AIP Conf. Proceedings vol. 359	1-56396-548-8

Other Related Titles from AIP Conference Proceedings

541 Theoretical High Energy Physics: MRST 2000
Edited by C. R. Hagen, November 2000, 1-56396-966-1

540 Particle Physics and Cosmology: Second Tropical Workshop
Edited by José F. Nieves, October 2000, 1-56396-965-3

539 Symmetries in Subatomic Physics: 3rd International Symposium
Edited by X.-H. Guo, A. W. Thomas, and A. G. Williams, October 2000, 1-56396-964-5

516 26th International Cosmic Ray Conference: ICRC XXVI, Invited, Rapporteur, and Highlight Papers
Edited by Brenda L. Dingus, David B. Kieda, and Michael H. Salamon
May 2000, 1-56396-939-4

515 GeV-TeV Gamma Ray Astrophysics Workshop: Towards a Major Atmospheric Cherenkov Detector VI
Edited by Brenda L. Dingus, Michael H. Salamon, and David B. Kieda,
May 2000, 1-56396-938-6

To learn more about these titles, or the AIP Conference Proceedings Series, please visit the webpage **http://www.aip.org/catalog/aboutconf.html**

PARTICLES AND FIELDS

Ninth Mexican School

Metepec, Puebla, México 9–19 August 2000

EDITORS

Gerardo Herrera Corral

CINVESTAV, México

Lukas Nellen

UNAM, México

AMERICAN INSTITUTE OF PHYSICS

Melville, New York, 2001

AIP CONFERENCE PROCEEDINGS ■ VOLUME 562

Editors:

Gerardo Herrera Corral
CINVESTAV
Departamento de Física
Apdo. Postal 14-740
07000 México, D. F.
MEXICO

E-mail: gerardo.herrera@fis.cinvestav.mx

Lukas Nellen
Instituto de Ciencias Nucleares, UNAM
Circuito Exterior, Cd. Universitaria
Apdo. Postal 70-543
04510 México, D. F.
MEXICO

E-mail: lukas@nuclecu.unam.mx

L.C. Catalog Card No. 2001088382
ISBN 1-56396-998-X
ISSN 0094-243X
Printed in the United States of America

CONTENTS

PREFACE

The IX Mexican School on Particles and Fields was held in Metepec, Puebla, from 9-19 August 2000. Metepec is a nice town near Puebla City and is located in a picturesque setting with a view of the volcano Popocatepetl. Almost 80 participants registered and about half of these were students from Mexico and Latin America.

This time, the workshop was organized together with the International Workshop on Observing Ultra-High Energy Cosmic Rays from Space and Earth. Several talks were shared, giving us the possibility of complementing the lecture courses of the school with review talks on cosmic rays.

The Medal 2000 of the Division of Particles and Fields of the Mexican Physical Society was awarded to Professor James Cronin from the University of Chicago for his contribution to the development of particle physics and astrophysics in Mexico. As a leader of the Auger project, he supported Mexican involvement. this is the second time that the medal has been awarded. The first was to Leon Lederman in 1999. We would like to mention that in its first two editions the medal was awarded to experimentalists. This should be seen as an expression of our community in supporting the development of experimental high energy physics in Mexico.

Gerardo Herrera Corral
President of the Division of Particles and Fields
Mexican Physical Society

Lukas Nellen
ICN-UNAM, Mexico City

ACKNOWLEDGEMENTS

We would like to thank our colleagues in the Local Organizing Committee

Isidro Aranda (EFM-UMSNH),
Lorenzo Díaz (IF-BUAP),
Arturo Fernández (FCFM-BUAP),
Mario Maya (FCFM-BUAP),
Myriam Mondragón (IF-UNAM),
Mauro Napsuciale (IFUG),
Miguel A. Pérez (CINVESTAV),
Alfonso Rosado (IF-BUAP),
Humberto Salazar (FCFM-BUAP),
Manuel Torres (IF-UNAM),
Luis Villaseñor (IF-UMSNH)

and to all others whose effort did so much for the success of the Workshop. In particular, we would like to acknowledge Cupatitzio Ramírez (FCFM-BUAP) and the members of the Organizing Committee of the International Workshop on Observing Ultra-High Energy Cosmic Rays From Space and Earth: Arnulfo Zepeda (CINVESTAV), Rebeca López (FCFM-BUAP), Enrique García (CINVESTAV) and Jacinta Grajales (FCFM-BUAP).

Special thanks to Pilar Menezes and Gabriela Murguía for their invaluable secretarial assistance and dedication.

It would have been impossible to organize the IX Mexican School on Particles and Fields without the financial support from the following institutions: CINVESTAV, CLAF - Rio de Janeiro, CLAF - Mexico, CONACyT and UNAM.

Finally we would like to thank to all the speakers and participants for their interest and excellent presentations, which where essential for making the IX Mexican School a successful event.

Gerardo Herrera Corral
CINVESTAV, Mexico City
Lukas Nellen
ICN-UNAM, Mexico City

SESSION IN HONOR OF JAMES W. CRONIN

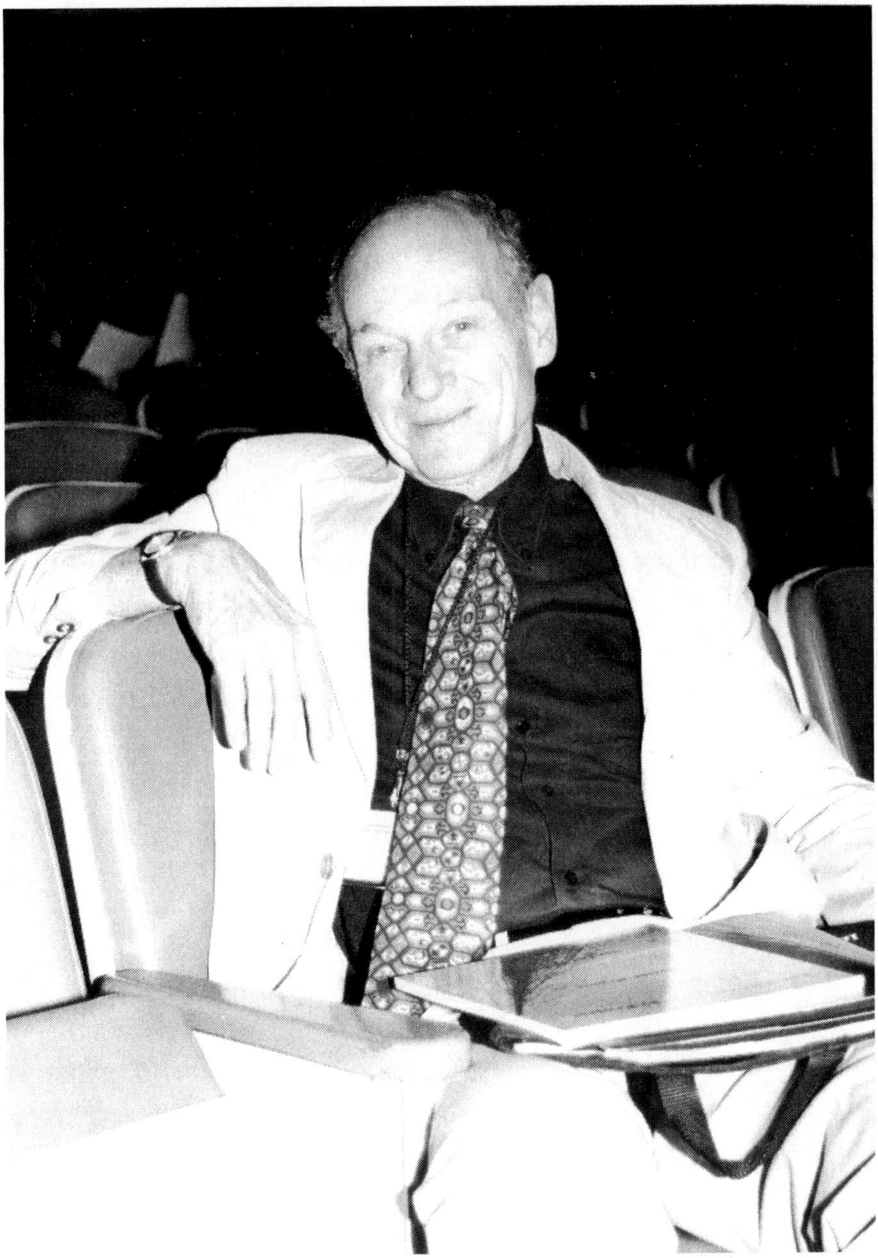

James Cronin and the Physics in Mexico

Gerardo Herrera Corral
President of the Division of Particles and Fields
Mexican Physical Society

Physics Department
CINVESTAV
Apdo. Postal 14 740, Mexico 07300, D.F.

Last year the Medal 99 of the Division of Particles and Fields was given to Prof. Leon Lederman, Nobel Laureate 1988.

This year the Medal 2000 is awarded to Prof. James Cronin, who is not stranger to winning awards, in 1980 he won the Nobel Prize in physics for the discovery of CP violation in the decay of neutral K-mesons.

At this point I want to say that being a Nobel Laureate is not a requisite to receive the Medal of the Division of Particles and Fields. In fact it does not help...

The Medal of the Division of Particles and Fields of the Mexican Physical Society is offered to national or foreign scientists which have contributed significantly to the development of particle physics in Mexico or to Mexican physicists, based in Mexico or abroad, who are recognized internationally for their contributions in particle physics.

In 1963, Prof. Cronin and Prof. Fitch observed CP violation in nature. In an experiment at Brookhaven National Laboratory they observed that two different kind of particles known as K mesons may decay into the same products. The ramifications of this discovery stretched far beyond the neutral K mesons and become a fundamental part of our present knowledge of nature. The paper published in 1964 by Physical Review Letters is nowadays considered a clasic in physics.

The conventional wisdom of particle physicists assumed that matter and antimatter were identical in every respect, except that a particle of antimatter is a mirror reversal of all the dimensions of its matter counterpart.

CP562, *Particles and Fields: Ninth Mexican School,* edited by G. Herrera Corral and L. Nellen

Dr. Cronin and Dr. Fitch performed the experiment that conclusively demonstrated a striking violation of these symmetry principles, and provided an explanation for the seemingly improbable existence of our universe, which is made entirely of matter, without any antimatter.

A particle of matter that encounters an antimatter particle causes the annihilation of both particles and the creation of gamma rays that race away at the speed of light.

We believe that the Big Bang created matter and antimatter in precisely equal amounts. But if that were so, why had not the primordial matter and antimatter canceled each other out completely, leaving only an invisible flash of energy ?

The answer, Dr. Cronin and Dr. Fitch learned from their particle-beam experiment at Brookhaven, was that nature is even less even-handed than had been supposed. They discovered that the "charge conjugation" and "parity" observed in the decay of unstable particles called neutral K-2 mesons are slightly violated by the "weak" nuclear force. In effect, nature is slightly biased by the weak nuclear force in favor of left-handed particle interactions.

The discovery of CP-violation at the microscopic particle level offered a possible explanation of why the universe was not annihilated soon after its birth. The amounts of matter and antimatter created by the Big Bang must have been equal but soon the fundamental forces of nature started to operate giving rise to a small difference. May be the miracle of this small effect was the final touch of creation.

Nowadays, Prof. Cronin is embarked in a new project. In a few minutes he will tell us something about the Auger Project and the new ways of looking at the sky and about the wonders of this.

As one of the spokespersons of the Pierre Auger Project, Prof. Cronin contributed significantly to the development of research in the area of ultra high energy cosmic ray physics in our country. This lead to the formation of a strong, national group, consisting of four institutions (Universidad de Puebla BUAP, Centro de Inv. y de Est. Avanzados CINVESTAV, Univ. Mich. San Nicolas Hidalgo and the Universidad Nacional Autónoma de México, UNAM). This group collaborate in the Pierre Auger Project and is making an impact in the development of the experiment.

Prof. Cronin has been in Mexico several times and has been always very supportive and ready to talk to those who are associated with science policy in our country.

The Medal is not only a recognition, it is also a symbol of the mexicans respect for the things Prof. Cronin is doing and therefore for the activities that my colleagues do here in Mexico.

The particle physics comunity in Mexico has a tradition in theoretical physics. It goes back to Manuel Sandoval Vallarta who by the way was very interested in cosmic ray research. Nowadays this comunity is becoming aware of the importance and the value of exploring nature experimentally. The fact that the Medal in its first two editions go to experimental physicists is a clear gesture of this, and I am very happy about it.

I would like to ask Prof. Cronin to come forward to receive the medal 2000 of the Division of Particles and Fields of the Mexican Physical Society.

LECTURE COURSES

Supersymmetric Dark Matter

John Ellis

Theoretical Physics Division, CERN,
CH-1211 Geneva 23, Switzerland

Abstract. Motivations for expecting supersymmetry to appear at energies $\lesssim 1$ TeV are reviewed, and it is emphasized that the lightest supersymmetric particle is an ideal candidate for cold dark matter. Experimental and cosmological constraints on supersymmetric dark matter are reviewed, and the prospects for indirect or direct detection are discussed. Finally, the potential implications of a Higgs boson weighing about 115 GeV are summarized.

I WHY SUPERSYMMETRY?

Theorists have been attracted to supersymmetry for many different reasons: because it is there, because it is beautiful, to unify theories of matter and interactions, etc. However, these arguments did not fix the energy scale where supersymmetry should appear. An argument for fixing this scale is provided by the chance offered by supersymmetry of making a large hierarchy of mass scales more natural, if there are spartners of the known particles weighing $\lesssim 1$ TeV [1].

In any model without supersymmetry, such as the Standard Model, there are many loop diagrams contributing to the renormalization of the Higgs and hence the W mass that are individually quadratically divergent:

$$\delta m_W^2 = \frac{g^2}{16\pi^2}\int^{\Lambda}\frac{d^4k}{k^2} \simeq \frac{\alpha}{4\pi}\Lambda^2 \tag{1}$$

where $\Lambda \gg m_W$ is a cutoff where the Standard Model ceases to be valid, or at least becomes incomplete. If $\Lambda \gg m_W/\sqrt{\alpha/4\pi} \simeq 1$ TeV, the radiative correction (1) renders a light W boson unnatural, in the sense that fine tuning is needed to keep it light. Essentially, one has to choose the bare value m_{W_0} very carefully, so that $m_W^2 = m_{W_0}^2 + \delta m_W^2$ is much smaller in magnitude than either the bare or the loop contribution. This is not impossible, nor in contradiction with the mathematical principle of renormalization, but it does seem rather unreasonable. For example, if $\Lambda \simeq m_P \simeq 10^{19}$ GeV, the cancellation must be accurate to some 36 decimal places, and readjusted in each higher order of perturbation theory [1].

CP562, *Particles and Fields: Ninth Mexican School,* edited by G. Herrera Corral and L. Nellen

On the other hand, a light W boson may be made natural in some theory with new physics appearing before the TeV scale. One such example is technicolour [2], which postulates that the Higgs boson is composite, with a size $R_H \simeq 1/1$ TeV. However, technicolour theories are disfavoured by the small magnitudes of flavour-changing neutral interactions [3] and by precision electroweak measurements, and are no longer popular.

The only extant alternative is to notice that the contributions (1) have opposite signs for diagrams with internal fermion and boson loops:

$$\delta m_W^2 = \frac{g_B^2}{16\pi^2}\int^\Lambda \frac{d^4k}{k^2} - \frac{g_F^2}{16\pi^2}\int^\Lambda \frac{d^4k}{k^2} \simeq \frac{\alpha}{4\pi} m_{susy}^2 \tag{2}$$

Thus there may be a cancellation in a supersymmetric theory [4], in which there are equal numbers of bosons and fermions, and they have equal couplings $g_B = g_F$, leaving a residuum related to the supersymmetry-breaking mass scale m_{susy}, as shown in (2). Having a small value of M_W becomes technically natural if $m_{susy} \lesssim 1$ TeV [1].

There are, moreover, two circumstantial experimental hints in favour of supersymmetry. One is the agreement of the gauge coupling strengths measured at LEP and elsewhere with supersymmetric grand unification, if the spartners of the Standard Model particles weigh ~ 1 TeV [5], and the other indication is provided by precision electroweak data, that favour a light Higgs boson [6], as predicted in the MSSM [7] [1].

There is a third, astrophysical, motivation for TeV-scale supersymmetry, namely the cold dark matter for which the lightest supersymmetric particle is an ideal candidate. Recall that conventional (baryonic) matter constitutes $\lesssim 5$ % of the critical density, according to analyses of Big-Bang nucleosynthesis [9] and the cosmic microwave background (CMB) radiation [10]. There are both astrophysical and particle-experimental reasons to believe that massive neutrinos could provide at most a similar percentage of the critical density [11]. However, analyses of the CMB and large-scale structure data, as well as theories of structure formation, suggest the presence of cold dark matter, in the form of massive weakly-interacting particles at the level of about 30 % of the critical density, far more than could be provided by either baryons or neutrinos [10]. The remaining 65 % of the critical density, as required in inflationary cosmology, indicated by data on high-redshift supernovae [12] and supported by the CMB data [10], is presumably provided by some form of vacuum energy.

Massive weakly-interacting particles that were originally in thermal equilibrium could have interesting cosmological densities if they have either (a) masses in the eV region (such as neutrinos?), (b) masses of a few GeV (now excluded by LEP), or (c) masses in the range $m_Z/2$ to about 1 TeV. It is the latter possibility that may be realized with supersymmetry.

---

1) At the end of this lecture, the possibility that direct Higgs searches may be providing more than an indication is also discussed [8].

After discussing in a general way why the lightest supersymmetric particle is such a good candidate for cold dark matter, and reviewing experimental constraints, I review the principal strategies for searching for supersymmetric dark matter. One of the tightest constraints on the mass of the lightest supersymmetric particle is provided by the Higgs search at LEP, so I append an update on its status.

II LIGHTEST SUPERSYMMETRIC PARTICLE

In many supersymmetric models, this is expected to be stable [13], and hence likely to be present today in the Universe as a cosmological relic from the Big Bang. It is stable because of a multiplicatively-conserved quantum number called R parity, which takes the value +1 for all conventional particles and -1 for all sparticles. Its conservation is linked to those of baryon and lepton numbers:

$$R = (-1)^{3B+L+2S} \tag{3}$$

where S is the spin. It is certainly possible to violate R by violating L either spontaneously or explicitly, but these options are limited by laboratory and cosmological constraints, and we discard them for the rest of this talk.

The conservation of R parity has the following three important consequences.
1) Sparticles are always produced in pairs, e.g., $pp \rightarrow \tilde{q}\tilde{q} + X$ or $e^+e^- \rightarrow \tilde{\mu}^+\tilde{\mu}^-$.
2) Heavier sparticles decay into lighter ones, e.g., $\tilde{q} \rightarrow q\tilde{q}$ or $\tilde{\mu} \rightarrow \mu\tilde{\gamma}$.
3) The lightest supersymmetric particle is stable, because it has no legal decay mode.

If the supersymmetric relic had either electric charge or strong interactions, it would have condensed into ordinary matter, and shown up as an anomalous heavy isotope. These have not been seen, and experiments impose [14]

$$\frac{n(relic)}{n(p)} \lesssim 10^{-15} \text{ to } 10^{-30} \tag{4}$$

for 1 GeV $\lesssim m_{relic} \lesssim$ 10 TeV, far below the expected supersymmetric relic abundance. We conclude that the supersymmetric relic is surely electrically neutral and weakly-interacting, e.g., it cannot be a gluino [15]. Possible scandidates are the sneutrinos of spin 0, neutralinos (i.e., $\tilde{\gamma}/\tilde{H}^0/\tilde{Z}$ mixtures) of spin 1/2, and the gravitino of spin 3/2. Sneutrinos are excluded by a combination of LEP and direct searches. The gravitino is the lightest supersymmetric particle in gauge-mediated models, in particular, but would not provide cold dark matter, so we do not discuss it further.

The lightest neutralino is therefore the favoured dark matter candidate in many supersymmetric models. At the tree level, it is characterized by 3 parameters: the unmixed gaugino mass, $m_{1/2}$, which we assume to be universal, the Higgs mixing parameter, μ, and the ratio of Higgs v.e.v.'s, $\tan\beta$ [15]. The neutralino composition simplifies in the limit $m_{1/2} \rightarrow 0$, where it becomes an almost pure photino [16], and in the limit $\mu \to 0$, where it becomes an almost pure Higgsino. However, both of these limits are excluded by LEP, which enforces $m_\chi \gtrsim 40$ GeV [17], as we discuss in more detail below.

One of the most exciting features of neutralino dark matter is that there are generic domains of parameter space where an 'interesting' cosmological relic density: $0.1 \lesssim \Omega_\chi h^2 \lesssim 0.3$ is possible for some suitable choice of the other supersymmetric model parameters [15,17], as also discussed later.

III EXPERIMENTAL AND COSMOLOGICAL CONSTRAINTS

LEP has established that *charginos* $\chi^\pm$ weigh $\gtrsim 100$ GeV, and has also established important limits by searching for the associatedc production of neutralinos: $e^+e^- \to \chi + \chi'$. LEP has also established a lower limit of about 95 GeV on the *selectron* mass. However, one of the most stringent sparticle limits comes indirectly from the *Higgs* search [17], since the Higgs mass is sensitive to their masses via radiative corrections [7]:

$$\delta m_h^2 \propto \frac{m_t^4}{m_W^2} \ln \left(\frac{m_{\tilde{t}}^2}{m_t^2} \right) \tag{5}$$

Because of the sensitivity to $m_{\tilde{t}}$ in (4), experimental constraints on *squarks*, in particular the lighter *stop*, also impact the neutralino limits.

Representative examples of the sensitivity of neutralino limits to Higgs limits are shown in Fig. 1 [17], and sample Higgs limits are shown in Fig. 2. These are calculated in a maximal mixing scenario, in which the ZZh coupling may be suppressed at large $\tan\beta$, relaxing the lower limit on m_h from its Standard Model value [2]. Fig. 1 also shows the impact of chargino and selectron searches, as well as the requirement that the lighter stau, $\tilde{\tau}_1$, not be the lightest supersymmetric particle [17]. Also shown, where relevant, is the constraint from the observed rate of $b \to s\gamma$ decay [18]. If one is nervous about the fate of our vacuum, one may also require that the effective potential have no charge- and colour-breaking (CCB) minimum [19], as also shown in Fig. 1.

The medium shaded region is that where $0.1 < \Omega_\chi h^2 < 0.3$. It has a tail extending up to $m/2 \sim 1400$ GeV, corresponding to $m_\chi \sim 600$ GeV, which is made possible by $\chi - \tilde{\tau}$ coannihilations [20]. The constraints from LEP and $b \to s\gamma$ impose a lower limit on m_χ within this cosmological region, as seen in Fig. 3. Its strength depends on $\tan\beta$ and theoretical assumptions, but we find in general that [17]

$$m_\chi \gtrsim 50 \text{ GeV} \quad \text{and} \quad \tan\beta \gtrsim 3 \tag{6}$$

[2] This does not happen in a constrained MSSM with universal soft supersymmetry-breaking scalar masses m_0.

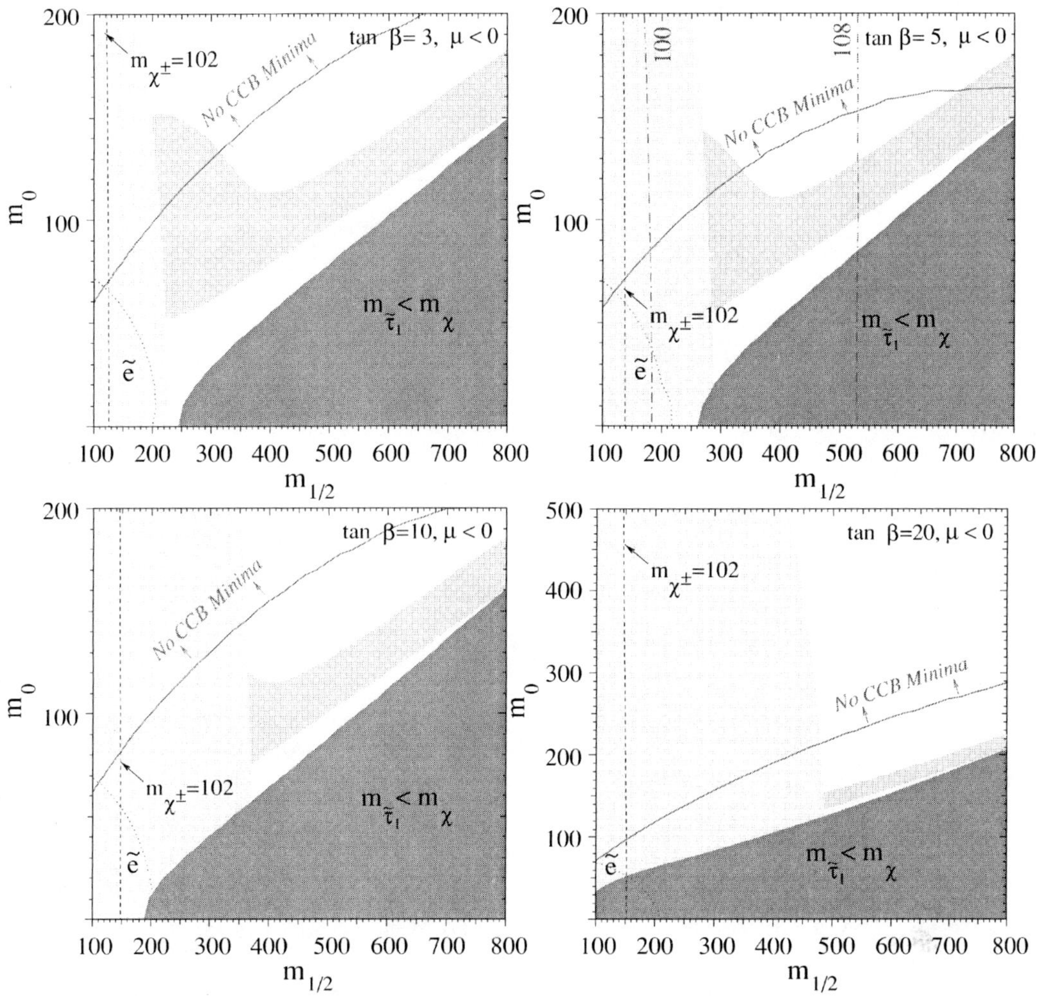

FIGURE 1. *Domains of the $m_0, m_{1/2}$ plane in the MSSM, showing the impacts of various experimental and cosmological constraints [17]. Vertical lines indicate Higgs mass contours, assuming universality of the soft supersymmetry-breaking scalar mass m_0. Other contours indicate the impacts of selectron and chargino searches at LEP. The dark shaded region is excluded because the lightest supersymmetric particle is charged, and the light shaded region is excluded by the measured $b \to s\gamma$ rate. Also shown is the potential CCB constiaint. Finally, the medium shaded region is that preferred by cosmology, where $0.1 < \Omega h^2 < 0.3$.*

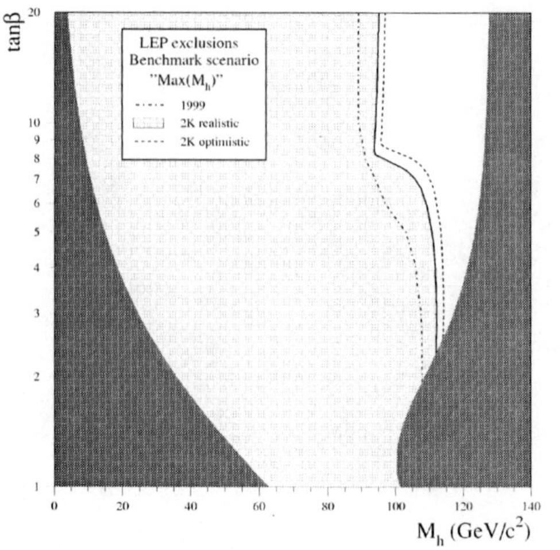

FIGURE 2. *Past, present and prospective constraints on MSSM Higgs bosons, in the $m_h, \tan\beta$ plane, assuming a maximal mixing scemario and neglecting CP violation [17].*

with the precise numbers depending on the sign of μ and different theoretical assumptions, for example whether or not the soft supersymmetry-breaking scalar masses are universal.

As seen in Fig. 1, at least within a universal framework, sparticle searches at Run II of the FNAL Tevatron collider will not be able to probe much of the parameter space not already excluded [21]. However, the LHC will be able to explore all the cosmological region [22]. Will the LHC get there before the searches for astrophysical dark matter?

IV SEARCHES FOR DARK MATTER

One strategy is to look for the annihilations of relic particles in the galactic halo [23], which may yield observable fluxes of (stable) particles such as $\bar{p}, \gamma$ and e^+. Measurements of the low-energy $\bar{p}$ flux already rule out some supersymmetric models [24]. On the other hand, γ searches do not yet rule out any models, unless the relic density in our galactic halo is strongly clumped [25]. There have been some reports of an excess of cosmic-ray positrons, but some of these have now been contradicted, and they cannot be taken as evidence for supersymmetric dark matter. Both of the $\bar{p}$ and positron searches may be interesting physics opportunities for the AMS experiment on the International Space Station.

Another search strategy is to look for annihilations of relic particles inside the Sun or Earth, following their capture after losing energy via elastic scattering [26].

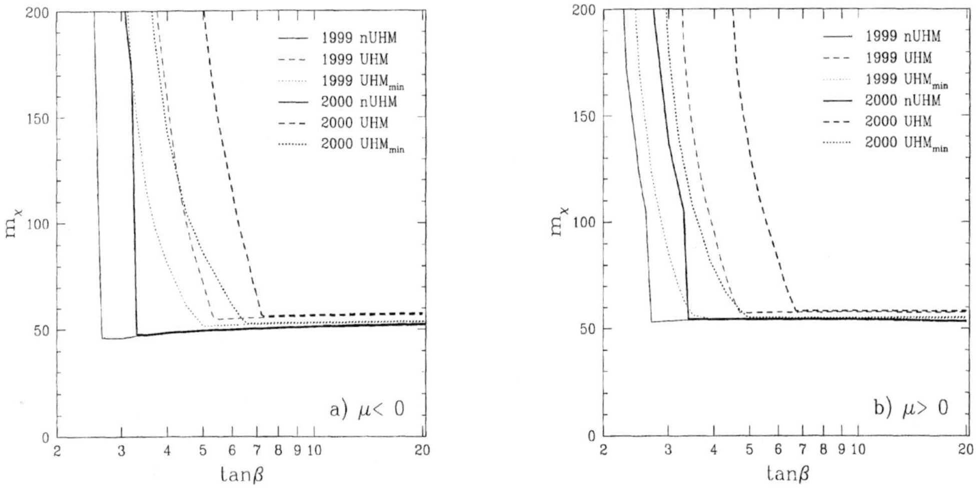

FIGURE 3. *Lower limits on the mass of the lightest supersymmetric particle. Different cases are shown, corresponding to the 1999 and 2000 data, to differing treatments of the CCB constraint, and to relaxing or retaining the universal scalar mass (UHM) constraint [17].*

The observable products of these annihilations are high-energy neutrinos, which may either be seen directly in underground detectors, or indirectly via ν collisions in rock that produce detectable muons. Searches for neutrino-induced muons from the Sun and Earth already exclude a significant number of supersymmetric models, and there are prospects for improving the current search sensitivities with a 1 km^2 (or 1 km^3) detector [27].

The third favoured strategy is to search directly for elastic dark matter scattering on nuclei in the laboratory [28]. Each scattering event would deposit typically an energy $E \sim m_\chi b^2/2 \sim$ tens of keV. There are two important types of interaction: spin-dependent, which is sensitive to the different quark contributions to the nucleon spin, and spin-independent, which is sensitive to the quark contributions to the nucleon mass. The former is likely to be more important for light nuclei, the latter for heavy nuclei.

One experiment reports annual modulation of the rate of energy deposition in their detector which may be interpreted as a possible signal for dark matter scattering [29]. We recently re-evaluated the rate of elastic scattering in a constrained supersymmetric model, in which all the soft scalar supersymmetry-breaking masses are assumed equal at the input unification scale [30]. As seen in Fig. 4, we found an elastic scattering cross-section below the region of experimental interest, at least for $\tan\beta \leq 10$. Subsequently, a second experiment published an upper limit that excluded most of the interesting region [31]. We have looked again at the elastic scattering rate, this time relaxing the assumption of universal scalar masses [32]. Although the rates could be somewhat higher than in the universal case, they still fell short of the region of experimental interest, at least for $\tan\beta \leq 10$ [33].

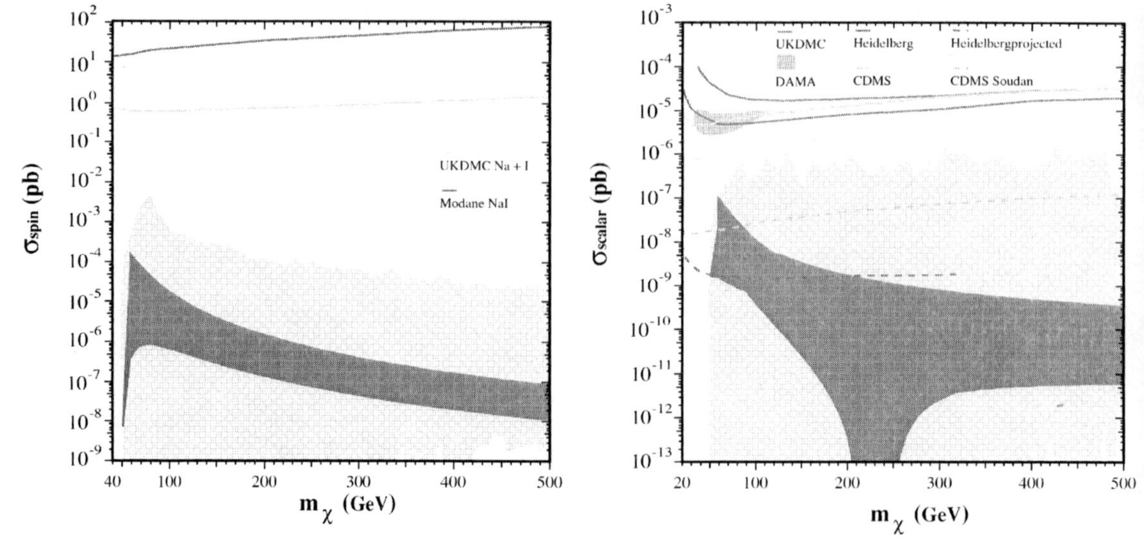

FIGURE 4. *Compilations of ranges of the elastic cross sections found in a sampling of supersymmetric models for* $\tan\beta \leq 10$*, compared with the sensitivity of the DAMA [29] and other experiments, for both spin-dependent and -independent scattering. The dark (light) shaded regions correspond to models with (without) universal soft scalar supersymmetry-breaking masses [32].*

V WHAT IF THE HIGGS WEIGHS 115 GEV?

On September 5th, the ALEPH and DELPHI experiments at LEP reported apparent excesses of Higgs candidates in the reaction $e^+e^- \to (H \to \bar{b}b) + (Z \to \bar{q}q)$. Two ALEPH events and one DELPHI event were rated as very good candidates that were quite unlikely to be due to $e^+e^- \to Z + Z$, the most dangerous background reaction. However, this background cannot be completely removed, and the significance of the reported signal only amounted to about 2.2 standard deviations [34]. Nevertheless, one may speculate on the interpretation of any signal of a Higgs boson weighing around 115 GeV [8].

The first implication is that the Standard Model must break down at some scale $\lesssim 10^6$ GeV, with the appearance of some new bosonic particles. This is because the Standard Model Higgs potential becomes unstable at a scale 10^6 GeV, with such a light Higgs boson and a relatively heavy top quark. What sort of new physics may appear? Technicolour and other models with strongly-interacting Higgs bosons are excluded yet again, because they generically predict a heavier Higgs boson: 300

GeV $\lesssim m_H \lesssim 1$ TeV. On the other hand, such a light Higgs boson is completely consistent with supersymmetry, which predicts that the lightest Higgs boson weighs $\lesssim 130$ GeV [7].

In fact, as seen in Figs. 1 and 5, a measurement of a Higgs mass around 115 GeV could be used to estimate the value of $m_{1/2}$ required in the radiative correction (5). Assuming that $m_t = 175 \pm 5$ MeV (which is the principal uncertainty) we find $m_{1/2} \gtrsim 240$ GeV, and hence [8]

$$m_\chi \gtrsim 95 \text{ GeV} \tag{7}$$

in the constrained minimal supersymmetric model with universal scalar masses, for $\tan\beta \lesssim 25$, as seen in Fig. 6. The limit (7) will survive, even if the present 'signal' eventually evaporates and becomes merely a lower limit of 115 GeV.

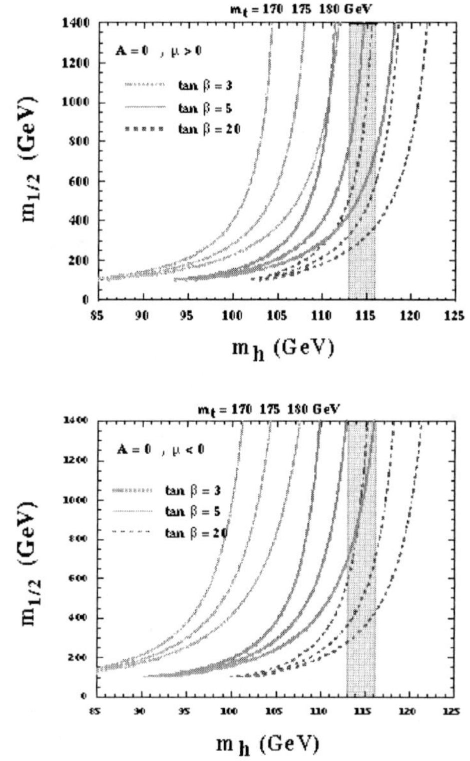

FIGURE 5. *Illustrations of the sensitivity of m_h to the universal soft supersymmetry-breaking parameter $m_{1/2}$, for different values of m_t and $\tan\beta$ [8].*

As seen in Fig. 6, this value of m_h is compatible with $m_{1/2} \lesssim 1400$ GeV, as favoured by supersymmetric dark matter. It is possible to strengthen the upper

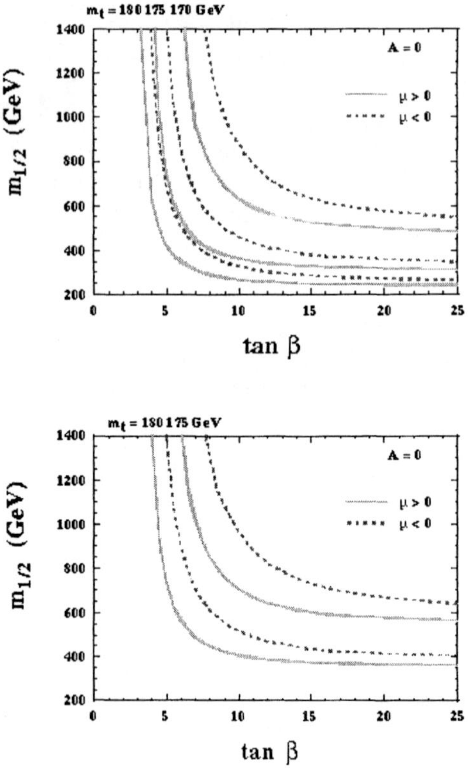

FIGURE 6. *Lower and upper limits on $m_{1/2}$ obtained by requiring 113 GeV$\leq m_h \leq$116 GeV [8].*

limit on $m_{1/2}$ if $m_t = 175$ or 180 GeV and $\tan\beta$ is sufficiently large, but not if $m_t = 170$ GeV. Of course, these upper limits would evaporate with the Higgs signal.

An interim update of the Higgs search was presented on October 10th. Some new Higgs candidates were reported, and the overall significance of the 'signal' increased to about 2.5 standard deviations. This was in line with the expectations if there really is a Higgs weighing 115 GeV [35].

The final 2000 data analysis will be presented on November 3rd. If the 'signal' strengthens significantly, CERN will be faced with the difficult choice whether to run LEP again in 2001 for a Higgs odyssey.

REFERENCES

1. L. Maiani, *Proceedings of the 1979 Gif-sur-Yvette Summer School On Particle Physics*, 1; G. 't Hooft, in *Recent Developments In Gauge Theories, Proceedings of the Nato Advanced Study Institute, Cargese, 1979*, eds. G. 't Hooft *et al.*, (Plenum Press, NY, 1980); E. Witten, Phys. Lett. **B105** (1981) 267.
2. E. Farhi and L. Susskind, Phys. Rept. **74** (1981) 277.
3. S. Dimopoulos and J. Ellis, Nucl. Phys. **B182** (1982) 505.
4. For reviews, see: H.P. Nilles, Phys. Rep. **110** (1984) 1; H.E. Haber and G.L. Kane, Phys. Rep. **117** (1995) 75.
5. J. Ellis, S. Kelley and D. V. Nanopoulos, Phys. Lett. **B249** (1990) 441 and Phys. Lett. **B260** (1991) 131; U. Amaldi, W. de Boer and H. Furstenau, Phys. Lett. **B260** (1991) 447; C. Giunti, C. W. Kim and U. W. Lee, Mod. Phys. Lett. **A6** (1991) 1745; P. Langacker and M. Luo, Phys. Rev. **D44** (1991) 817.
6. LEP and SLD experiments, *A Combination of Preliminary Electroweak Measurements and Constraints on the Standard Model*, CERN EP/2000-016, as updated by B. Pietrzyk, Talk at XXXth International Conference on High Energy Physics, 27 July - 2 August, 2000, Osaka, Japan: `http://lepewwg.web.cern.ch/LEPEWWG/lepww/talks_notes/welcome.html`.
7. J. Ellis, G. Ridolfi and F. Zwirner, Phys. Lett. **B257** (1991) 83; M.S. Berger, Phys. Rev. **D41** (1990) 225; Y. Okada, M. Yamaguchi and T. Yanagida, Prog. Theor. Phys. **85** (1991) 1; Phys. Lett. **B262** (1991) 54; H.E. Haber and R. Hempfling, Phys. Rev. Lett. **66** (1991) 1815.
8. J. Ellis, G. Ganis, D. V. Nanopoulos and K. A. Olive, hep-ph/0009355.
9. K. A. Olive, G. Steigman and T. P. Walker, Phys. Rept. **333-334** (2000) 389.
10. A. E. Lange *et al.*, astro-ph/0005004.
11. J. Ellis, summary talk at the *XIXth International Conference on Neutrino Physics and Astrophysics*, Sudbury, Canada, June 2000: hep-ph/0008334 and references therein.
12. See, for example, N. Bahcall, J. P. Ostriker, S. Perlmutter and P. J. Steinhardt, Science **284** (1999) 1481.
13. P. Fayet, in *Proceedings of the Europhysics Study Conf. on Unification of Fundamental Interactions*, Erice, Italy, Mar 17-24, 1980.

14. See, for example, J. L. Basdevant, R. Mochkovitch, J. Rich, M. Spiro and A. Vidal-Madjar, Phys. Lett. **B234** (1990) 395.
15. J. Ellis, J.S. Hagelin, D.V. Nanopoulos, K.A. Olive and M. Srednicki, Nucl. Phys. **B238** (1984) 453.
16. H. Goldberg, Phys. Rev. Lett. **50** (1983) 1419.
17. J. Ellis, T. Falk, G. Ganis and K. A. Olive, hep-ph/0004169.
18. CLEO Collaboration, M.S. Alam et al., Phys.Rev.Lett. **74** (1995) 2885; S. Ahmed et al., `CLEO CONF 99-10`; ALEPH Collaboration, R. Barate *et al.*, Phys. Lett. **B429** (1998) 169; G. Degrassi, P. Gambino and G. F. Giudice, hep-ph/0009337; M. Carena, D. Garcia, U. Nierste and C. E. Wagner, hep-ph/0010003.
19. J. A. Casas, A. Lleyda and C. Munoz, Nucl. Phys. **B471** (1996) 3; H. Baer, M. Brhlik and D. Castano, Phys. Rev. **D54** 6944 1996; S. Abel and T. Falk, Phys. Lett. **B444** (1998) 427.
20. J. Ellis, T. Falk and K. A. Olive, Phys. Lett. **B444**, 3; J. Ellis, T. Falk, K. A. Olive and M. Srednicki, Astropart. Phys. **13** (2000) 181.
21. See S. Abel *et al.*, Tevatron SUGRA Working Group Collaboration, hep-ph/0003154, and references therein.
22. ATLAS Collaboration, Detector and Physics Performance Technical Design Report, `http://atlasinfo.cern.ch/Atlas/GROUPS/PHYSICS/TDR/access.html`;
CMS Collaboration, Technical Proposal,
`http://cmsinfo.cern.ch/TP/TP.html`.
23. J. Silk and M. Srednicki, Phys. Rev. Lett. **53** (1984) 624.
24. L. Bergstrom, J. Edsjo and P. Ullio, astro-ph/9902012.
25. L. Bergstrom, J. Edsjo, P. Gondolo and P. Ullio, Phys. Rev. **D59** (1999) 043506.
26. J. Silk, K. Olive and M. Srednicki, Phys. Rev. Lett. **55** (1985) 257.
27. P. Gondolo, talk at the *XIXth International Conference on Neutrino Physics and Astrophysics*, Sudbury, Canada, June 2000:
`http://nu2000.sno.laurentian.ca/`.
28. M. W. Goodman and E. Witten, Phys. Rev. **D31** (1985) 3059.
29. R. Bernabei *et al.*, DAMA Collaboration, Phys. Lett. **B480** (2000) 23.
30. J. Ellis, A. Ferstl and K. A. Olive, Phys. Lett. **B481** (2000) 304.
31. R. Abusaidi *et al.*, CDMS Collaboration, Nucl. Instrum. Meth. **A444** (2000) 345.
32. J. Ellis, A. Ferstl and K. A. Olive, hep-ph/0007113.
33. For another view and earlier references, see: A. Bottino, F. Donato, N. Fornengo, S. Scopel, hep-ph/0010203.
34. Presentations at the open LEPC seminar, Sept. 5th, 2000:
ALEPH Collaboration, W.-D. Schlatter,
`http://alephwww.cern.ch/`;
DELPHI Collaboration, T. Camporesi,
`http://delphiwww.cern.ch/~offline/physics_links/lepc.html`;
L3 Collaboration, J.-J. Blaising,
`http://l3www.cern.ch/analysis/latestresults.html`;
OPAL Collaboration, C. Rembser,
`http://opal.web.cern.ch/Opal/PPwelcome.html`;
C. Tully, for the LEP Working Group on Higgs boson searches,

http://lephiggs.web.cern.ch/LEPHIGGS/talks/index.html.
35. Presentations at the LEPFest, Oct. 10th, 2000:
ALEPH Collaboration, F. Cerutti,
http://alephwww.cern.ch/;
DELPHI Collaboration, V. Hedberg,
http://delphiwww.cern.ch/~offline/physics_links/lepc.html;
L3 Collaboration, M. Kienzle,
http://l3www.cern.ch/analysis/latestresults.html;
OPAL Collaboration, R. Hemingway,
http://opal.web.cern.ch/Opal/PPwelcome.html;
T. Junk, for the LEP Working Group on Higgs boson searches,
http://lephiggs.web.cern.ch/LEPHIGGS/talks/index.html.

INTRODUCTION TO SUPERSYMMETRY

CUPATITZIO RAMIREZ ROMERO

Facultad de Ciencias Físico Matemáticas
Universidad Autónoma de Puebla
Puebla, Pue. México

Abstract. The goal of these lectures is to provide an insight on topics going from supersymmetry representations to supergravity, stressing on techniques useful to construct and understand phenomenological theories. Historical aspects are not considered. Supergravity is shortly touched, giving before an account of the vierbein formalism of gravity.

Outline of the lectures

1. Introduction
2. Supersymmetry algebra, representations
3. Supersymmetry transformations, superspace
4. Superfields, scalar and vector superfields
5. Supersymmetric matter actions
6. Supersymmetric gauge theories
7. Spontaneous supersymmetry breaking
8. Supergravity

I INTRODUCTION

There are almost 30 years that supersymmetry has been discovered, since then, in spite of the lack of unambiguous experimental evidence, it has become the dominant framework for formulating physics beyond the standard model. The reasons are the variety of situations in which supersymmetry improve the behavior of theories, sometimes dramatically. In these lectures I will focus exclusively on the technical side. Phenomenological aspects will be treated in other lectures. Some good reviews

CP562, *Particles and Fields: Ninth Mexican School,* edited by G. Herrera Corral and L. Nellen

are Wess and Bagger, [1] West, [2] Sohnius [3] and Weinberg, [4]. There are also many reviews, see for example Lykken [5]

II SUPERSYMMETRY ALGEBRA, REPRESENTATIONS

A The 4-dimensional supersymmetry algebra

Supersymmetry is a group of transformations which act on bosons transforming them into fermions and vice versa, in a way consistent with Poincaré symmetry. It can be said that it is the only possible nontrivial extension of the known space-time symmetries of particle physics, fact made possible by the inclusion of anticommuting transformation generators.

Let us take a bosonic field, A, and transform it in such a way that at the r.h.s. a fermion field appears,

$$\delta_\xi A = \xi^\alpha \psi_\alpha,$$

where ξ^α are the transformation infinitesimal parameters. As far as, in order to satisfy Fermi statistics, fermions are anticommuting fields, this rule is meaningful only if ξ^α are anticommuting quantities.

$$\{\xi^\alpha, \xi^\beta\} \equiv \xi^\alpha \xi^\beta + \xi^\beta \xi^\alpha = 0 \tag{1}$$

Anticommuting quantities generate a Grassmann algebra, whose elements are power expansions of these quantities. These expansions are always finite because of the nilpotent character of anticommuting quantities, i.e.

$$(\xi)^2 = 0,$$

which results from their anticommuting property. Thus, supersymmetry transformations have anticommuting parameters and consequently, the supersymmetry generators Q_α, $\delta_\xi A = \xi^\alpha Q_\alpha A$, will have also anticommuting character.

The most general 4-dimensional supersymmetry algebra is given by

$$\begin{aligned}
\{Q^A_\alpha, \overline{Q}_{\dot\beta B}\} &= 2\sigma^m_{\alpha\dot\beta} P_m \delta^A_B \\
\{Q^A_\alpha, Q^B_\beta\} &= \epsilon_{\alpha\beta} a^{\ell AB} B_\ell \\
\{\overline{Q}_{\dot\alpha A}, \overline{Q}_{\dot\beta B}\} &= -\epsilon_{\dot\alpha\dot\beta} a^*_{\ell AB} B^\ell \\
[Q^A_\alpha, P_m] &= [\overline{Q}^A_{\dot\alpha}, P_m] = 0 \\
[Q^A_\alpha, M_{mn}] &= \sigma_{mn\alpha}{}^\beta Q^A_\beta \\
[\overline{Q}^{\dot\alpha}_A, M_{mn}] &= \bar\sigma^{\dot\alpha}_{mn\dot\beta} \overline{Q}^{\dot\beta}_A \\
[P_m, P_n] &= 0 \\
[M_{mn}, P_p] &= i(\eta_{np} P_m - \eta_{mp} P_n)
\end{aligned} \tag{2}$$

$$
\begin{aligned}
[M_{mn}, M_{pq}] &= -i(\eta_{mp}M_{nq} - \eta_{mq}M_{np} - \eta_{np}M_{mq} + \eta_{nq}M_{mp}) \\
[Q_\alpha^A, B_\ell] &= S_{\ell\ B}^{A} Q_\alpha^B \\
[\overline{Q}_{\dot\alpha A}, B_\ell] &= -S_{\ell A}^{*\ \ B} \overline{Q}_{\dot\alpha B} \\
[B_\ell, B_k] &= iC_{\ell k}{}^j B_j \\
[P_m, B_\ell] &= [M_{mn}, B_\ell] = 0
\end{aligned}
$$

where P_m and M_{mn} are the generators of translations and Lorentz transformations, Q_α^A and $\bar{Q}_{\dot\alpha}^A$ are the $2N$ supersymmetry generators ($A = 1, \ldots, N$) in *four*-dimensional Weyl spinor notation, and B_ℓ are the generators of internal symmetry group $\mathcal{G}$, of dimension N, which are Lorentz scalars. Thus the supersymmetric charges Q carry representations of the Lorentz group, as well as representations of $\mathcal{G}$. When $N = 1$, it is called simple supersymmetry and when $N > 1$, it is is called N-extended supersymmetry. In the first case the internal group is abelian, with a global $U(1)$ symmetry with charge R, and it can be written as

$$
\begin{aligned}
[Q_\alpha, R] &= Q_\alpha \\
[\overline{Q}_{\dot\alpha}, R] &= -\overline{Q}_{\dot\alpha} \quad .
\end{aligned} \tag{3}
$$

B Lorentz Weyl spinor representations

The Weyl spinor notation relies on the decomposition of Dirac spinors in two two−component spinors as follows

$$
\psi = \begin{pmatrix} \psi_\alpha \\ \bar\chi^{\dot\alpha} \end{pmatrix} \tag{4}
$$

Let us consider the space of hermitic 2×2 matrices, a basis in this space is given by the matrices $\sigma^m = (\sigma^0, \sigma^i)$, where $\sigma^0 \equiv 1$ and σ^i are the Pauli matrices,

$$
\sigma^1 = \begin{pmatrix} 0 & 1 \\ 1 & 0 \end{pmatrix}, \qquad \sigma^2 = \begin{pmatrix} 0 & -i \\ i & 0 \end{pmatrix}, \qquad \sigma^3 = \begin{pmatrix} 1 & 0 \\ 0 & -1 \end{pmatrix} \tag{5}
$$

Thus the matrices in this space can always be written as expansions $\mathcal{P} = P_m \sigma^m$. Obviously this space is invariant under the transformations $\mathcal{P} \to M\mathcal{P}M^\dagger$, where M are complex matrices. If moreover M are elements of $SL(2, C)$, then also

$$
\det \mathcal{P} = \det \begin{pmatrix} P_0 + P_3 & P_1 - iP_2 \\ P_1 + iP_2 & P_0 + P_3 \end{pmatrix} = P_0^2 - P_1^2 - P_2^2 - P_3^2 \tag{6}
$$

is invariant. These transformations thus induce Lorentz transformations on P_m. Therefore, M as well as $M^\dagger$, elements of $SL(2, C)$, carry two dimensional representations of the Lorentz group, similar to the case of $SU(2)$ with $SO(3)$.

These representations act as

$$\psi'_\alpha = M_\alpha{}^\beta \psi_\beta, \qquad \bar{\psi}'_{\dot\alpha} = M^{*\dot\beta}_{\dot\alpha} \bar{\psi}_{\dot\beta} = \bar{\psi}_{\dot\beta} M^{\dagger\dot\beta}{}_{\dot\alpha} \tag{7}$$

i.e. $\bar{\psi}_{\dot\alpha}$ are complex conjugates of ψ_α.

In order to write invariants, the contragredient representations, defined as

$$\psi'^{\ \alpha} = \psi^\beta M^{-1}{}_\beta{}^\alpha, \qquad \bar{\psi}'^{\ \dot\alpha} = \bar{\psi}^{\dot\beta} (M^*)^{-1}{}_{\dot\beta}{}^{\dot\alpha} = (M^\dagger)^{-1}{}^{\dot\alpha}{}_{\dot\beta} \bar{\psi}^{\dot\beta} \tag{8}$$

are needed. Thus, $\psi\chi \equiv \psi^\alpha \chi_\alpha$ and $\bar{\psi}\bar{\chi} \equiv \bar{\psi}_{\dot\alpha} \bar{\chi}^{\dot\alpha}$ are invariants. We see also that the $\sigma-$matrices have the index structure $\sigma^m_{\alpha\dot\alpha}$, from which it turns out that $\psi\sigma^m\bar{\chi} \equiv \psi^\alpha \sigma^m_{\alpha\dot\alpha} \bar{\chi}^{\dot\alpha}$ transforms as a $four$-vector.

It is well known that the contragredient representations are equivalent to the other by the relations

$$\psi^\alpha = \varepsilon^{\alpha\beta} \psi_\beta, \qquad \bar{\psi}^{\dot\alpha} = \varepsilon^{\dot\alpha\dot\beta} \bar{\psi}_{\dot\beta}, \tag{9}$$

where the invariant matrices $\varepsilon^{\alpha\beta}$ and $\varepsilon^{\dot\alpha\dot\beta}$, which act as metrics, are antisymmetric and are defined by $\varepsilon^{12} = \varepsilon^{\dot1\dot2} = -1$.

C The representations of supersymmetry

In order to study supersymmetry representations, we need to know first the Casimir operators. For the Poincaré algebra, the Casimir operators are given by the mass operator $P^2 = P_m P^m$, with eigenvalues m^2, and the square W^2 of the Pauli-Lubanski vector

$$W_m = \frac{1}{2} \epsilon_{mnpq} P^n M^{pq} \quad . \tag{10}$$

with eigenvalues $-m^2 s(s+1)$, s=0, $\frac{1}{2}, 1, \ldots$ for massive states. For massless states $W_m = \lambda P_m$, where λ is the helicity.

For N=1 supersymmetry, P^2 is still a Casimir, but W^2 is not because supersymmetry charges carry spin. The corresponding Casimir is $C^2 = C_{mn} C^{mn}$, where

$$\begin{aligned} C_{mn} &= B_m P_n - B_n P_m \quad , \\ B_m &= W_m - \frac{1}{4} \bar{Q}_{\dot\alpha} \bar{\sigma}_m^{\dot\alpha\beta} Q_\beta \quad . \end{aligned} \tag{11}$$

Thus supersymmetry representations are characterized by mass m and spin s or helicity λ. The anticommutative charges Q, because of their nilpotent nature, contribute to the discrete part of the spectrum. In the following of this chapter, it will be shown how the irreducible representations of supersymmetry on asymptotic (on-shell) one particle physical states are constructed.

First of all, let us remark that the vanishing of the commutator of supersymmetric charges with the translations operator P_m means that all states created by the action of supersymmetry transformations have the same mass, that is, supersymmetric multiplets, supermultiplets, are characterized by only one mass value.

D N=1 massive states

The study of massive states is done as usual, the rest frame $P_m = (m, \vec{0})$ is chosen, such that

$$C^2 = 2m^4 J_i J^i \quad ,$$
$$J_i \equiv S_i - \frac{1}{4m}\overline{Q}\bar{\sigma}_i Q \quad , \tag{12}$$

where S_i is the spin operator of the Lorentz algebra ($i = 1, 2, 3$).

It turns out that

$$[J_i, J_j] = i\epsilon_{ijk}J_k \quad , \qquad [J_i, Q_\alpha] = [J_i, \bar{Q}_{\dot{\alpha}}] = 0 \tag{13}$$

i.e. J^2 has eigenvalues $j(j+1)$, $j = \frac{n}{2}$, $n = 0, 1, 2, \ldots$ and any states created by the action of Q or $\bar{Q}$ will have the same j value.

Let us define

$$a_{1,2} = \frac{1}{\sqrt{2m}}Q_{1,2} \quad ,$$
$$a^\dagger_{1,2} = \frac{1}{\sqrt{2m}}\overline{Q}_{\dot{1},\dot{2}} \quad , \tag{14}$$

which are nilpotent and form a Clifford algebra,

$$\{a_\alpha, a^\dagger_\beta\} = \sigma^0_{\alpha\dot{\beta}} = 2m\begin{pmatrix} 1 & 0 \\ 0 & 1 \end{pmatrix} \quad . \tag{15}$$

Thus, $a^\dagger_\alpha$ are 'creation' and a_α 'annihilation' operators acting on the 'vacuum' state defined by $|\Omega> = a_1 a_2|m, j>$, with $2j+1$ components. These operators $a^\dagger_\alpha$ fill out the N=1 massive supersymmetry representation of fixed m and j as follows:

$$|\Omega>$$
$$a^\dagger_1|\Omega> \tag{16}$$
$$a^\dagger_2|\Omega>$$
$$\frac{1}{\sqrt{2}}a^\dagger_1 a^\dagger_2|\Omega> \; = -\frac{1}{\sqrt{2}}a^\dagger_2 a^\dagger_1|\Omega>$$

With a total of $4(2j+1)$ of states. As an example, consider the $j = 0$, $N = 1$ massive representation. Since $|\Omega>$ has spin zero there are a total of four states in the representation and it is irreducible, with spin projections $s = 0, -\frac{1}{2}, \frac{1}{2}$, and 0, respectively. Since the parity operation interchanges $a^\dagger_1$ with $a^\dagger_2$, one of the spin zero states is a scalar and the other pseudoscalar. Thus these four massive states correspond to one Weyl fermion, one real scalar, and one real pseudoscalar.

E N=1 supersymmetry, massless states

In this case we take the light-like reference frame $P_m = (E, 0, 0, E)$ and it is easy to see that the Casimir operator is,

$$C^2 = -2E^2(B_0 - B_3)^2 = -\frac{1}{2}E^2\overline{Q}_{\dot{2}}Q_2\overline{Q}_{\dot{2}}Q_2 = 0 \quad . \tag{17}$$

Also we have,

$$\begin{aligned} \{Q_{\dot{1}}, \overline{Q}_{\dot{1}}\} &= 4E \quad , \\ \{Q_{\dot{2}}, \overline{Q}_{\dot{2}}\} &= 0 \quad . \end{aligned} \tag{18}$$

Therefore,

$$< \Omega|Q_2\overline{Q}_{\dot{2}}|\Omega >= 0 \quad . \tag{19}$$

i.e. $\overline{Q}_{\dot{2}}|\Omega >$ has zero norma and $\overline{Q}_{\dot{2}}$ can be set equal to zero. Thus, there is just one pair of creation/annihilation operators:

$$a = \frac{1}{2\sqrt{E}}Q_1 \quad , \quad a^\dagger = \frac{1}{2\sqrt{E}}\overline{Q}_{\dot{1}} \quad . \tag{20}$$

The vacuum state $|\Omega >$ is nondegenerate and has definite helicity λ. The creation operator $a^\dagger$ increases helicity by 1/2 and the representation contains two states:

$$\begin{aligned} |\Omega > &\text{ helicity } \lambda \quad , \\ a^\dagger|\Omega > &\text{ helicity } \lambda + \frac{1}{2} \quad . \end{aligned} \tag{21}$$

In general, this is not a CPT eigenstate, requiring that we pair two representations to obtain four states with helicities λ, $\lambda+\frac{1}{2}$, $-\lambda-\frac{1}{2}$, and $-\lambda$.

F N>1 supersymmetry, no central charges

In this case there are N copies of the supersymmetry charges and the analysis can be done for each of these copies as in the preceding section. Thus, in the massless case, there are N creation operators $a_A^\dagger$. and in the massive case, there are 2N creation operators $(a_\alpha^A)^\dagger$. In order to see how many states are generated, let us consider how many different powers of anticommuting quantities a_i, $(i = 1 \ldots n)$ there are. The general term is $a_{i_1}a_{i_2}\ldots a_{i_k}$, with all indices different, no matter the order, and $k \leq n$. There are $\begin{pmatrix} n \\ k \end{pmatrix}$ different such states and a total of

$$\sum_{k=1}^{n} \begin{pmatrix} n \\ k \end{pmatrix} = 2^n. \tag{22}$$

Therefore in the massless case there are 2^N states and in the massive case there are $2^{2N}(2j+1)$ states, where the vacuum spin is considered. In order to determine the spin values of these states it must be taken into account that the internal group indices are 'scalar'. In the massless case the helicity is incremented by each 'creation' operator acting on the vacuum state by $\frac{1}{2}$, in such a way that if the vacuum has helicity$=\lambda$, then the states of the representation will have helicities $\lambda, \lambda+\frac{1}{2}, \ldots, \lambda+\frac{N}{2}$, with multiplicity $\begin{pmatrix} N \\ k \end{pmatrix}$ for the state of helicity $\lambda+\frac{k}{2}$.

In the massive case the spin values are obtained from the the composition of the angular momentum, $s=\frac{1}{2}$, of the creation operators, that is from the irreducible representations corresponding to configurations of definite symmetry of the spinorial indices.

For example, in the fundamental N=2 massive representation there are a total of 16 states in a supermultiplet as follows,

1 state :	$\vert\Omega>$	1 spin 0 state
4 states :	$(a_\alpha^A)^\dagger\vert\Omega>$	4 spin $\frac{1}{2}$ states
6 states :	$(a_{\alpha_1}^{A_1})^\dagger(a_{\alpha_2}^{A_2})^\dagger\vert\Omega>$	3 spin 1 and 3 spin 0 states
4 states :	$(a_{\alpha_1}^{A_1})^\dagger(a_{\alpha_2}^{A_2})^\dagger(a_{\alpha_3}^{A_3})^\dagger\vert\Omega>$	4 spin $\frac{1}{2}$ states
1 state :	$(a_{\alpha_1}^{A_1})^\dagger(a_{\alpha_2}^{A_2})^\dagger(a_{\alpha_3}^{A_3})^\dagger(a_{\alpha_4}^{A_4})^\dagger\vert\Omega>$	1 spin 0 state

If the spinorial indices are symmetric, then the corresponding internal indices are antisymmetric and vice versa, so that the global anticommutativity is kept.

G N>1 supersymmetry, with central charges

In the $N>1$ case, central charges can appear. In the massive case, in the rest frame the algebra can be written as,

$$\begin{aligned} \{Q_\alpha, \overline{Q}_{\dot\beta}\} &= 2m\sigma^0_{\alpha\dot\beta}\delta^a_b\delta^L_M \quad , \\ \{Q^{aL}_\alpha, Q^{bM}_\beta\} &= \epsilon_{\alpha\beta}\epsilon^{ab}\delta^{LM}Z_M \quad , \\ \{\overline{Q}_{\dot\alpha aL}, \overline{Q}_{\dot\beta bM}\} &= -\epsilon_{\dot\alpha\dot\beta}\epsilon_{ab}\delta_{LM}Z_M \quad , \end{aligned} \tag{23}$$

where the internal indices A, B have now been replaced by the index pairs (a, L), (b, M), with $a, b = 1, 2$ and $L, M = 1, 2, \ldots [\frac{N}{2}]$.

Thus, there are 2N pairs of creation/annihilation operators:

$$\begin{aligned} a^L_\alpha &= \frac{1}{\sqrt{2}}\left[Q^{1L}_\alpha + \epsilon_{\alpha\beta}\overline{Q}_{\dot\gamma 2L}\bar\sigma^{0\dot\gamma\beta}\right] \quad , \\ (a^L_\alpha)^\dagger &= \frac{1}{\sqrt{2}}\left[\overline{Q}_{\dot\alpha 1L} + \epsilon_{\dot\alpha\dot\beta}\bar\sigma^{0\dot\beta\gamma}Q^{2L}_\gamma\right] \quad , \\ b^L_\alpha &= \frac{1}{\sqrt{2}}\left[Q^{1L}_\alpha - \epsilon_{\alpha\beta}\overline{Q}_{\dot\gamma 2L}\bar\sigma^{0\dot\gamma\beta}\right] \quad , \\ (b^L_\alpha)^\dagger &= \frac{1}{\sqrt{2}}\left[\overline{Q}_{\dot\alpha 1L} - \epsilon_{\dot\alpha\dot\beta}\bar\sigma^{0\dot\beta\gamma}Q^{2L}_\gamma\right] \quad . \end{aligned} \tag{24}$$

with anticommutation relations:

$$\{a_\alpha^L, (a_\beta^M)^\dagger\} = (2m + Z_M)\sigma^0_{\alpha\dot\beta}\delta^L_M \quad ,$$
$$\{b_\alpha^L, (b_\beta^M)^\dagger\} = (2m - Z_M)\sigma^0_{\alpha\dot\beta}\delta^L_M \quad . \tag{25}$$

Since $\{a, a^\dagger\}$ and $\{b, b^\dagger\}$ are positive definite operators, and since the Z_M are nonnegative, then $Z_M \leq 2m$ and there are two cases:

- When $Z_M < 2m$, the situation is very similar to the one without central charges, indeed, there are $2N$ creation operators $(2m + Z_M)^{-1/2}(a_\alpha^M)^\dagger$ and $(2m - Z_M)^{-1/2}(b_\alpha^M)^\dagger$. As well, the multiplicities are the same as for the case with no central charges.

- When $Z_M = 2m$ for some or all Z_M, it is said that the bound is saturated. In this case, the corresponding b_α^M operators can be set to zero and as in the massless case, there are only half of the corresponding creation operators. If all Z_M saturate the bound, then this representation has only $2^N(2j+1)$ states instead of $2^{2N}(2j+1)$ states.

These states are often referred to as BPS-saturated states, because of the connection to BPS monopoles in supersymmetric gauge theories.

An important property of supersymmetry representations is the fact that the number of bosonic degrees of freedom equals already the number of fermionic degrees of freedom. This can be shown as follows, consider any representation and represent the supersymmetric charges by complex matrices, fact any time possible because Clifford algebras can be represented by Dirac matrices. Further define the fermionic number $(-1)^{N_F}$, where N_F counts the number of fermions. $(-1)^{N_F}$ anticommutes with supersymmetric charges and commutes with bosonic ones. Thus we can write

$$Tr\left[(-1)^{N_F}\{Q_\alpha^A, \bar{Q}_{\dot\beta\ B}\}\right] = Tr\left[-Q_\alpha(-1)^{N_F}\bar{Q}_{\dot\beta\ B} + \bar{Q}_{\dot\beta\ B}(-1)^{N_F}Q_\alpha\right] = 0 \tag{26}$$

which vanishes because of the cyclic property of the trace. The left-hand side of this expression gives also

$$2\sigma^m_{\alpha\dot\beta}\delta^A_B Tr\left[(-1)^{N_F}P_m\right] = 0 \tag{27}$$

which taking into account that supersymmetric states have all the same mass gives $Tr(-1)^{N_F} = 0$, that is there is an equal number of bosonic and of fermionic states. This result gives a very useful criterion to decide if a supermultiplet is complete or if there are missing degrees of freedom. Note that this demonstration supposes that the operator P_m acts one to one, so that it is nonsingular. There are examples, e.g. in two dimensions, in which it does not work.

III SUPERSYMMETRY TRANSFORMATIONS, SUPERSPACE

1 Supersymmetry transformations

Supersymmetry acts on fields changing their character, from bosonic to fermionic and vice versa. In this way, off-shell supermultiplets, with not fixed mass value, can be created by successive action of supersymmetry transformations, taking into account supersymmetry algebra. In general these transformations define new fields, and if applied successively on the new fields, if dimensionally possible, they can give derivatives of the other fields. The action of these transformations must be consistent, that is they must satisfy the supersymmetry algebra,

$$[\,\xi\, Q, \bar{\eta}\,\overline{Q}\,] = 2\xi\sigma^m\bar{\eta}P_m. \tag{28}$$

Let us start with $N = 1$ supersymmetry and a bosonic field A,

$$\delta_\xi A \equiv (\xi Q + \bar{\xi}\bar{Q})A = \sqrt{2}\xi^\alpha\psi_\alpha, \tag{29}$$

further we wish to know how ψ itself transforms, it can be through a derivative of A (translation), or by a new bosonic field F,

$$\delta_\eta\delta_\xi A \equiv \sqrt{2}\xi^\alpha\delta_\eta\psi_\alpha = ai\xi\sigma^m\bar{\eta}\partial_m A + b\xi\eta F. \tag{30}$$

Consistency of this product of transformations with (28) fixes the coefficients $a = b = 2$. A further application of supersymmetry must close the algebra on ψ as well. It turns out that it closes with no need of new fields if

$$\delta_\xi F = i\sqrt{2}\bar{\xi}\sigma^m\partial_m\psi \tag{31}$$

Thus we get a supermultiplet $\{A, \psi, F\}$, called scalar or chiral supermultiplet. Following this procedure, under observation of supersymmetry algebra, all off-shell representations can be generated. However it can be that supersymmetry algebra cannot be closed, unless certain field equations are satisfied. In these cases on-shell supermultiplets are obtained.

A Superspace formalism

Superspace is a formalism that allows a tensorial type construction of representations of supersymmetry. It is similar to relativistic quantum field theory, which relies upon the fact that the space-time coordinates x^m parameterize the group of translations. As far as supersymmetry transformations can be seen as the "square root" of translations, we can expect that a similar scheme for supersymmetry can be found. For simplicity we will discuss the case of N=1 supersymmetry.

The first step is to introduce constant Grassmann (anticommuting) spinors θ^α, $\overline{\theta}_{\dot\alpha}$, with the properties of the parameters of the supersymmetry transformations:

$$\{\theta^\alpha, \theta^\beta\} = \{\overline{\theta}_{\dot\alpha}, \overline{\theta}_{\dot\beta}\} = \{\theta^\alpha, \overline{\theta}_{\dot\beta}\} = 0 \quad . \tag{32}$$

This allows us to replace the anticommutators in the N=1 supersymmetry algebra with commutators:

$$\begin{aligned} &[\,\theta\, Q, \overline{\theta}\,\overline{Q}\,] = 2\theta\sigma^m\overline{\theta}P_m \quad , \\ &[\,\theta\, Q, \theta\, Q\,] = [\,\overline{\theta}\,\overline{Q}, \overline{\theta}\,\overline{Q}\,] = 0 \quad . \end{aligned} \tag{33}$$

Thus, operators θQ and $\bar\theta\bar Q$, together with translations, constitute a Lie algebra which can be exponentiated to get the group element:

$$G(x, \theta, \overline{\theta}) = \mathrm{e}^{i[-x^m P_m + \theta\, Q + \overline{\theta}\,\overline{Q}]} \quad , \tag{34}$$

which is unitary since $(\theta Q)^\dagger = \overline{\theta}\,\overline{Q}$. The product law for this group element can be obtained using the Baker–Campbell–Hausdorff formula for the product of exponentials of operators, $e^A e^B = e^{A+B+\frac{1}{2}[A,B]+\cdots}$, where the dots represent higher commutators which in our case vanish. Hence,

$$G(y, \xi, \bar\xi)G(x, \theta, \overline{\theta}) = G(x + y + i\theta\sigma^m\bar\xi - i\xi\sigma^m\bar\theta, \theta + \xi, \overline{\theta} + \bar\xi). \tag{35}$$

The r.h.s. of this equations shows how superspace, whose elements are the parameters of the supersymmetry transformations $(x, \theta, \overline{\theta})$, transforms under supersymmetry. This transformation law allows constructing representations of supersymmetry on superfields, functions on superspace, by the group left multiplication law. Consider any linear representation of supersymmetry on elements Φ_0 and define the superfield $\Phi(x, \theta, \overline{\theta}) \equiv G(x, \theta, \overline{\theta})\Phi_0$, then

$$\Phi'(x, \theta, \overline{\theta}) = G(y, \xi, \bar\xi)\Phi(x, \theta, \overline{\theta}) = \Phi(x + y + i\theta\sigma^m\bar\xi - i\xi\sigma^m\bar\theta, \theta + \xi, \overline{\theta} + \bar\xi) \tag{36}$$

defines a linear representation of supersymmetry on arbitrary superfields. As far as the product of functions is a new function, the product of superfields will be a superfield and will transform as (36).

The nilpotency property of anticommuting variables does not allow the usual definition of a measure in superspace and the usual notions of continuity, differentiability, etc., cannot be given. Superfields are defined as formal power expansions on the anticommuting variables. In N=1, four dimensions we have,

$$\begin{aligned} &\Phi(x, \theta, \overline{\theta}) = f(x) + \theta\phi(x) + \overline{\theta}\overline{\chi}(x) + \theta\theta m(x) + \overline{\theta}\overline{\theta} n(x) \\ &+\theta\sigma^m\overline{\theta}v_m(x) + (\theta\theta)\overline{\theta}\overline{\lambda}(x) + (\overline{\theta}\overline{\theta})\theta\psi(x) + (\theta\theta)(\overline{\theta}\overline{\theta})d(x) \quad . \end{aligned} \tag{37}$$

Any higher power of the θ variables will vanish because at least one of the components will appear repeated.

The Grassmann derivatives are defined as anticommuting operators on superfields by,

$$\partial_\alpha \theta^\beta = \delta_\alpha^\beta \quad , \qquad \overline{\partial}^{\dot\alpha}\overline{\theta}_{\dot\beta} = \delta_{\dot\beta}^{\dot\alpha}, \tag{38}$$

their action on any other quantity vanishes. The derivative index is raised by $-\epsilon^{\alpha\beta}$, where the minus sign is needed such that $\partial^\alpha\theta_\beta = \delta_\beta^\alpha$, similarly for the dotted derivative.

Opposite to the derivative, superspace integration cannot be done following the bosonic integration rules because the bosonic integral increases the power of variables and we would obtain almost always zero. Instead there is a definition which has many of the properties of bosonic integrals over all space, this definition essentially coincides with the definition of superspace derivation,

$$\int d\theta_\alpha\, \theta^\beta = \delta_\alpha^\beta, \qquad \int d\overline{\theta}^{\dot\alpha}\, \overline{\theta}_{\dot\beta} = \delta_{\dot\beta}^{\dot\alpha} \tag{39}$$

$$\int d\theta_\alpha \quad = 0, \qquad \int d\overline{\theta}_{\dot\alpha} \quad = 0 \tag{40}$$

In this case we can define the whole Grassmann space integration, taking into account that the indices of the differentials are raised by the usual rule, as

$$\int d^4\theta \Phi(\theta, \bar\theta) \equiv \frac{1}{16}\int d\theta^\alpha d\theta_\alpha \;\; d\bar\theta_{\dot\alpha} d\bar\theta^{\dot\alpha} \;\; \Phi(\theta, \bar\theta) = d(x), \tag{41}$$

where $d(x)$ is the highest component of the superfield (37).

B Supersymmetry generators

The infinitesimal supersymmetry transformations of superfields (36) can be written as:

$$\delta_\xi \phi(x, \theta, \overline{\theta}) = (\xi^m P_m + \xi^\alpha Q_\alpha + \bar\xi^{\dot\alpha} \bar Q_{\dot\alpha})\phi(x, \theta, \overline{\theta}) \tag{42}$$

where P, Q and $\bar Q$ are differential operators. Due to the fact that supersymmetry acts on superspace by translations, it can be easily seen that

$$\begin{aligned} P_m &: i\partial_m \quad , \\ Q_\alpha &: \partial_\alpha - i\sigma^m_{\alpha\dot\beta}\overline{\theta}^{\dot\beta}\partial_m \quad , \\ \overline{Q}_{\dot\alpha} &: \overline{\partial}_{\dot\alpha} + i\theta^\beta \sigma^m_{\beta\dot\alpha}\partial_m \quad . \end{aligned} \tag{43}$$

If we use the definitions of Grassmann derivatives and compute the commutators, anticommutators of these operators the supersymmetry algebra is recovered,

$$\{Q_\alpha, Q_{\dot\beta}\} = 2\sigma^m_{\alpha\dot\beta} P_m, \tag{44}$$

and all other (anti)commutators vanishing.

From (35), we see that superspace transformation rules are of local nature, because they depend on the anticommuting variables, although they are rigid with respect to space-time variables. This means that space-time derivatives are always covariant, but Grassmann derivatives are not. In order to construct supersymmetric expressions, we need derivatives that keep the superfield transformation properties, i.e. covariance. They are given by,

$$\begin{aligned} D_\alpha &= \partial_\alpha + i\sigma^m_{\alpha\dot\beta}\overline{\theta}^{\dot\beta}\partial_m \quad , \\ \overline{D}_{\dot\alpha} &= -\overline{\partial}_{\dot\alpha} - i\theta^\beta\sigma^m_{\beta\dot\alpha}\partial_m \quad . \end{aligned} \tag{45}$$

With the properties

$$\{D_\alpha, Q_\beta\} = \{D_\alpha, \bar{Q}_{\dot\beta}\} = \{\bar{D}_{\dot\alpha}, Q_\beta\} = \{\bar{D}_{\dot\alpha}, \bar{Q}_{\dot\beta}\} = 0, \tag{46}$$

which insure that $D_\alpha\Phi$ and $\bar{D}_{\dot\alpha}\Phi$ transform as superfields.

The covariant derivatives satisfy also,

$$\begin{aligned} \{D_\alpha, \bar{D}_{\dot\beta}\} &= -2\sigma^m_{\alpha\dot\beta}P_m \\ \{D_\alpha, D_\beta\} &= \{\bar{D}_{\dot\alpha}, \bar{D}_{\dot\beta}\} = 0. \end{aligned} \tag{47}$$

Therefore, starting with a given set of superfields, we can construct new ones by making linear combinations, products, or by acting on them with space-time derivatives or covariant Grassmann derivatives. In the following we will show how to construct invariants by means of Grassmann integration.

C Component transformations

The supersymmetry transformations of the superfield components are computed by the use of (36) and comparing the coefficients of the l.h.s. with the coefficients of the r.h.s. That is, for infinitesimal transformations, the coefficients of the different θ powers of

$$\begin{aligned} \delta_\xi\Phi(x,\theta,\overline{\theta}) &= \delta_\xi f(x) + \theta\delta_\xi\phi(x) + \overline{\theta}\delta_\xi\overline{\chi}(x) + \theta\theta\delta_\xi m(x) + \overline{\theta\theta}\delta_\xi n(x) \\ &+\theta\sigma^m\overline{\theta}\delta_\xi v_m(x) + (\theta\theta)\overline{\theta}\delta_\xi\overline{\lambda}(x) + (\overline{\theta\theta})\theta\delta_\xi\psi(x) + (\theta\theta)(\overline{\theta\theta})\delta_\xi d(x) \quad . \end{aligned} \tag{48}$$

must be equated to the corresponding coefficients of

$$\delta_\xi\Phi(x,\theta,\bar\theta) = (\xi Q + \overline{\xi Q})\Phi(x,\theta,\bar\theta), \tag{49}$$

where Q and $\bar{Q}$ are given by (43). After algebraic manipulations, the component fields of Φ can be seen to transform as:

$$
\begin{aligned}
\delta_\xi f &= \xi\phi + \bar{\xi}\bar{\chi} \quad , \\
\delta_\xi \phi_\alpha &= 2\xi_\alpha m + \sigma^m_{\alpha\dot{\beta}}\bar{\xi}^{\dot{\beta}}[i\partial_m f + v_m] \quad , \\
\delta_\xi \bar{\chi}^{\dot{\alpha}} &= 2\bar{\xi}^{\dot{\alpha}} n + \xi^\beta \sigma^m_{\beta\dot{\gamma}}\epsilon^{\dot{\gamma}\dot{\alpha}}[i\partial_m f - v_m] \quad , \\
\delta_\xi m &= \bar{\xi}\bar{\lambda} - \frac{i}{2}\partial_m \phi \sigma^m \bar{\xi} \quad , \\
\delta_\xi n &= \xi\psi + \frac{i}{2}\xi\sigma^m \partial_m \bar{\chi} \quad , \\
\delta_\xi v_m &= \xi\sigma_m \bar{\lambda} + \psi\sigma_m \bar{\xi} + \frac{i}{2}\xi\sigma_m \phi - \frac{i}{2}\partial_m \bar{\chi}\bar{\xi} \quad , \\
\delta_\xi \bar{\lambda}^{\dot{\alpha}} &= 2\bar{\xi}^{\dot{\alpha}} d + \frac{i}{2}\bar{\xi}^{\dot{\alpha}}\partial^m v_m + i(\xi\sigma^m \epsilon)^{\dot{\alpha}}\partial_m m \quad , \\
\delta_\xi \psi_\alpha &= 2\xi_\alpha d - \frac{i}{2}\xi_\alpha \partial^m v_m + i(\sigma^m \bar{\xi})_\alpha \partial_m n \quad , \\
\delta_\xi d &= \frac{i}{2}\partial_m \left[\psi\sigma^m \bar{\xi} + \xi\sigma^m \bar{\lambda}\right] \quad .
\end{aligned}
\tag{50}
$$

This representation is however reducible, take for example the components $\{f, \phi, m\}$ and all other equal to zero, then their transformations close,

$$
\begin{aligned}
\delta_\xi f &= \xi\phi \quad , \\
\delta_\xi \phi_\alpha &= 2\xi_\alpha m + i\sigma^m_{\alpha\dot{\beta}}\bar{\xi}^{\dot{\beta}}\partial_m f \quad , \\
\delta_\xi m &= -\frac{i}{2}\partial_m \phi \sigma^m \bar{\xi} \quad ,
\end{aligned}
\tag{51}
$$

as well for $\{f, \bar{\chi}, n\}$,

$$
\begin{aligned}
\delta_\xi f &= \bar{\xi}\bar{\chi} \quad , \\
\delta_\xi \bar{\chi}^{\dot{\alpha}} &= 2\bar{\xi}^{\dot{\alpha}} n + i\xi^\beta \sigma^m_{\beta\dot{\gamma}}\epsilon^{\dot{\gamma}\dot{\alpha}} i\partial_m f \quad , \\
\delta_\xi n &= \frac{i}{2}\xi\sigma^m \partial_m \bar{\chi} \quad ,
\end{aligned}
\tag{52}
$$

Similar results are obtained for $\{n, \psi, d\}$ or $\{m, \bar{\lambda}, d\}$. The first and the third representations transform as the scalar supermultiplet, the others as its complex conjugated.

From these transformation laws an important fact arises, the highest ($\theta-$power) component transforms always as a divergence and its space-time integral can be used as an invariant action. This is the way to obtain supersymmetry invariants. More, we have seen that the Grassmann integral delivers precisely this highest component (41). Therefore supersymmetric invariants can be obtained if we integrate any superfield on superspace. In this way supersymmetric actions can be constructed.

D N=1 chiral superfields

Covariant derivatives can be used for the imposition of covariant constraints on superfields, by means of which irreducible superfields can be obtained.

The simplest condition is to impose the "left chirality" condition

$$\overline{D}_{\dot{\alpha}}\Phi = 0 \quad . \tag{53}$$

This condition can be solved observing that

$$\bar{D}_{\dot{\alpha}}y^m = 0, \qquad \bar{D}_{\dot{\alpha}}\theta_\alpha = 0, \tag{54}$$

where

$$y^m = x^m + i\theta\sigma^m\overline{\theta}. \tag{55}$$

Thus, the solution to (53) is given by the chiral superfield, in the so called chiral base,

$$\Phi(y,\theta) = A(y) + \sqrt{2}\theta\psi(y) + \theta\theta F(y). \tag{56}$$

If the variables y are substituted by their value, the usual base expansion is obtained,

$$\begin{aligned}\Phi(y,\theta) = A(x) + \sqrt{2}\theta\psi(x) + \theta\theta F(x) +\\ i\theta\sigma^m\overline{\theta}\partial_m A(x) + \frac{i}{\sqrt{2}}(\theta\theta)\partial_m\psi(x)\sigma^m\overline{\theta} - \frac{1}{4}(\theta\theta)(\overline{\theta}\overline{\theta})\Box A(x) \quad .\end{aligned} \tag{57}$$

There are 4+4 = 8 real off-shell field components; this is twice the number in the on-shell fundamental N=1 massive irreducible representations. The supersymmetry transformations obtained from this superfield coincide with the ones of the scalar supermultiplet.

From this expression of the chiral superfield, it can be seen that in the new variables $(y, \theta, \overline{\theta})$, the covariant derivatives have a particularly simple expression,

$$\begin{aligned} D_\alpha &= \partial_\alpha + 2i\sigma^m_{\alpha\dot{\beta}}\overline{\theta}^{\dot{\beta}}\partial_m \quad , \\ \overline{D}_{\dot{\alpha}} &= -\overline{\partial}_{\dot{\alpha}} \quad , \end{aligned} \tag{58}$$

where of course here ∂_m is a partial derivative with respect to y^m rather than x^m.

E N=1 vector superfields

Another way to obtain an irreducible representation is by means of the reality condition

$$\Phi(x,\theta,\bar{\theta}) = \Phi^\dagger(x,\theta,\bar{\theta}) \quad , \tag{59}$$

or, in components:

$$\begin{aligned} f = f^* \quad , \qquad \bar{\chi} = \phi^* \quad , \qquad m = n^* \quad , \\ v_m = v_m^* \quad , \qquad \bar{\lambda} = \psi^* \quad , \qquad d = d^* \quad . \end{aligned} \tag{60}$$

this is the vector superfield, called $V(x, \theta, \overline{\theta})$.

IV SUPERSYMMETRIC MATTER ACTIONS

One of the characteristics of supersymmetry is that, for each matter field a complete chiral superfield must be introduced. It is not possible to have a big supermultiplet, which contains the whole particle zoo inside. This has the consequence of the well-known proliferation of s-particles, the supersymmetric partners of known matter particles, which constitutes one of the major phenomenological shortcomings of supersymmetry. On the other side, this seeming problem is one of the theoretical great advantages of supersymmetry, from which an underlining geometrical structure arises, as well as important holomorphy properties.

Let us consider the superfield

$$\begin{aligned}\Phi(y,\theta) = {} & A(x) + \sqrt{2}\theta\psi(x) + \theta\theta F(x) + \\ & i\theta\sigma^m\bar{\theta}\partial_m A(x) + \frac{i}{\sqrt{2}}(\theta\theta)\partial_m\psi(x)\sigma^m\bar{\theta} - \frac{1}{4}(\theta\theta)(\bar{\theta}\bar{\theta})\Box A(x) \quad . \end{aligned} \tag{61}$$

and its complex conjugated

$$\begin{aligned}\Phi^\dagger(y^\dagger,\bar{\theta}) = {} & A^\dagger(x) + \sqrt{2}\bar{\theta}\bar{\psi}(x) + \bar{\theta}\bar{\theta}F^\dagger(x) - \\ & i\theta\sigma^m\bar{\theta}\partial_m A^\dagger(x) + \frac{i}{\sqrt{2}}(\bar{\theta}\bar{\theta})\partial_m\bar{\psi}(x)\bar{\sigma}^m\theta - \frac{1}{4}(\theta\theta)(\bar{\theta}\bar{\theta})\Box A^\dagger(x) \quad . \end{aligned} \tag{62}$$

Thus, the coefficient of the highest θ power of $\Phi\Phi^\dagger$ will be given by

$$\Phi^\dagger\Phi\Big|_{\theta\theta\bar{\theta}\bar{\theta}} = -\frac{1}{4}\Box A^* A - \frac{1}{4}A^*\Box A + F^*F + \frac{1}{2}\partial^m A^*\partial_m A + \frac{i}{2}\psi\sigma^m\partial_m\bar{\psi} - \frac{i}{2}\bar{\psi}\bar{\sigma}^m\partial_m\psi, \tag{63}$$

which contains the usual kinetic terms for $A(x)$, ψ and $\bar{\psi}$, the field $F(x)$ appears without derivatives. Further, it is easy to see that the θ^2 term of any power of $\Phi(y,\theta)$, will be a function of $A(y)$, $\psi(y)$ and $F(y)$, without y derivatives.

The action for a single massive chiral superfield with Yukawa interactions, the Wess-Zumino action, is given by

$$\int d^4x \int d^4\theta\ \Phi^\dagger\Phi - \int d^4y \left[\int d^2\theta\ (\frac{1}{2}m\Phi^2 + \frac{1}{3}g\Phi^3) + \text{h.c.}\right] \quad , \tag{64}$$

which after the fermionic variables integration contains the right kinetic terms for the fields A, ψ and $\bar{\psi}$ and the potential terms

$$\int d^4x \left[F^*F - (mAF + gA^2F - \frac{1}{2}m\psi\psi - g\psi\psi A + \text{h.c.})\right] \tag{65}$$

Which contains no derivatives acting on $F(x)$, i.e. $F(x)$ is an auxiliary field which can be eliminated by its equations of motion,

$$\frac{\delta \mathcal{L}}{\delta F} = F^* - mA - gA^2 = 0 \quad ,$$
$$\frac{\delta \mathcal{L}}{\delta F^*} = F - mA^* - gA^{*2} = 0 \quad . \tag{66}$$

Introducing the solution of these equations back into (64), the bosonic part of the Wess-Zumino action is just

$$\int d^4x \, [\partial^m A^* \partial_m A - V(A, A^*)] \quad , \tag{67}$$

where the scalar potential $V(A, A^*)$ is given by:

$$V(A, A^*) = |F|^2 = (mA^* + gA^{*2})(mA + gA^2) \quad . \tag{68}$$

The Wess-Zumino action is the most general (power counting) renormalizable action with one massive (Majorana) fermionic field.

A more general action is given by

$$\int d^4x \int d^4\theta \; \Phi^\dagger \Phi - \int d^4x \left[\int d^2\theta \; P(\Phi) + \text{h.c.} \right] \quad , \tag{69}$$

where the superpotential $P(\Phi)$ is a holomorphic function of Φ, i.e. it depends only on Φ, not on $\Phi^\dagger$. This fact insures, as mentioned, that there are no space-time derivatives.

The θ expansion of this superfield can be easily seen to be,

$$P(\Phi) = P(A) + \sqrt{2}\psi \frac{\partial P(A)}{\partial A} + \theta\theta \left[F \frac{\partial P(A)}{\partial A} - \frac{1}{2}\psi\psi \frac{\partial^2 P(A)}{\partial A^2} \right] , \tag{70}$$

where $P(A) \equiv P(\Phi = A)$.

The kinetic term is the same and the auxiliary field F can be eliminated in a similar way as in the Wess-Zumino model, resulting the positive definite scalar potential

$$V_F(A, A^*) = |F|^2 = \left| \frac{\partial P(A)}{\partial A} \right|^2 \quad . \tag{71}$$

V SUPERSYMMETRIC GAUGE THEORIES

A Abelian gauge transformations

Let $\Lambda(y, \theta)$ be a chiral superfield, then $\Lambda + \Lambda^\dagger$ and $i(\Lambda - \Lambda^\dagger)$ are special cases of vector superfields. For the second one, in components we have:

$$i(\Lambda - \Lambda^\dagger) = i(A - A^*) + i\sqrt{2}\theta\psi - i\sqrt{2}\bar{\theta}\bar{\psi} + i\theta\theta F - i\bar{\theta}\bar{\theta}F^* - \theta\sigma^m\bar{\theta}\partial_m(A + A^*)$$
$$-\frac{1}{\sqrt{2}}(\theta\theta)\bar{\theta}\bar{\sigma}^m\partial_m\psi + \frac{1}{\sqrt{2}}(\bar{\theta}\bar{\theta})\theta\sigma^m\partial_m\bar{\psi} - i\frac{1}{4}(\theta\theta)(\bar{\theta}\bar{\theta})\Box(A - A^*) \quad . \tag{72}$$

If we compare it with the vector superfield, we see that it contains the correct term for an (abelian) gauge transformation for the vector component,

$$\begin{aligned} v_m &\to v_m + \partial_m\alpha \quad ; \\ \alpha &= -(A + A^*) \quad . \end{aligned} \tag{73}$$

Thus, we see that we can define the superfield analog of an infinitesimal abelian gauge transformation to be

$$V \to V + i(\Lambda - \Lambda^\dagger) \quad , \tag{74}$$

from which the infinitesimal transformations for the other components (after some field redefinitions) can be obtained,

$$\begin{aligned} f' &= f + i(A - A^*) \\ \chi' &= \chi + i\psi, \quad \bar{\chi}' = \bar{\chi} - i\bar{\psi} \\ m' &= m + iF, \quad m^{*\prime} = m^* - iF^* \\ \psi' &= \psi, \quad \bar{\psi}' = \bar{\psi} \\ d' &= d. \end{aligned}$$

We see that the fields f, χ, $\bar{\chi}$ and m can be eliminated by a gauge transformation, called Wess-Zumino gauge. In this case the vector superfield reduces to

$$V_{WZ} = -\theta\sigma^m\bar{\theta}v_m + i(\theta\theta)\bar{\theta}\bar{\lambda} - i(\bar{\theta}\bar{\theta})\theta\lambda + \frac{1}{2}(\theta\theta)(\bar{\theta}\bar{\theta})D \quad . \tag{75}$$

This gauge fixing uses only the imaginary part of the field A and lets as a remnant the abelian gauge freedom of the vector (gauge) component $v_m \to v_m - (A + A^*)$. It is clear that this gauge fixing is not respected by supersymmetry transformations, although field dependent supersymmetry transformations can be defined, which preserve it.

The construction of gauge invariant supersymmetric actions, needs of gauge invariant superfields with the right kinetic terms. The right superfields are the gauge invariant spinorial chiral field strength superfield W_α, and its hermitic conjugated $\overline{W}_{\dot{\alpha}}$, given by

$$W_\alpha = -\frac{1}{4}(\overline{D}\overline{D})D_\alpha V(x, \theta, \bar{\theta}) \qquad \overline{W}_{\dot{\alpha}} = -\frac{1}{4}(DD)\overline{D}_{\dot{\alpha}}V(x, \theta, \bar{\theta}). \tag{76}$$

Their gauge invariance turns out from the fact that the covariant derivatives in (76) kill the gauge transformation superfields. Indeed,

$$(\overline{DD})D(\Lambda + \Lambda^\dagger) = (\overline{DD})D_\alpha\Lambda = \overline{D}^{\dot\beta}\{\overline{D}_{\dot\beta}, D_\alpha\}\Lambda = 0. \tag{77}$$

The superfield strengths have the important property to be (anti)chiral, as can be seen from the fact that the covariant derivatives $\overline{D}_{\dot\alpha}$ (or D_α) anticommute among them (47) and their third power always vanishes.

The fact that the superfields (76) originate both from the same real superfield, has as a consequence the reality condition,

$$\overline{D}_{\dot\alpha}\overline{W}^{\dot\alpha} = D^\alpha W_\alpha \quad , \tag{78}$$

which can be geometrically interpreted as a Bianchi identity.

The gauge invariance of the field strength superfield, allows to simplify the computation of its component expansion, which can be done in the WZ gauge. In the chiral base it it turns out that,

$$W_\alpha = -i\lambda_\alpha + \theta_\alpha D - \frac{i}{2}(\sigma^m\bar\sigma^n\theta)_\alpha(\partial_m v_n - \partial_n v_m) + (\theta\theta)\sigma^m_{\alpha\dot\beta}\partial_m\bar\lambda^{\dot\beta} \quad , \tag{79}$$

which is an irreducible off-shell multiplet with 8 real components.

With this data, the supersymmetric abelian gauge action can immediately written as

$$\frac{1}{4}\int d^4x d^4\theta(WW + \bar W\bar W) = \int d^4x(\frac{1}{2}d^2 - \frac{1}{4}F^{mn}F_{mn} - i\lambda\sigma^m\partial_m\bar\lambda) \tag{80}$$

For the study of topological properties the knowledge of the finite form of the transformation law (74) is needed, and is given by,

$$\mathrm{e}^V \to \mathrm{e}^{-i\Lambda^\dagger}\mathrm{e}^V\mathrm{e}^{i\Lambda} \quad , \tag{81}$$

This allows us to write an equivalent form for the field strengths,

$$W_\alpha = -\frac{1}{8}(\overline{DD})\mathrm{e}^{-2V}D_\alpha\mathrm{e}^{2V} \quad , \qquad \overline{W}_{\dot\alpha} = \frac{1}{8}(DD)\mathrm{e}^{2V}\overline{D}_{\dot\alpha}\mathrm{e}^{-2V} \quad . \tag{82}$$

B Nonabelian case

Gauge fields can be usually written as Lie algebra valued fields, $A = A_a T^a$, where T^a are the generators of the Lie algebra of the symmetry (compact) group which can be parameterized to satisfy

$$[T^a, T^b] = if^{abc}\, T^c \quad , \qquad \mathrm{tr}\, T^a T^b = \delta^{ab} \quad . \tag{83}$$

Thus, we write $V = V_a T^a$, and the abelian transformation can be generalized to

$$\mathrm{e}^V \to \mathrm{e}^{-i\Lambda_a^\dagger T^a}\mathrm{e}^V\mathrm{e}^{i\Lambda_a T^a} \quad , \tag{84}$$

where Λ_a are chiral superfields put together to a Lie algebra valued chiral field $\Lambda \equiv \Lambda_a T^a$

To find the infinitesimal nonabelian transformation, the Baker-Campbell-Hausdorff formula can be applied. However it is better to observe that the first terms of the transformation law are the same as the ones of the abelian case,

$$V' = V + i(\Lambda - \Lambda^\dagger) + \cdots \tag{85}$$

Thus the Wess Zumino gauge can still be applied, in which the vector superfield has the form

$$V_{WZ} = -\theta\sigma^m\overline{\theta}v_m + i(\theta\theta)\overline{\theta}\overline{\lambda} - i(\overline{\theta}\overline{\theta})\theta\lambda + \frac{1}{2}(\theta\theta)(\overline{\theta}\overline{\theta})D \quad , \tag{86}$$

where all components are Lie algebra valued, i.e. they transform in the adjoint representation. Thus $V^3 = 0$, and the transformation law can be evaluate to give

$$\delta V = i(\Lambda - \Lambda^\dagger) - \frac{i}{2}[(\Lambda + \Lambda^\dagger), V] \quad . \tag{87}$$

This implies the usual nonabelian gauge transformation for the gauge field $v_m(x)$. The fields $\lambda(x)$, and $D(x)$ are not anymore invariant, they transform in the adjoint representation.

The field strengths W_α and $\overline{W}_{\dot{\alpha}}$ are given by the same expressions as in the abelian case (82). However they are not invariant, a substitution of the transformation law of the vector superfield (84) gives the transformation laws

$$W_\alpha \to e^{-i2\Lambda}W_\alpha e^{i2\Lambda} \quad , \qquad \overline{W}_{\dot{\alpha}} \to e^{-i2\Lambda^\dagger}\overline{W}_{\dot{\alpha}}e^{i2\Lambda^\dagger} \quad . \tag{88}$$

The computation of W_α and $\bar{W}_{\dot{\alpha}}$ can be done in the Wess Zumino gauge (86), with the result

$$W_\alpha = -\frac{1}{4}(\overline{D}\overline{D})D_\alpha V_{WZ} + \frac{1}{2}(\overline{D}\overline{D})V_{WZ}D_\alpha V_{WZ} - \frac{1}{4}(\overline{D}\overline{D})D_\alpha, V_{WZ}^2 \tag{89}$$

with the component expansion,

$$W_\alpha = -i\lambda_\alpha(y) + \theta_\alpha D(y) - \sigma_\alpha^{mn\beta}\theta_\beta F_{mn}(y) + (\theta\theta)\sigma^m_{\alpha\dot{\beta}}\nabla_m \bar{\lambda}^{\dot{\beta}}(y) \quad , \tag{90}$$

where

$$F_{mn} = \partial_m v_n - \partial_n v_m + i[v_m, v_n], \tag{91}$$

is the Yang-Mills field strength and

$$\nabla_m \bar{\lambda}^{\dot{\beta}} = \partial_m \bar{\lambda}^{\dot{\beta}} + i[v_m, \bar{\lambda}^{\dot{\beta}}]. \tag{92}$$

C N=1 supersymmetric Yang-Mills theory

The Yang-Mills action can be written observing that the W_α superfield transforms in the adjoint representation, so that an invariant can be obtained taking the trace,

$$\begin{aligned}&\frac{1}{2}\int d^4x \int d^2\theta \, \mathrm{tr}\, W^\alpha W_\alpha \\ &= \int d^4x \, \mathrm{tr}\left[-\frac{1}{4}F_{mn}F^{mn} - \frac{i}{8}\epsilon^{mnpq}F_{mn}F_{pq} - i\lambda\sigma^m \nabla_m \bar{\lambda} + \frac{1}{2}D^2\right].\end{aligned} \tag{93}$$

For matter couplings, we have that the chiral matter superfields transform under the usual transformation law,

$$\Phi \to \mathrm{e}^{-i\Lambda}\Phi \tag{94}$$

where the transformation parameters are Lie algebra valued chiral superfields. Opposite to the usual case, where the parameters of the group are real, here they are given by complex quantities. Thus, the exponential on the r.h.s. of (86) is not unitary and the product $\Phi^\dagger\Phi$ is not gauge invariant. However from the transformation law of the vector superfield,

$$\mathrm{e}^V \to \mathrm{e}^{-i\Lambda^\dagger}\mathrm{e}^V\mathrm{e}^{i\Lambda} \quad , \tag{95}$$

we see that under a gauge transformation,

$$\Phi^\dagger\mathrm{e}^V \to \left(\Phi^\dagger\mathrm{e}^V\right)\mathrm{e}^{i\Lambda} \tag{96}$$

has the correct transformation properties to make an invariant when multiplied to the right with Φ. This product gives us the generalized kinetic term for the nonabelian gauged version of the Wess-Zumino action:

$$\begin{aligned}&\int d^4x \int d^4\theta \; \Phi^\dagger \mathrm{e}^{2V}\Phi \\ &-\int d^4x \left[\int d^2\theta \, (\frac{1}{2}m_{ij}\Phi^i\Phi^j + \frac{1}{3}g_{ijk}\Phi^i\Phi^j\Phi^k) + \text{h.c.}\right] \quad ,\end{aligned} \tag{97}$$

where Φ^i are the components of Φ, and m_{ij} and g_{ijk} are supposed to be tensors in such a way that the contractions in (97) are singlets and the action is gauge invariant.

Let us now consider the sum of the actions (93) and (97). The kinetic term in this action contains the terms trA^*DA and trD^2. There is no more dependence on the fields D, which thus are auxiliary and can be eliminated by their equations of motion, which turn out to be,

$$D^a = -\frac{1}{2}if^{abc}A_bA_c^* = -\frac{1}{2}[A, A^*]^a \quad , \tag{98}$$

This implies that there is a new contribution to the scalar potential in the coupled Yang-Mills-Wess-Zumino model,

$$V_D = \frac{1}{2} tr D^2 = \frac{1}{8} tr \left([A, A^*]\right)^2 \quad . \tag{99}$$

So the complete scalar potential is the sum of positive definite F and D-term contributions:

$$V(A^i, A^{i*}) = V_F + V_D = |F|^2 + \frac{1}{2} tr D^2 \tag{100}$$

If we forget about renormalizablity, the most general N=1 supersymmetric gauge invariant action is:

$$\begin{aligned} \mathcal{L} = & \int d^4\theta \; K(\Phi^i, \Phi^{j\dagger} e^{2V}) - \left[\int d^2\theta \; P(\Phi^i) + \text{h.c.} \right] \\ & + \frac{1}{8\pi} \text{Re} \left[\tau \int d^4x \int d^4\theta \; \text{tr} \, f(\Phi^i) W^\alpha W_\alpha \right] \quad , \end{aligned} \tag{101}$$

where $f(\Phi^i)$ is a new holomorphic function called the gauge kinetic function and τ is a complex parameter.

VI SPONTANEOUS SUPERSYMMETRY BREAKING

The spontaneous breaking of supersymmetry is studied in general along the same lines as for usual symmetries, however supersymmetry imposes additional conditions.

From the anticommutators

$$\{Q_\alpha, \bar{Q}_{\dot{\alpha}}\} = 2\sigma^m P_m = 2 \begin{pmatrix} P_0 + P_3 & P_1 - iP_2 \\ P_1 + iP_2 & P_0 - P_3 \end{pmatrix}, \tag{102}$$

we see that

$$H = \frac{1}{4}(\{Q_1, \bar{Q}_1\} + \{Q_3, \bar{Q}_3\}) = \frac{1}{4}(Q_1\bar{Q}_1 + \bar{Q}_1 Q_1 + Q_3\bar{Q}_3 + \bar{Q}_3 Q_3). \tag{103}$$

Therefore

$$< \Psi | H | \Psi > \geq 0. \tag{104}$$

If supersymmetry is spontaneously broken, the vacuum is degenerated under the action of supersymmetry, that is $Q|0> \neq 0$. Thus in this case the vacuum energy cannot be zero and it must be positive. Opposite, if the vacuum energy is nonzero, then from (103) we see that there is a degenerated vacuum and supersymmetry is spontaneously broken.

The computations to determine spontaneous breaking of supersymmetry are done with the scalar potential, i.e. the part of the potential that contains only scalar fields. The reason is that only scalars can have nonzero vaccum expectation value because of Lorenz invariance of the vacuum

$$\delta_{Lorenz} < \psi > = < J\psi - \psi J > = 0. \tag{105}$$

Let us consider a model of the type considered above, with n scalar superfields Φ_i

$$I = \int d^4x \int d^4\theta \; \Phi_i^\dagger \Phi_i + \int d^4x \int d^4\theta \; [P(\Phi_i) + h.c.] \,, \tag{106}$$

whose potential part can be obtained from the expansion (70)

$$\int d^4x \left[F_i^* F_i + F_i \frac{\partial P(A)}{\partial A_i} - \frac{1}{2} \frac{\partial^2 P(A)}{\partial A_i \partial A_j} \psi_i \psi_j + h.c. \right] . \tag{107}$$

The scalar potential is given by the first two terms and the last term will give the fermion masses. The equations of the auxiliary fields are

$$F_i^* + \frac{\partial P(A)}{\partial A_i} = 0 \quad , \qquad F_i + \frac{\partial P^*(A^*)}{\partial A_i^*} = 0 \tag{108}$$

which substituted back into the action, give the positive definite scalar field potential

$$V(\Phi) = \sum_i |F_i|^2 = \sum_i \left| \frac{\partial P(A)}{\partial A_i} \right|^2 \tag{109}$$

This potential vanishes if and only if there is a solution for the system of equations for the vacuum expectation values $a_i = < A_i >$,

$$\frac{\partial P(a)}{\partial a_i} = 0. \tag{110}$$

As there are as many equations as independent variables, in general we can expect that there is a solution. However, there are potentials that do not have solution. Let us take

$$P(a, b) = \sum b_i P_i(a_n), \tag{111}$$

in this case the equations (110) are given by

$$P_i(a_n) = 0 \quad , \qquad \sum_n b_i \frac{\partial P_i}{\partial a_n} = 0. \tag{112}$$

The second equations can be always solved by $b_i = 0$. In this case, if the number of variables a_n is smaller than the number of b_i, then the first system of equations is overdetermined and in general it has no solution and supersymmetry is broken. This is a generalization of the model proposed by O'Rafeartaigh, which gives the simplest version of it. Basically, let us take three superfields, A, B_1 and B_2, and

$$P_1(A) = A^2 - a_0, \qquad P_2(A) = A. \tag{113}$$

Then, after setting $B_1 = B_2 = 0$, the two equations to be solved are

$$\begin{aligned} A^2 - a_0 &= 0, \\ A &= 0. \end{aligned}$$

Obviously there is no solution unless $a_0 = 0$, and supersymmetry is broken.

There is a general characteristic of spontaneous symmetry breaking of bosonic symmetries, given by the existence of massless bosons as indicated by the Goldstone theorem. In the case of supersymmetry, there is a generalization of this theorem. Let us look for the minimum of the scalar potential,

$$\sum_n \frac{\partial^2 P(a)}{\partial a_m \partial a_n} \frac{\partial P^*(a^*)}{\partial a_n^*} + \sum_n \frac{\partial^2 P^*(a^*)}{\partial a_m^* \partial a_n^*} \frac{\partial P(a)}{\partial a_n} = 0, \tag{114}$$

which, due to the independence of the variables a and a^*, reduce to

$$\sum_n \frac{\partial^2 P(a)}{\partial a_m \partial a_n} \frac{\partial P^*(a^*)}{\partial a_n^*} = 0. \tag{115}$$

If there is supersymmetry breaking, then this equation means that the matrix

$$\mathcal{M} = \frac{\partial^2 P(a)}{\partial a_m \partial a_n} \tag{116}$$

has at least one eigenvector with zero eigenvalue, which tells us, as can be seen from the action (107), that there is at least one massless fermion, called Goldstino.

Further, if we consider the transformation law of fermions in this model, they are of the type

$$\delta_\xi \psi_n = i\sqrt{2}\sigma^m \bar{\xi} \partial_m A_n + \sqrt{2}\xi F_n \tag{117}$$

if we take the vacuum expectation value of this expression, the first term of the r.h.s. vanises and we have

$$< \delta_\xi \psi_n > = \sqrt{2}\xi < F_n >, \tag{118}$$

which gives the other distinctive characteristic of Goldstinos, and of Goldstones, that their transformation laws are inhomogeneous.

A Gauge theories

In the case of abelian gauge theories, the highest component d of the vector superfield is gauge invariant and can be added to the gauge invariant action as the superspace integral of the vector superfield V,

$$\int d^4x \int d^4\theta \, \frac{1}{4} \mathrm{tr} \left(W^\alpha W_\alpha + \bar{W}^\alpha \bar{W}_\alpha \right) + \int d^4x \int d^4\theta \left(\Phi^{i\dagger} e^{2eV} \Phi^i + 2\kappa V \right) + \int d^4x \, P(\Phi_i) \quad . \tag{119}$$

In this case the bosonic potential will be given by

$$\int d^4x \left[\frac{1}{2} d^2 + \kappa d + \frac{e_i}{2} d A_i^* A_i + F_i^* F_i + F_i \frac{\partial P(A)}{\partial A_i} + h.c. \right] \tag{120}$$

where e_i are the charges of A_i. The field equations for the auxiliary fields are

$$d + \kappa + e_i A_i^* A_i = 0$$
$$F_i^* + \frac{\partial P(A)}{\partial A_i} = 0 \, , \quad F_i + \frac{\partial P^*(A^*)}{\partial A_i^*} = 0 \tag{121}$$

and the scalar potential will be

$$V = \frac{1}{2} d^2 + F_i^* F_i, \tag{122}$$

where d and F_i are substituted from their equations. This is a sum of positive quantities and it vanishes if each of them vanish. Thus the supersymmetry breaking conditions are in this case

$$\kappa + e_i A_i^* A_i = 0 \quad , \quad \frac{\partial P(A)}{\partial A_i} = 0 \tag{123}$$

For example, if we take $P = m\Phi_1\Phi_2$, then the second equations above are $mA_i = 0$ and the system has no solution. More general, it can be shown that if the second equations have a solution, then the first equation does not have unless $\kappa = 0$. Therefore, in this case supersymmetry breaking is induced by a nonvanishing v.e.v. $< d >= \kappa$.

The Goldstone theorem is valid here as well, in this case the Goldstino is given by the λ spinor of the vector superfield as can be seen from its transformation law

$$\delta_\xi \lambda = i\xi d + \sigma^{mn} v_{mn}. \tag{124}$$

Thus, if d has nonvanishing v.e.v., the field λ gets the inhomogeneous transformation law of Goldstone particles.

VII $N = 1$ SUPERGRAVITY

Supergravity is obtained by gauging the supersymmetry transformations, by making them local, $\xi_m \to \xi_m(x)$, $\xi_\alpha \to \xi_\alpha(x)$. Of course suitable gauge fields have to be introduced. This formulation requires permanent consistency checks in order to have supersymmetry. Superspace allows formulating supergravity insuring automatically this consistency, in this case superspace locality is required for the transformation parameters. In order to understand how supergravity can be formulated in this way, we will show how its work for general relativity.

A Vierbein formulation of general relativity

Let us take a scalar field $\phi(x)$, on a D dimensional manifold $\mathcal{M}$. Under infinitesimal general coordinate transformations, i.e. local translations, $x \to x + \xi$, it transforms as,

$$\delta_\xi \phi = -\xi^m \partial_m \phi. \tag{125}$$

Derivatives are not covariant under this transformation, thus covariant derivatives $D_m \equiv \partial_m + h_m{}^n \partial_n$ are required, where $h_m{}^n$ are the translational gauge fields. It is easy to see that covariance

$$\delta_\xi D_m \phi = -\xi^n \partial_n D_m \phi, \tag{126}$$

requires that the gauge field transforms as

$$\delta_\xi h_m{}^n = -\xi^r \partial_r h_m{}^n + h_m{}^r \partial_r \xi^n + \partial_m \xi^n. \tag{127}$$

Let us now define the inverse vierbein field $e_m{}^n = \delta_m^n + h_m{}^n$, it transforms as

$$\delta_\xi e_m{}^n = -\xi^r \partial_r e_m{}^n + e_m^r \partial_r \xi^n. \tag{128}$$

This is the transformation law of a contravariant vector with respect to the upper index n, the lower index m does not transform and for this reason is usually denoted by a different type of character. We will write $e_a{}^m$, where the lower indices take as well the values $a = 0, 1, 2, 3$, but have a 'scalar' character under coordinate reparametrizations, these indices are called local indices. Thus, covariant derivatives will be written as $D_a \phi = e_a{}^m \partial_m \phi$, consistently with the transformations (126). The vierbein field $e_m{}^a$ is defined by $e_m{}^a e_a{}^n = \delta_m^n$ and $e_a{}^m e_m{}^b = \delta_a^b$. Its transformation law corresponds to the transformation law of a covariant vector under the lower index m,

$$\delta_\xi e_m{}^a = -e_m{}^b \delta_\xi e_b{}^n e_n{}^a = -\xi^r \partial_r e_m{}^a - \partial_m \xi^n e_n{}^a. \tag{129}$$

or

$$e'_m{}^a(x') = \frac{\partial x^n}{\partial x'^m} e_n{}^a(x). \tag{130}$$

Covariant vectors are elements of the tangent space[1] to the manifold $\mathcal{M}$. The tangent space can be seen as the space generated on each point of $\mathcal{M}$ by the tangent vectors to the curves in $\mathcal{M}$. A basis on it is given by the differentials dx^m. This basis can be transformed by the vierbein to a new basis $e^a = dx^m e_m{}^a$, which can be chosen to be orthogonal on each point of $\mathcal{M}$. In this case a local flat metric η_{ab} can be introduced on the tangent space, and Lorentz transformations on its elements can be defined in such a way that the scalars $u^a v_a$ are Lorentz invariant. Thus, Lorentz transformations are local and act on local indices

$$v'_a(x) = \Lambda_a{}^b(x) v_b(x) \tag{131}$$

By means of the vierbein, any contravariant tensor can be transformed into a Lorentz vector, $v^m \to v^a = v^n e_n{}^a$, as well as for covariant tensors $v_m \to v_a = e_a{}^n v_n$. As well, the flat metric is transformed by the vierbein into the manifold metric:

$$g_{mn} = e_m{}^a e_n{}^b \eta_{ab} \tag{132}$$

The local Lorentz transformations require the introduction of new gauge fields in order to get covariant derivatives for local Lorentz tensors

$$D_m v_a = \partial_m v_a - \omega_{ma}{}^b v_b, \tag{133}$$

where $\omega_{ma}{}^b = \omega_m{}^{cd}(L_{cd})_a{}^b$ and $(L_{cd})_a{}^b = \frac{1}{2}(\eta_{ca}\delta^b_d - \delta^b_c \eta_{da})$ are the generators of the Lorentz transformations, antisymmetric separately in cd and in ab. Thus, the covariant derivatives under coordinate reparametrizations and Lorentz transformations will be given by $D_a v_b = e_a{}^m D_m v_b$. Consistency requires that the covariant derivatives of world vectors are defined by $D_m v_n = e_n{}^a D_m v_a$, which can be also written in terms of the affine connection,

$$\begin{aligned} D_m v_n &= e_n{}^a(\partial_m v_a - \omega_{ma}{}^b v_b) \\ &= e_n{}^a[\partial_m(e_a{}^p v_p) - \omega_{ma}{}^b e_b{}^p v_p] \\ &= \partial_m v_n - [-e_n{}^a \partial_m e_a{}^p + \omega_{mn}{}^p] v_p = \partial_m v_n - \Gamma_{mn}{}^p v_p. \end{aligned} \tag{134}$$

If the torsion vanishes, as is the case for Einstein theory, this relation coincides with the usual definition of the affine connection. In fact, the torsion is the tensor part of the affine connection and can be defined by the commutator of two covariant derivatives on a scalar field,

$$[D_m, D_n]\,\phi = -(\Gamma_{mn}{}^p - \Gamma_{nm}{}^p) D_p \phi = -T_{mn}{}^p D_p \phi. \tag{135}$$

---

[1] In fact, in the mathematical literature it is called cotangent space, the tangent space is the one of contravariant vectors

If this commutator is applied on a vector field, besides the torsion field, the curvature tensor arises

$$[D_m, D_n]\, v_a = -T_{mn}{}^b D_b v_a - R_{mna}{}^b v_b. \tag{136}$$

The curvature tensor can be also seen as the field strength of the Lorentz or spin connection $\omega_{ma}{}^b$,

$$R_{mn}{}^{ab} = \partial_m \omega_{na}{}^b - \partial_n \omega_{ma}{}^b - \omega_{ma}{}^c \omega_{nc}{}^b + \omega_{na}{}^c \omega_{mc}{}^b. \tag{137}$$

The Jacobi identities that commutators satisfy, originate identities among the curvature and torsion fields, called Bianchi identities,

$$\sum_{cyclic\{abc\}} \left(R_{abc}{}^d - D_a T_{bc}{}^d - T_{ab}{}^e T_{ec}{}^d\right) = 0$$
$$\sum_{cyclic\{abc\}} \left(D_a R_{bcd}{}^e - T_{ab}{}^f R_{fcd}{}^e\right) = 0. \tag{138}$$

Now we can write invariant actions under general coordinate reparametrizations and local Lorenz transformations. The Lagrangian must be an invariant density, which transforms, under general reparametrizations and local Lorentz transformations as

$$\mathcal{L}'(x') = det\left(\frac{\partial x}{\partial x'}\right) \mathcal{L}(x), \tag{139}$$

in such a way that the integral over it is invariant. An invariant density is the determinant of the vierbein $e = det(e_m{}^a) = \sqrt{-det(g_{mn})}$, whose transformation law can be obtained from (130). Thus, lagrangians are obtained multiplying e by any scalar field, ensuring that the result is an invariant density again. Einstein gravity, in the Palatini formulation, is obtained in this way, $L_E = e\mathcal{R}$, where $\mathcal{R} = R_{ab}{}^{ab}$ is the curvature scalar. In this formulation the spin connection is an independent field and if the action is variated with respect to it, the resulting equation is that torsion vanishes. The solution to this equation gives the spin connection in terms of the vierbein,

$$\omega_{mnp} = e_{ma}(\partial_n e_p{}^a - \partial_p e_n{}^a) - e_{na}(\partial_m e_p{}^a - \partial_p e_m{}^a) - e_{pa}(\partial_m e_n{}^a - \partial_n e_m{}^a). \tag{140}$$

which, when substituted in the afinne connection from (134), gives the usual result in terms of the metric.

B Local supersymmetry

In this and the next sections, a very short account of some basic topics of supergravity will be given, for a thorough account and for notations, we refer the reader to [1].

As mentioned, supergravity can be obtained requiring invariance under local superspace transformations, $\xi^A \to \xi^A(z^M) \equiv \xi^A(x^m, \theta, \bar{\theta})$, where $\xi^A = (\xi^a, \xi^\alpha, \bar{\xi}_{\dot{\alpha}})$ are the supersymmetry parameters and $z^M = (x^m, \theta_\mu, \bar{\theta}^{\dot{\mu}})$ are superspace coordinates. Thus there are general transformations for the fermionic coordinates, which are treated on the same footing as the bosonic ones. The tangent space will contain as well fermionic coordinates, which will transform under the spin-1/2 Lorentz representations of chapter II. Thus, the vierbein (four legs in german) is substituted by the vielbein (many legs) $E_M{}^A(z)$, and the local Lorentz indices A transform under local Lorentz vector and spinor representations, $L_A{}^B = \{L_a{}^b, L_\alpha{}^\beta, L^{\dot{\alpha}}{}_{\dot{\beta}}\}$.

$$E'_M{}^A(z) = E_M{}^B(z) L_B{}^A(z). \tag{141}$$

The covariant derivatives are,

$$\mathcal{D}_A \Phi_B = E_A{}^M (\partial_M \Phi_B - \omega_{MB}{}^C \Phi_C). \tag{142}$$

The commutation properties of superspace can be written as

$$z^M z^N = (-1)^{mn} z^N z^M \tag{143}$$

, where $m = 0$ if M is bosonic and $m = 1$ if M is fermionic. In this case we have, for example for a two tensor,

$$\mathcal{D}_A \Phi_{BC} = E_A{}^M \left[\partial_M \Phi_{BC} - \omega_{MB}{}^D \Phi_{DC} - (-1)^{b(m+c+d)} \omega_{MC}{}^D \Phi_{BD}\right], \tag{144}$$

because in the last term of the r.h.s. the connection $\omega_{MC}{}^D$ has to 'go through' the index B.

Superspace invariant densities are obtained from the vielbein. The new fact is that the Jacobian corresponding to superspace coordinate reparametrizations, is given by the the superdeterminant or Berezinian [6] of $\partial z'^M / \partial z^N$. For a matrix

$$\mathcal{M} = \left(M_A{}^B\right) = \begin{pmatrix} M_a{}^b & M_a{}^\beta \\ M_\alpha{}^b & M_\alpha{}^\beta \end{pmatrix}, \tag{145}$$

the superdeterminant is defined by

$$Sdet(\mathcal{M}_A{}^B) = \frac{det(M_a{}^b - M_A{}^\alpha M_\alpha^{-1\beta} M_\beta{}^b)}{det(M_\alpha{}^\beta)}. \tag{146}$$

Thus, the invariant density is given by the superdeterminant of the vielbein. In order to obtain the supergravity action, we need a suitable scalar superfield, which cannot be obtained as easy as for Einstein theory. In order to get it, a couple of steps are in order.

C Minimal supergravity

Supersymmetry is characterized by the introduction of a number of new fields. In our case the vierbein and the spin connection are the lowest components of their corresponding superfields and all other components are new fields, which in particular can have too high spin values. These representations are highly reducible and their reduction is achieved under the imposition of suitable constrictions. These constrictions could be imposed on the vielbein and spin connections superfields, but in order to keep manifest covariance, superfield tensorial constraints must be imposed. The simplest nontrivial tensors constructed from these quantities are torsion and curvature, which however, satisfy generalized Bianchi identities, similar to the bosonic ones (138). The constraints can be guessed by an analysis of the components of the mentioned superfields, which should be eliminated. It turns out that, the constraints which lead to the minimal version of supergravity are,

$$\begin{aligned} &T_{\underline{\alpha}\underline{\beta}}^{\underline{\gamma}} = 0, \qquad T_{\alpha\beta}^{c} = T_{\dot{\alpha}\dot{\beta}}^{c} = 0, \qquad T_{\underline{\alpha}b}^{c} = 0, \qquad T_{ab}^{c} = 0 \\ &T_{\alpha\dot{\beta}}^{c} = 2i\sigma^{c}_{\alpha\dot{\beta}}, \end{aligned} \tag{147}$$

where $\underline{\alpha} = (\alpha, \dot{\alpha})$. These constraints must be compatible with the Bianchi identities, which are solved algebraically to give a total of three superfields R, G_a and $W_{\alpha\beta\gamma}$, satisfying remnant conditions. As a consequence, the torsion and curvature tensors and their covariant derivatives, with local Lorentz indices, can be written in terms of these superfields, the covariant derivatives are obtained by means of the Bianchi identities. From (147), we see that the bosonic torsion vanishes and as mentioned in the preceding section, it has as a consequence that the spin connection can be written in terms of the vierbein, here plus fermionic contributions.

As well as the superfields, the transformation parameters have here a lot of higher components, reflected in a corresponding number of pure gauge degrees of freedom. It is convenient to eliminate as many as possible of these gauge degrees, keeping supersymmetry, and general space-time and Lorentz covariance.

From the generalization of the bosonic transformation laws

$$\begin{aligned} \delta E_M{}^A &= -\partial_M \xi^N E_N{}^A - \xi^L \partial_L E_M{}^A + E_M{}^B L_B{}^A \\ \delta \omega_{MA}{}^B &= -\partial_M L_A{}^B + \omega_{MA}{}^C L_C{}^B - L_A{}^C \omega_{MC}{}^B - \partial_M \xi^N \omega_{NA}{}^B - \xi^L \partial_L \omega_{MA}{}^B, \end{aligned} \tag{148}$$

it can be seen that the first terms can be used to fix the gauge. In the case of the vielbein, the parameter which must be used to fix the gauge is ξ^A, in such a way that the first term in the transformation law, given by $\partial_M \xi^A$, is inhomogeneous. All the higher θ-components of ξ^A and $L_A{}^B$ are used for this gauge fixing, remaining still the gauge freedom of the transformations by $\xi^A(x)$ and $L_A{}^B(x)$. By this gauge fixing the $\theta = 0$ components of $E_\mu{}^a$ and $\omega_{\mu A}{}^B$ are set to zero and $E_\mu{}^\alpha = \delta_\mu^\alpha$. For example, we have $\delta E_\mu{}^a = -\partial_\mu \xi^a + \cdots$, which means that the first θ power of ξ^a can be used to eliminate $E_\mu{}^a(x)$. It turns out, that after the gauge fixing, the only remaining degrees of freedom are $e_m{}^a = E_m{}^a|_{\theta=0}$, $\psi_m{}^\alpha = 2E_m{}^\alpha|_{\theta=0}$, $\bar{\psi}_{m\dot{\alpha}} = 2E_{m\dot{\alpha}}|_{\theta=0}$ and

$\omega_{ma}{}^b(x) = \omega_{ma}{}^b|_{\theta=0}$. Further, from the gauge fixing and the solution of the Bianchi identities, it can be seen that only the $\theta = 0$ components $M(x) = R|_{\theta=0}$ and $b_a(x) = G_a|_{\theta=0}$ are independent. It can be shown that all higher θ components of the vielbein, the spin connection and the superfields R and G_a, and the whole superfield $W_{\alpha\beta\gamma}$, can be written in terms of the graviton $e_m{}^a(x)$, the gravitino $(\psi_m{}^\alpha(x), \bar{\psi}_{m\dot{\alpha}}(x))$, the scalar $M(x)$ and the vector field $b_a(x)$, and their space-time derivatives. These fields constitute the minimal supergravity supermultiplet, which turns out to close under the remaining local supersymmetry transformations. One important result is that the highest component of the superfield R is given by,

$$\begin{aligned} R^{(2)} = & \frac{1}{3} e_a{}^m e_b{}^n R_{mn}{}^{ab} + \frac{2}{3} \bar{\psi}^m \bar{\sigma}^n \psi_{mn} + \frac{1}{12} \varepsilon^{klmn} (\bar{\psi}_k \sigma_l \psi_{mn} + \psi_k \sigma_l \bar{\psi}_{mn}) \\ & - \frac{2}{3} i e_a{}^m \mathcal{D}_m b^a + \frac{4}{9} M M^* + \frac{2}{9} b^2 + \frac{1}{3} \bar{\psi}^2 M - \frac{1}{3} \psi_m \sigma^m \bar{\psi}_n b^n . \end{aligned} \tag{149}$$

Further, the invariant density can be computed and it turns out that to be given by[2]

$$\mathcal{E} = e \left[1 + \frac{i}{2} \theta \sigma^m \bar{\psi}_m - \frac{1}{2} \theta^2 (M^* + \bar{\psi}_m \bar{\sigma}^{mn} \bar{\psi}_n \right] . \tag{150}$$

Therefore, the minimal supergravity action can be written as

$$\begin{aligned} \mathcal{L} &= \int d^4x d^4\theta \mathcal{E} \mathcal{R} \\ &= e \left[-\frac{1}{2} \mathcal{R} - \frac{1}{3} M M^* + \frac{1}{3} b^2 + \frac{1}{2} \varepsilon^{klmn} (\bar{\psi}_k \bar{\sigma}_l \tilde{\mathcal{D}}_m \psi_n - \psi_k \sigma_l \tilde{\mathcal{D}}_m \bar{\psi}_n) \right] . \end{aligned} \tag{151}$$

If the auxiliary fields M and b_a are eliminated, it coincides with the supergravity action of Freedman et.al. [8].

This formalism allows the construction of any matter coupling to supergravity. In order to do that, it is enough to take the global supersymmetric action and to proceed as in the case of gauge theories, substituting the global supersymmetric covariant derivatives by local supersymmetric and Lorentz covariant derivatives and then add it to the supergravity action (151).

REFERENCES

1. J. Wess and J. Bagger, Supersymmetry and Supergravity, 2nd edition, Princeton University Press, Princeton NJ, 1992.
2. Peter West, Introduction to Supersymmetry and Supergravity, 2nd edition, World Scientific, Singapore, 1990.

---

[2] In fact, many of these results are best obtained in the formalism of the 'new Θ-variables', based on a reparametrization of superfields in which the superfield components are obtained from their fermionic covariant derivatives, and which gives a natural framework for a gauge covariant Wess-Zumino gauge, keeping the whole geometric formulation of superspace. [1,7]

3. M. Sohnius, Introducing Supersymmetry, Phys. Rep., 128, 39, (1985).
4. S. Weinberg, The quantum theory of fields: Supersymmetry, Cambridge Univ. Press 2000.
5. Joseph D. Lykken, Introduction to supersymmetry, Tasi Proceedings 1996.
6. B.S. De Witt, Supermanifolds, Cambridge Univ. Press, 1984.
7. Cupatitzio Ramirez, Ann. Phys. N.Y., 186(1988)43.
8. D.Z. Freedman, S. Ferrara and P. Van Nieuwenhuizen, Phys. Rev. D13(1976)3214.

Theories in More than Four Dimensions

Abdel Pérez-Lorenzana

Department of Physics, University of Maryland, College Park, Maryland 20742, USA
Departamento de Física, Centro de Investigación y de Estudios Avanzados del I.P.N.
Apdo. Post. 14-740, 07000, México, D.F., México.

Abstract.

Particle physics models where there are large hidden extra dimensions are currently on the focus of an intense activity. The main reason is that these large extra dimensions may come with a TeV scale for quantum gravity (or string theory) which leads to a plethora of new observable phenomena in colliders as well in other areas of particle physics. Those new dimensions could be as large as millimeters implying deviations of the Newton's law of gravity at these scales. Intending to provide a basic introduction to this fast developing area, we present a general overview of theories with large extra dimensions. We center our discussion on models for neutrino masses, high dimensional extensions of the Standard Model and gauge coupling unification. We discuss the recently proposed technic of splitting fermion wave functions on a tick brane which may solve the problem of a fast proton decay and produce fermion mass hierarchies without invoking extra global symmetries. Randall-Sundrum model and some current trends are also commented.

I INTRODUCTION

New extra dimensions beyond the four of our world could possible exist in Nature. This idea is as old as Kaluza and Klein's work dating back to the 1920's [1]. In modern terms, the idea arose again with the advent of string theories. However, it was conventional to assume that such extra dimensions were compactified to manifolds of small radii with a size of the order of the inverse Planck scale, $\ell_P = M_{P\ell}^{-1} = G_N^{1/2}$ or so, such that they would remain hidden to low energy physics considerations. Thus, beyond the interest of a small community, the study of theories with extra dimensions was almost far away from the scope of many particle physicists.

During the last two years of the XX century, the work on theories in more than four dimensions has increased almost exponentially. The intriguing fact that strongly motivates this renewed interest is the realization of the possibility that extra dimensions as large as millimeters [2] could exist and yet being hidden to the experiments [3–7], but with new effects not so far of being observed. Among the experimental signatures are deviation on the gravitational Newton's law at small

CP562, *Particles and Fields: Ninth Mexican School,* edited by G. Herrera Corral and L. Nellen

distances and a rich phenomenology for collider physics. One of the attractive features of these theories is that they may provide a natural solution to the hierarchy problem. In this scenarios, it is believed that we live on a hypersurface (3-brane) embedded in a higher dimensional world (the bulk). Although it is fair to say that similar ideas were proposed on the 80's by several authors [8], they were missed by some time, until recent developments on string theory provided an independent realization of such models [9–11], given them certain credibility.

Our goal for the present notes is to give a general overview of this field. Among the scenarios presented until now in the literature, we will focus on studying mainly the case of large (millimetric) extra dimensions. This, in turn, will provide us the insight to extend our study to other more elaborated models.

To motivate the ideas we will start with the discussion of the origin of the long standing hierarchy problem and how extra dimensions offer a new way to understand it by providing a new low fundamental scale, M, which, based on the phenomenological and experimental constrains, could be just at the TeV range. At this point we shall assume that the Standard Model particles are attached to the brane and that only gravity propagates on the bulk. As we will see, in such a theory the weakness of gravity is related to the large size of the volume of the extra space.

As it is clear, with a low fundamental scale all the particle physics phenomena that invoke high energy scales will not work any more. Then, problems as neutrino masses and mixings, gauge coupling unification and proton decay should be reviewed under the light of this new theories. This will be the central issue along our discussion, and in order to address it we will modify the above model accordingly.

First, after discussing some general aspects of the brane-bulk theories, we will explore neutrino oscillations phenomena. Here, we shall show how an isosinglet bulk neutrino that couples to the standard neutrinos may help to solve the neutrino puzzles [12–22]. Next we will analize the possibility that the standard model particles propagate in the extra dimensions and develop Kaluza-Klein excitations [23–26]. The contribution of this exited modes to standard processes will set bounds to the bulk radius and provide collider signatures [7]. Those modes will also modify the profile of the renormalization group equations that govern the running of the gauge coupling constants. The net effect is a power law running that accelerates their meeting, which now may occur at very low energies [26–32], even at the TeV scale. We comment on the accuracy of this low energy gauge coupling unification.

Next, we shall discuss a slightly modified scenario that may account for the explanation of fermion mass hierarchies and the suppression of proton decay by introducing a splitting of the wave functions along the extra dimension [33,34]. Finally, we will present the Randall-Sundrum model [35] which provides an alternative explanation for the hierarchy problem based on small extra dimensions and a non factorizable bulk geometry. We will also mention some of the current trends on the study of this class of models [36–45].

II HIERARCHY PROBLEM AND LARGE EXTRA DIMENSIONS

A Radiative corrections and the Higgs mass

Our starting point is the scalar sector of the Standard Model (SM). This is perhaps the most peculiar sector in the theory. The scalar Higgs field, H, realizes the spontaneous symmetry breaking and gives masses to all other particles, fermions and gauge bosons, and yet, its own mass is introduced as a free parameter on the theory. The Higgs is the only particle of the SM which remains to be observed. It is also the only field that have self interactions that are only constrained by gauge invariance and the renormalizability condition which leave a λH^4 term, with λ a free parameter. And, what is more important for us, it is the only field for which the quantum corrections require a large fine tunning on the mass parameter. While the self energy diagrams of all other fields, evaluated in the $\overline{MS}$ scheme, develop logarithmic divergences that depend on their own bare mass, the scalar field develops quadratic divergences that are independent of its bare mass. For instance, at one loop order one gets

$$\delta m_H^2 = \frac{1}{8\pi^2}\left(\lambda_H^2 - \lambda_F^2\right)\Lambda^2 + (\text{log. div.}) + \text{finite terms.} \tag{2.1}$$

where λ_H is the self-couplings of H, λ_F is the Yukawa coupling to fermions, and Λ is the physical cut off of the theory, which is usually believed to be the Planck scale, $M_{P\ell} \sim 10^{19}\ GeV$, or the GUT scale, $M_{GUT} \sim 10^{16}\ GeV$.

Nevertheless, in order to keep the WW scattering cross section from violating unitarity, the physical Higgs boson mass, m_H, must be less than about 1 TeV. Then we get the unpleasant result, $m_H^2 = m_{H,0}^2 + \delta m_H^2 + \text{counterterm}$, where the counterterm must be adjusted to a precision of roughly 1 part in 10^{15} in order to cancel the quadratically divergent contributions to δm_H^2. Moreover, this adjustment must be made at each order in perturbation theory. This large fine tunning is what is known as the hierarchy problem.

Of course, the quadratic divergence can be renormalized away in exactly the same manner as is done for logarithmic divergences, and in principle, there is nothing formally wrong with this fine tuning. In fact if this calculation is performed in the dimensional regularization scheme, DR, one obtains only $1/\epsilon$ singularities which are absorbed into the definitions of the counterterms, as usual. Hence, the problem of quadratic divergences does not become apparent there. It arises only when one attempts to give a physical significance to the cut-off Λ. In other words, if the SM were a fundamental theory then the using of DR would be justified. However, most theorists believe that the final theory should also include gravity, then a cut-off must be introduced in the SM. Hence, we regard this fine tunning as unattractive.

B A low fundamental scale from new dimensions

Explaining the hierarchy problem has been a leading motivations to explore new physics during the last twenty years. Supersymmetry [46] and compositeness [47] are two of the main proposals that solve this problem. Supersymmetry predicts the existence of new particles, the super-partners, at the TeV scale. They belong to the same dimensional representation than those of the SM, but they differ each other on the spin. Their couplings are symmetric such that their contribution to the quadratic divergences get balanced and nullify each other. This has been, so far, the most popular extension of the SM. Compositeness, on the other hand, assumes that the Higgs is not a fundamental field, but a quantum condensate of other fields. Both theories reflect a common idea on the structure of physics beyond the SM: A new effective field theory will be revealed at the weak scale, $m_{EW} \sim 1\ TeV$, stabilizing and perhaps explaining the origin of the hierarchy, $m_{EW}/M_{P\ell}$, while, a desert among those scales will remain until the Planck scale, which is assumed to be the fundamental scale where gravity becomes as strong as the gauge interactions and where the quantum theory of gravity is revealed. Eventually the desert could be populated by other effective theories, responsible from explaining other phenomena (as fermion masses) or from triggering dynamical symmetry breakings, and a big deal of work has been dedicated to study those pictures.

It has been realized recently that if there are more than four dimensions, as it is suggested by string theory, the above scenario could be drastically modified. To explain this let us assume as in Ref. [2] that there are δ extra space-like dimension which are compact. Compactification will be in principle a natural ingredient that would explain why we see only four dimensions. Let us also imaging that all (some) of these dimensions are large with a common radius R. Then, if two test particles with masses m_1 and m_2 were separated each other by a distance $r \gg R$, they would feel the usual gravitational potential

$$U(r) = G_N \frac{m_1 m_2}{r}. \qquad (2.2)$$

However, if $r \ll R$, the potential between each other should be

$$U(r) = G \frac{m_1 m_2}{r^{\delta+1}}, \qquad (2.3)$$

where $G^{-1} = M^{\delta+2}$, is the coupling constant of gravity in $\delta + 4$ dimensions which defines the *fundamental scale* M where gravity becomes strong. Therefore, if we take the last relationship as the fundamental one, then Eq. (2.2) will imply that our *effective* four dimensional gravity scale is given by [2]

$$M_{P\ell}^2 = M^{\delta+2} V_\delta, \qquad (2.4)$$

where V_δ is the volume of the extra space. It is worth saying that this same relationship arises if we dimensional reduce to four dimensions the gravity action in $4 + \delta$ dimensions, assuming the space-time to be $\mathcal{R}^4 \times \mathcal{M}_\delta$, where $\mathcal{M}_\delta$ is an δ dimensional compact manifold of volume V_δ. If the volume were large enough, then the fundamental scale could be as low as m_{EW}, and the hierarchy would be naturally removed. Of course, the price one has to pay is to explain why the extra dimensions are so large. For $\delta = 2$, we get for $M \sim 1\ TeV$ that R is less than 1 millimeter. This case is highly interesting since it is on the current limits on the low distance gravity experiments [3] that should detect the predicted deviation of the inverse squared law of gravity at distances, $r \sim V_\delta^{1/\delta}$. Nevertheless, more than two extra dimensions could be expected (strings predicts six more), and their size do not have to be the same. More complex scenarios with a hierarchical distributions of the sizes could be natural. Then, for the investigation of the model, we will use a single large extra dimension, implicitly assuming other smaller dimensions.

Now well, while submillimeter dimensions remain untested for gravity, the SM gauge forces have certainly been accurately measured up to weak scale distances. Therefore, the SM particles can not freely propagate in these large extra dimensions, but must be constrained to a four dimensional submanifold. Then the scenario we have in mind is one where we live in a four dimensional surface embedded in a higher dimensional space. This picture is similar to the D-brane models [48], as in the Horava-Witten theory [10]. We may also imagine our world as a domain wall of size M^{-1} where the SM fields are trapped by some dynamical mechanism [2]. As this framework solves the hierarchy problem, supersymmetry is no longer needed. However, we should notice that it may still be crucial for the self-consistency of the theory of quantum gravity above the M scale, although such theory is yet unknown. It could be superstring theory, but it may also be something yet to be discovered. In any case, supersymmetry may coexist with large extra dimensions.

C Experimental bounds

There are a number of dramatic experimental consequences of large extra dimensions. First, as already mentioned, there is the deviation of the inverse squared law on gravity at submillimeter distances. The current experiments are just at this limit, so $R < 0.2\ mm$ [3]. Also, as gravity becomes comparable in strength to the gauge interactions at energies $M \sim$ TeV, the nature of the quantum theory of gravity would become accessible to LHC and NLC. With gravity freely propagating on the bulk, the effect of the gravitational couplings will be mostly of two types: missing energy, that goes into the bulk, and corrections to the standard cross sections from graviton exchange. The Feynman rules were calculated from the linearized bulk gravity theory, and are given in Ref. [4]. From the effective four dimensional point of view, the graviton develops Kaluza Klein (KK) excitations of masses n/R, with n an integer number (see next section). The coupling of each one of those modes to the SM particles is suppressed by the Planck scale, but the overall coupling that considers all the KK excitations become suppressed just by M.

TABLE 1. Collider limits for the fundamental scale M. Graviton Production.

Process	Background	M limit	Collider
$e^+e^- \to \gamma G$	$e^+e^- \to \gamma\bar{\nu}\nu$	1 TeV	L3
$e^+e^- \to ZG$	$e^+e^- \to Z\bar{\nu}\nu$	$\left\{ \begin{array}{l} 515\ GeV \\ 600\ GeV \end{array} \right.$	LEPII L3
$Z \to \bar{f}fG$	$Z \to \bar{f}f\bar{\nu}\nu$	0.4 TeV	LEP

TABLE 2. Collider limits for the fundamental scale M. Virtual Graviton exchange

Process	M limit	Collider
$e^+e^- \to ff$	0.94 TeV	Tevatron & HERA
$e^+e^- \to \gamma\gamma, WW, ZZ$	$\left\{ \begin{array}{l} 0.7 - -1\ TeV \\ 0.8\ TeV \end{array} \right.$	LEP L3
All above	1 TeV	L3
Bhabha scattering	1.4 TeV	LEP
$\left. \begin{array}{l} q\bar{q} \to \gamma\gamma \\ gg \to \gamma\gamma \end{array} \right\}$	0.9 TeV	CDF

A long number of studies on this topic have appeared already [4,5], some nice and short reviews of collider signatures are given in [49]. We summarize some of the current bounds in tables 1 and 2. At e^+e^- colliders (LEP,LEPII, L3), the best signals would be the production of gravitons with Z, γ or fermion pairs $\bar{f}f$. There is also the interesting monojet production [4] at hadron colliders (CDF, LHC) which is yet untested.

The virtual exchange of gravitons either leads to modifications of the SM cross sections and asymmetries, or to new processes not allowed in the SM at tree level. The amplitude for exchange of the entire tower naively diverges when $\delta > 1$ and has to be regularized, typically by a cut off at M, which performs the replacement of the sum suppressed by $M_{P\ell}$ into the simple suppression by M. An interesting channel is $\gamma\gamma$ scattering, which appears at tree level, and may surpasses the SM background at $s = 0.5$ TeV for $M = 4$ TeV. Bi-boson productions of $\gamma\gamma$, WW and ZZ has been already analized [4,5]. Some experimental limits, most of them based on existing data, are given in Table 2. The upcoming experiments will easily overpass those limits.

Notice that all collider limits are about 1 TeV. A more stringent constraint comes from SN1987A and astrophysics [6], which gives $M > 30 - 100$ TeV.

III AN INTRODUCTION TO BRANE-BULK MODELS

Before getting into the discussion of models with large extra dimensions in more detail, let us make some useful remarks on the generical properties of brane-bulk models.

As we already mentioned, from the four dimensional effective theory point of view, a bulk field, as the graviton, will appear as an infinite tower of KK excitations. The appreciation of the impact of this KK excitations will depend in the relevant energy of the experiment, and on the compactification scale $\frac{1}{R}$. Roughly speaking, if $E \ll \frac{1}{R}$ the theory behaves purely four dimensional. However, for energies above $\frac{1}{R}$, a large number of KK excitations, $\sim (ER)^\delta$, becomes kinematically accessible making physics looks $4+\delta$ dimensional. To clarify this matters let us consider a five dimensional model where the fifth dimension has been compactified on a circle of circumference $2\pi R$. The generalization of these results is straightforward. Let ϕ be a bulk scalar field for which the action has the form

$$S_\phi = \frac{1}{2}\int d^4x\, dy\, \left(\partial^A\phi\partial_A\phi - m^2\phi^2\right); \tag{3.1}$$

where $A = 1,\ldots,5$, and y denotes the fifth dimension. Demanding periodicity on the extra compact dimension under $y \to y + 2\pi R$, the field may be Fourier expanded as

$$\phi(x,y) = \frac{1}{\sqrt{2\pi R}}\phi_0(x) + \sum_{n=1}^{\infty}\frac{1}{\sqrt{\pi R}}\left[\phi_n(x)\cos\left(\frac{ny}{R}\right) + \hat{\phi}_n(x)\sin\left(\frac{ny}{R}\right)\right]. \tag{3.2}$$

Notice the different normalization of the excited modes, ϕ_n and $\hat{\phi}_n$, with respect to the zero mode, ϕ_0. By introducing the last equation into the action and integrating over the extra dimension we get

$$S_\phi = \sum_{n=0}^{\infty}\frac{1}{2}\int d^4x\left(\partial^\mu\phi_n\partial_\mu\phi_n - m_n^2\phi_n^2\right) + \sum_{n=1}^{\infty}\frac{1}{2}\int d^4x\left(\partial^\mu\hat{\phi}_n\partial_\mu\hat{\phi}_n - m_n^2\hat{\phi}_n^2\right), \tag{3.3}$$

where the KK mass is given as $m_n^2 = m^2 + \frac{n^2}{R^2}$. Therefore, in the effective theory, the higher dimensional field looks like an infinite tower of fields with masses m_n. These are the so called KK modes. Notice that all these modes are fields with the same spin, and quantum numbers as ϕ. But they differ in the KK number n, associated with the fifth component of the momentum. For $m = 0$ it is clear that for energies below $\frac{1}{R}$ only the massless zero mode will be kinematically accessible, making the theory looks four dimensional. As the energy increases, once it surpasses the threshold of the first exited level, the manifestation of the KK modes will evidence the higher dimensional nature of the theory.

The five dimensional field ϕ has mass dimension $\frac{3}{2}$, while all modes are dimension one. In general for δ extra dimensions we will get $[\phi] = d_4 + \frac{\delta}{2}$, where d_4 is the natural mass dimension of the field in four dimensions. Because this change on the dimensionality of ϕ, the higher order operators (beyond the mass term) will all have dimensionful couplings. To keep them dimensionless a mass parameter should be introduced to correct the dimensions. As usual, the natural choice for this parameter is the cut off of the theory. For instance, let us consider the quartic couplings of ϕ. Since all potential terms should be of dimension five, we should write down $\frac{\lambda}{M}\phi^4$.

After integrating the fifth dimension, this operator will generate quartic couplings among all KK modes. Four normalization factors containing $1/\sqrt{R}$ appear in the expansion of ϕ^4. Two of them will be removed by the integration, thus, we left with the effective coupling λ/MR. By introducing Eq. (2.4) we observe that the effective couplings have the form

$$\lambda\left(\frac{M}{M_{P\ell}}\right)^2 \phi_k\phi_l\phi_m\phi_{k+l+m}. \tag{3.4}$$

where the indices are arranged to respect the conservation of the fifth momentum. From the last expression we conclude that in the low energy theory ($E < M$), the effective coupling appears suppressed. Thus, the effective four dimensional theory is weaker interacting compared with the bulk theory. Something similar happens to gravity on the bulk, where the coupling constant is stronger than in the brane, because it is also suppressed by the volume of the extra space as given in Eq. (2.4). By the naturalness principle, we must assume that all dimensionless coupling constants are of order one.

Let us now consider the brane fields represented by a fermion $\psi(x)$. The theory describing this brane fermion is purely four dimensional. Indeed, its action is localized by a delta function in the complete theory in the form

$$S_\psi = \int d^4x\; dy\; \mathcal{L}\left(\psi\right)\delta(y-y_0), \tag{3.5}$$

where y_0 is the position of the brane. The coupling of ψ to the bulk scalar field only may take place at the brane, where ψ lives. For simplicity we will assume that the brane is located at the position $y_0 = 0$, which in the case of orbifolds corresponds to a fixed point. So, the part of the action that describes the brane-bulk coupling is

$$S_{int} = \int d^4x\; dy\; \mathcal{L}_{int}\left(\phi,\psi\right)\delta(y). \tag{3.6}$$

Lets choose for instance the term

$$\begin{aligned} S_{\phi\psi} &= \int d^4x\; dy\; \frac{f}{\sqrt{M}}\bar{\psi}(x)\psi(x)\phi(x,y=0)\;\delta(y) \\ &= \int d^4x \frac{M}{M_{P\ell}} f\cdot\bar{\psi}\psi\left(\phi_0+\sqrt{2}\sum_{n=1}^{\infty}\phi_n\right). \end{aligned} \tag{3.7}$$

Here the Yukawa coupling constant f is dimensionless. On the right hand side we have used the expansion (3.2) and Eq. (2.4). From here, we notice that the coupling of brane to bulk fields is generically suppressed by the ratio $\frac{M}{M_{P\ell}}$. Also, notice that the modes $\hat{\phi}_n$ decouple from the brane. Then, to get ride of those (harmless) fields in the effective theory we could assume that ϕ is an even field under the transformation $y \to -y$, thus, imposing a $\mathcal{Z}_2$ symmetry on the theory, though the theory is consistent even without this extra assumption. That symmetry, on the other hand, is characteristic of the orbifolds and domain walls. By picking up an explicit parity for ϕ we are projecting out one half of the KK excitations, it means, the expansion (3.2) will involve either the cosine or the sine part, but not both.

Let us stress that the couplings in (3.7) do not conserve the KK number. This reflects the fact that the brane breaks the translational symmetry along the extra dimension. Nevertheless, it is worth noticing that the four dimensional theory is still Lorentz invariant. Physically this means that when the interactions among the brane fields reach enough energy to produce real emission of KK modes, part of the energy of the brane is released into the bulk. This is also the case of gravity.

Let us mention, parenthetically, that in general the linear perturbations of the metric lead to particles of spin two, one and zero; however, only the spin two graviton, $G_{\mu\nu}$, and one scalar field, b, the radion; couple to matter at the weak field limit. The Feynman rules are given in [4]. Briefly speaking, they come from the couplings to the energy momentum tensor

$$\mathcal{L} = -\frac{1}{M_{p\ell}}\sum_n \left[G^{(n)\mu\nu} - \frac{\kappa}{3}b^{(n)}\eta^{\mu\nu}\right]T_{\mu\nu}. \tag{3.8}$$

Here κ is a parameter of order one. Notice that $G^{(0)\mu\nu}$ is massless. That is the source of long range four dimensional gravity. It is worth saying that $b^{(0)}$ is not actually massless, it get a mass from the stabilization mechanism. From supernova constrains such a mass should be larger than 10^{-3} eV. Experimental bounds for graviton production were given in the previous section.

Next, let us consider the scattering process among brane fermions: $\psi\psi \to \psi\psi$. In the present toy model this interaction is mediated by all the KK excitations of ϕ, then the typical amplitude will receive the contribution

$$\mathcal{M} = \hat{f}^2\left(\frac{1}{q^2-m^2} + 2\sum_{n=1}\frac{1}{q^2-m_n^2}\right)D(q^2), \tag{3.9}$$

where $\hat{f}$ represents the effective coupling, and $D(q^2)$ is an operator that depends on the Feynman rules and is usually independent of the index n, as a consequence of the universal coupling manifested on Eq. (3.7). In more than five dimensions the equivalent to the above sum usually diverges and has to be regularized by introducing a cut off at the fundamental scale. Roughly speaking, at high energies, $qR \gg 1$, the overall factor becomes $\hat{f}^2N$, where N is the number of KK modes until the cut off. It is $N = MR = M_{P\ell}^2/M^2$. In the case of the graviton, $\hat{f}$ is about $1/M_{P\ell}$, and therefore, the overall coupling becomes just $1/M$ [4,5]. At low energies, on the other hand, by assuming that $q^2 \ll m^2 \ll 1/R^2$ we may integrate out the KK excitations, and at the first order we get

$$\mathcal{M} = \frac{\hat{f}^2}{m^2}D(q^2)\left(1 + \frac{\pi^2}{3}m^2R^2\right). \tag{3.10}$$

The last term between parenthesis is a typical correction produced by the KK modes exchange to the pure four dimensional result.

Let us now mention on some characteristics of other bulk fields. A massless bulk fermion field is defined as the solution of the higher dimensional Dirac Equation $\left(i\partial^M\Gamma_M\right)\Psi = 0$, where the $\delta+4$ Dirac matrices satisfy the algebra $\{\Gamma_M, \Gamma_N\} = 2g_{MN}$. In five dimensions we use the Weyl basis

$$\Gamma_\mu = \gamma_\mu = \begin{pmatrix} 0 & \sigma_\mu \\ \bar{\sigma}_\mu & 0 \end{pmatrix}; \qquad \text{and} \qquad \Gamma_5 = \gamma_5 = -i\begin{pmatrix} \mathbf{1} & 0 \\ 0 & -\mathbf{1} \end{pmatrix}. \tag{3.11}$$

Therefore, a bulk fermion is necessarily a four component spinor. Thus, only vector-like theories are in principle possible. However, if the fifth dimension is orbifolded, the theory on the brane may look as a chiral theory. In this basis, Ψ is conveniently decomposed as

$$\Psi = \begin{pmatrix} \nu_R \\ \nu_L \end{pmatrix}; \tag{3.12}$$

with each component having its own set of KK excitations. The free action for the massless field may be expanded in terms of the KK modes, and it has the form

$$\begin{aligned} S_\Psi &= \int d^4x\; dy\; i\bar{\Psi}\Gamma^M\partial_M\Psi \\ &= \int d^4x\; \left(i\bar{\nu}_{nL}\bar{\sigma}^\mu\partial_\mu\nu_{nL} + i\bar{\nu}_{nR}\bar{\sigma}^\mu\partial_\mu\nu_{nR} - \frac{n}{R}\bar{\nu}_{nR}\nu_{nL} + h.c\right); \end{aligned} \tag{3.13}$$

where a sum over n is to be understood. Notice that all KK masses are Dirac masses.

Two different Lorentz invariant fermion bilinears are possible in five dimensions: Dirac mass terms $\bar{\Psi}\Psi$ and Majorana masses $\Psi^T C_5\Psi$, where $C_5 = \gamma^0\gamma^2\gamma^5$. However, the Dirac mass is an odd function under the orbifold symmetry, $y \to -y$, under which $\Psi \to \gamma^5\Psi$. So, if the theory is invariant under this symmetry that term will be zero.

Lets now consider the extension of a gauge field to five dimensions. For simplicity let us consider only the case of a free gauge abelian theory. The Lagrangian in five dimensions is given as

$$\mathcal{L}_{5D} = -\frac{1}{4}F_{MN}F^{MN}\ . \tag{3.14}$$

Upon integration over the extra dimension one gets [26]

$$\mathcal{L} = -\frac{1}{4}\sum_{n=0}^{\infty} F_{\mu\nu}^{(n)}F^{(n)\mu\nu} - \frac{1}{2}\sum_{n=0}^{\infty}\left[\partial_\mu A_5^{(n)} + \frac{n}{R}A_\mu^{(n)}\right]^2 \tag{3.15}$$

The gauge invariance of the theory can be now expressed in terms of the (expanded) gauge transformation of the KK modes

$$A_\mu^{(n)} \to A_\mu^{(n)} + \partial_\mu \theta^{(n)} \qquad \text{and} \qquad A_5^{(n)} \to A_5^{(n)} - \frac{n}{R}\theta^{(n)}. \tag{3.16}$$

Therefore, by fixing the gauge one may absorb the scalar field A_5 into A_μ. Hence, the only massless vector is the zero mode, all KK modes acquired mass by absorbing the Goldston bosons, $A_5^{(n)}$, associated to the spontaneous isometry breaking [50]. That keeps the gauge symmetry on the four dimensional effective theory untouched. We must stress that this new theory is essentially non renormalizable for the infinite number of fields that it involves. However, the truncated theory that only consider a finite number of KK modes is renormalizable. The cut off for the inclusion of the excited modes will be again the scale M. Other aspects of this theories will be discussed later on when we addressed the SM extensions.

IV NEUTRINO MASSES

Several experiments have provided conclusive evidence for a deficit on the expected flux of atmospheric and solar neutrinos, and there is also the direct observation of $\bar{\nu}_e$ in the $\bar{\nu}_\mu$ beam of the LSND experiment. A simple explanation of these anomalies arises if neutrinos are massive and a large amount of work is devoted nowadays to explain the origin of their small squared mass differences [12]. However, most of the four dimensional mechanisms invoke high energy physics with scales about 10^{12} GeV or higher. Obviously, with a fundamental TeV scale, understanding the small neutrino masses poses a theoretical challenge to the new theories.

A second possible problem is the enhancement of all non renormalizable operators, now suppressed only by powers of $1/M$. Among them, there are those which produce a dangerous fast proton decay and the operator

$$\frac{(LH)^2}{M} \tag{4.1}$$

which produces a large neutrino mass of the order $\langle H\rangle^2/M$. To exclude this operator one has to make the additional assumption that the theory respect lepton number symmetry (or more precisely $B-L$). Two class of models are then possible depending on whether $B-L$ is a global or local symmetry. We discuss both possibilities in this section.

A Models with global $B-L$ symmetry

In the context of models that have a global $U(1)_{B-L}$ symmetry, one can get small neutrino masses by introducing isosinglet neutrinos in the bulk [13] which carry lepton number. As this is a sterile neutrino, it comes natural to assume that it may propagate into the bulk as well as gravity, while the SM particles remain attached to the brane. These models are interesting since they lead to small neutrino masses without any extra assumptions.

Let $\nu_B(x^\mu, y)$ be a bulk neutrino which we take to be massless since the Majorana mass violates conservation of Lepton number and the five dimensional Dirac mass is forbidden by the orbifold symmetry, which we assume. This neutrino couples to the standard lepton doublet, L, and to the Higgs field, H, via $\frac{h}{\sqrt{M}}\bar{L}H\nu_{BR}\,\delta(y)$. Once the Higgs develops its vacuum, this coupling will generate the four dimensional Dirac mass terms

$$m\bar{\nu}_L\left(\nu_{0R} + \sqrt{2}\sum_{n=1}^{\infty}\nu_{nR}\right), \tag{4.2}$$

where the mass m is given by [14]

$$m = hv\frac{M}{M_{P\ell}} \sim 10^{-4}\ eV \times \frac{hM}{1\ TeV}. \tag{4.3}$$

Therefore, if $M \sim 1\ TeV$ we get just the right order of magnitude on the mass as required by the experiments. Moreover, even if the KK decouple for a small R, we will still get the same Dirac mass for ν_L and ν_{0R}, as far as M remains in the TeV range. After including the KK masses from Eq. (3.13), we may write down all mass terms in the compact form [15]

$$(\bar{\nu}_{eL}\bar{\nu}'_{BL})\begin{pmatrix} m & \sqrt{2}m \\ 0 & \partial_5 \end{pmatrix}\begin{pmatrix} \nu_{0B} \\ \nu'_{BR} \end{pmatrix}, \tag{4.4}$$

where the notation is as follows: ν'_B represents the KK excitations. The off diagonal term $\sqrt{2}m$ is actually an infinite row vector of the form $\sqrt{2}m(1, 1, \cdots)$ and the operator ∂_5 stands for the diagonal and infinite KK mass matrix whose n-th entrance is given by n/R. Notice that the left handed zero mode, ν_{0BL}, has decoupled from the spectra and will remain massless. Using this short hand notation it is straightforward to calculate the exact eigensystem for this mass matrix [16]. Simple algebra yields the characteristic equation $2\lambda_n = \pi\xi^2\cot(\pi\lambda_n)$, with $\lambda_n = m_nR$, $\xi = \sqrt{2}mR$, and where m_n is the mass eigenvalue [13,14]. The weak eigenstate is given in terms of the mass eigenstates, $\tilde{\nu}_{nL}$, as

$$\nu_L = \sum_{n=0}^{\infty}\frac{1}{N_n}\tilde{\nu}_{nL}, \tag{4.5}$$

where the mixing N_n is given by $N_n^2 = \left(\lambda_n^2 + f(\xi)\right)/\xi^2$, with $f(\xi) = \xi^2/2 + \pi^2\xi^4/4$ [16]. Therefore, ν_L is actually a coherent superposition of an infinite number of massive modes. As they evolve differently on time, the above equation will give rise to neutrino oscillations, $\nu \to \nu_B$, even though there is only one single flavour. This is a totally new effect that arise the possibility of new ways of understanding the neutrino anomalies. An analysis of the implications of the mixing profile in these models for solar neutrino deficit was presented in [14]. Implications for atmospheric neutrinos were discussed in [17], and some phenomenological bounds were given in [17,18]. A comprehensive analysis for three flavours is given in [19]. Here we summarize some of the main results.

B New patterns of neutrino oscillations

Small ξ. For $\xi \ll 1$, the mixing in Eq. (4.5) is given by $\tan\theta_n \sim \xi/n$ [14]. The masses are: m for $n = 0$ and n/R otherwise. Therefore, as expected, the main component of ν_L is the lightest mass eigenstate. The mixing, on the other hand, will induce a resonant conversion into bulk (sterile) modes. The survival probability has the form

$$P_{surv}(L) = 1 - \frac{4}{\eta^4}\xi^2 \sum_{n=1}^{\infty} \frac{\sin^2\left(\frac{n^2 L}{4ER^2}\right)}{n^2} - \frac{2}{\eta^4}\xi^4 \sum_{k,n=1}^{\infty} \frac{\sin^2\left[\frac{(n^2-k^2)L}{4ER^2}\right]}{n^2k^2}, \tag{4.6}$$

where $\eta = (1+\pi^2\xi^2/6)^{1/2}$. Thus, the oscillation length is given by $L_{osc} = 4\pi ER^2$. Clearly, at the low ξ order the above expresion becomes $P_{surv}(L) \approx 1 - 4\xi^2\sin^2(L/4ER^2)$, which mimics the former small mixing angle case. Therefore, if $1/R \sim 10^{-3}\ eV$, which is just $R \sim 0.2\ mm$, we get an explanation for solar data by introducing MSW effects as described in [14] that is consistent with other astrophysical constrains [14], though it has been suggested that SN1987A may impose more stringent bounds: $R < 1$ Å [17]. A similar explanation for the atmospheric anomaly seems disfavored [14,19], since it needs large mixing angle.

Large ξ. On the continuos approximation, the survival probability reads [17]

$$P_{surv}(z) = \left|1 - \operatorname{erf}\left(\frac{\pi}{2}\xi^2\sqrt{iz}\right)\right|^2 \tag{4.7}$$

where $z = L/2ER^2$. Notice that now the probability has no oscillatory nature in terms of L/E, this feature should distinguish this models form the conventional two neutrino oscillations. Indeed, it is easy to check the physical origin of this effect from the exact solutions [19]. For $\xi \gtrsim 1$, the eigenvalues λ_n start to deviate from the integral value n, and a large number ($N_\xi \sim \pi^2\xi^2/4$) of equally suppressed eigenstates contribute to (4.5). Then, once ν_L is released, the time evolution of the different components will most likely wash out the original coherent superposition and the initial ν_L will almost disappear, and the maximal probability will not be recovered away from the source [13]. The fast developing slope of P_{surv} is governed by the single parameter ξ^2/R. As solar and atmospheric have $\overline{P}_{surv} \approx 0.5$, in order to avoid a large deficit on P_{surv}, one must keep that parameter on a small range. For atmospheric one gets that $\xi^2/R \approx 10^{-2} eV$ [17,19]. That means $10^{-3}\ eV < 1/R < 10^{-2}\ eV$, and $1 < \xi^2 < 10$, and then, R should remain in the submillimeter range. An explanation for solar data is, on the other hand, not possible in this limit [19].

C Three flavour oscillations

The extension of this model to three brane generations, $\nu_{e,\mu,\tau}$, is straightforward. It was observed earlier [16] that to give masses to the three standard generations three bulk neutrinos are needed. This comes out from the fact that with a single bulk neutrino only one massless right handed neutrino is present (the zero mode), then, the coupling to brane fields will generate only one new massive Dirac neutrino. After introducing a rotation by an unitary matrix U on the weak sector, the most general Dirac mass terms with three flavours and arbitrary Yukawa couplings may be written down as

$$-\mathcal{L} = \sum_{\alpha=1}^{3} \left[m_\alpha \bar{\nu}_{\alpha L} \nu_{BR}^{\alpha}(y=0) + \int dy\, \bar{\nu}_{BL}^{\alpha} \partial_5 \nu_{BR}^{\alpha} + h.c. \right] , \tag{4.8}$$

where $\nu_{aL} = U_{a\alpha}\nu_{\alpha L}$, with $a = e, \mu, \tau$ and $\alpha = 1, 2, 3$. The mass parameters m_α are the eigenvalues of the Yukawa couplings matrix multiplied by the vacuum v, and as stated before are naturally of the order of eV or less. This reduces the analysis to considering three sets of mixings given as in the previous case. Each set (tower) of mass eigenstates is characterized by its own parameter $\xi_\alpha \equiv \sqrt{2} m_\alpha R$. Now, each weak eigenstate can be expressed as a coherent superposition of these three different towers by

$$\nu_a = \sum_{\alpha=1}^{3} \sum_{k=0}^{\infty} U_{a\alpha} \frac{1}{N_{\alpha k}} \tilde{\nu}_{\alpha k}; \tag{4.9}$$

with $\tilde{\nu}_{\alpha k}$ being the k-th mass eigenstates of the α-th tower. It is now clear that the three flavour oscillations will correspond to the oscillations among the three towers. In this regards, if the KK do not decouple, the explanation to neutrino puzzles will not be any more described in terms of three single neutrinos. Now, the transition probability is given by $P_{ab} = \sum_{\alpha\beta} U_{a\alpha}^{*} U_{b\alpha} U_{b\beta}^{*} U_{a\beta}\, p_{\alpha\beta}$, where we have introduced the partial transition probabilities $p_{\alpha\beta}$ defined as $p_{\alpha\beta}(L) \equiv \overline{\langle \nu_\alpha(L)|\nu_\alpha(0)\rangle} \langle \nu_\beta(L)|\nu_\beta(0)\rangle$. The diagonal $p_{\alpha\alpha}$ is interpreted as the survival probability of (the non standard) ν_α, and it has the form of (4.6) and (4.7) for small and large ξ_α respectively.

The analysis of Ref. [19] shows that there is not simultaneous explanation for solar, atmospheric and LSND data within this minimal model. Without LSND, we have the following possible scenarios: (i) $\xi_{1,2,3} \ll 1$; then, R is smaller than 1 μm. KK modes decouple. Solar and atmospheric neutrino data are understood as in the case of four dimensional models. (ii) $\xi_{1,2} \ll 1 \lesssim \xi_3$; solar neutrino data is provided as in the four dimensional models but atmospheric data is explained by ν_μ to ν_{bulk} oscillation as from Eq. (4.7). R is on the submillimeter range. (iii) $\xi_1 \ll 1 \lesssim \xi_{2,3}$. Both, solar and atmospheric data are explained by $\nu \to \nu_{bulk}$ oscillations. $1/R^2 \sim \Delta m_{sol}^2 \sim 10^{-3}\, eV$ with matter effects for solar and $\xi_{2,3}^2 \sim 10$. Finally, (iv) for $\xi_{1,2,3} \gg 1$ there is no explanation for solar neutrino data. This case is therefore ruled out.

D Models for Majorana masses

Some extended scenarios that consider the generation of Majorana masses from the breaking of lepton number either on the bulk or on a distant brane have been considered in Ref. [20,21] (the breaking of global symmetries at distant branes was first proposed in Ref. [51]). In these models a bulk scalar field χ that carries lepton number is introduced. It develops a small vacuum and gives mass to the neutrinos which are generically of the form

$$m_\nu \sim \frac{\langle H\rangle^2}{M}\frac{\langle\chi\rangle_B}{M^{3/2}}; \tag{4.10}$$

then, with M of the order of 10 TeV, we need $\langle\chi\rangle_B \sim (10\ MeV)^{3/2}$. Such a small vacuum is possible in both classes of models, though it usually needs a small mass for χ. Obviously, with Majorana masses, a bulk neutrino may not be needed but new physics must be invoked. We should notice the there is also a Majoron field associated to the spontaneous breaking of the lepton number symmetry. Its phenomenology depends on the details of the specific model. In the simplest scenario, the coupling $(LH)^2\chi$ is the one responsible for generating Majorana masses. It also gives an important contribution for neutrinoless double beta decay which is just right at the current experimental bounds [21].

E Models with local $B-L$ symmetry

We now proceed to consider the second class of models. For the case where $B-L$ is local, anomaly cancellation requires that right handed neutrinos must be present in the brane as in the models discussed in Ref. [15,16]. The simplest gauge model where this scenario is realized is the left-right symmetric model where the right handed symmetry is broken by the Higgs doublet $\chi_R(1,2,1)$, where the number inside the parenthesis correspond to the quantum numbers under $SU(2)_L\times SU(2)_R\times U(1)_{B-L}$. The model then contains the left and right handed brane leptons and a blind bulk neutrino. The relevant terms of the action for one generation are [15]

$$\mathcal{S} = \int d^4x[\kappa\bar{L}\chi_L\nu_B(y=0) + \kappa\bar{R}\chi_R\nu_B(y=0) + h\bar{L}\phi R] + \int d^4x dy \bar{\nu}_B\Gamma^5\partial_5\nu_B + h.c.$$

By setting $\langle\chi_R^0\rangle = v_R$ and $\langle\chi_L^0\rangle = 0$, the following Dirac neutrino mixing matrix is obtained

$$(\bar{\nu}_{eL}\ \bar{\nu}_{0BL}\ \bar{\nu}'_{BL})\begin{pmatrix} hv & 0 \\ \kappa v_R & 0 \\ \sqrt{2}\kappa v_R & \partial_5 \end{pmatrix}\begin{pmatrix} \nu_{eR} \\ \nu'_{BR} \end{pmatrix}. \tag{4.11}$$

Note that in general the effect of $\langle\chi_L^0\rangle = 0$ may also be produced if the bulk neutrino breaks explicitly the parity symmetry [22]. Now, a massless field, ν_0, which is predominantly the electron neutrino, provided that $\kappa v_R \gg hv \simeq$ few MeV, appears. Since $\kappa \simeq \frac{M}{M_{Pl}}$, this constraint implies that M must be as large as 10^8 GeV or so, however R may remain in the submillimeters. Oscillations into bulk neutrino will now result. The profile of the oscillations in the present case is quite different from the case of models with global $B - L$. Now, the mass eigenstates obey the characteristic equation $\lambda_n = \pi\kappa^2 v_R^2 R^2 \cot(\pi\lambda_n)$. For the weak eigenstate we found $\nu_e = \cos\theta\ \nu_0 + \sin\theta\ \tilde{\nu}_0$, where $\tan\theta = \frac{hv}{\kappa v_R}$; and $\tilde{\nu}_0$ is given in terms of the mass eigenstates as $\tilde{\nu}_0(t) = \sum \frac{1}{\eta_n}\nu_n$, with the mixing factors given by $\eta_n^2 = 2(\lambda_n^2/\zeta^2)(\lambda_n^2 + f(\zeta))$, with $\zeta = \sqrt{2}\kappa v_R R$ and $f(\zeta)$ as before. Thus, in this case the KK contributions enter in ν_e trough the universal mixing angle θ, in contrast with Eq. (4.5). Now, the survival probability after the neutrino traverses a distance L in vacuum reads $P_{surv}(L) = \cos\theta^4 + \sin\theta^4\ |\langle\tilde{\nu}_0|\tilde{\nu}_0(L)\rangle|^2 + 2\ \cos\theta^2\sin\theta^2\ Re\langle\tilde{\nu}_0|\tilde{\nu}_0(L)\rangle$. The averaged probability has the form $\overline{P_{surv}} = \cos\theta^4 + \frac{2}{3}\sin\theta^4$, which is smaller than the two neutrino case with the same mixing angle, though, for small mixing it approaches the former result.

Further analysis extending the present model to three brane generations was presented in Ref. [16]. There, seesaw Majorana terms were included, and a single bulk neutrino plays the role of a sterile neutrino with its lightness associated to the geometry of the extra dimension. The possible scenario that explain the neutrino anomalies could be as follows: solar data is given by $\nu_e \to \nu_B$ oscillations and matter effects, this implies a submillimeter radius. Atmospheric data, is provided by the usual $\nu_\mu \to \nu_\tau$ oscillations thanks to a natural decoupling of this sector from the KK modes. Finally, LSND is explained by a small $\nu_e - \nu_\mu$ mixing.

V STANDARD MODEL IN EXTRA DIMENSIONS

Considering the possibility that the SM particles propagate on the extra dimensions drives the models back to the former KK theories, nevertheless, besides the new possibility of having large extra dimensions, the fact that some particles could be still attached to the branes makes this scenario quite different from the former one. Before considering any possible case one must notice that the conservation of the charges, that is, the consistency of the local gauge symmetry, implies that the first natural candidates to propagate in the bulk are the gauge bosons. Once they are promoted to be bulk fields we will think on the SM fields as the zero modes of higher dimensional fields. However, as there is not experimental evidence of light copies of Z, W, etc., we lead to the conclusion that this models can not have a too large extra dimension. The current experimental data provide lower bounds for the size of R just at the TeV scale, suggesting that these extra dimensions may show up in the near future at the colliders. The whole scenario could be as follows: There are several extra dimensions. The SM particles are free to propagate within one (or more) p-brane(s), where $p > 3$ and where the largest extra compact dimensions are about TeV's, while gravity lives in a higher dimensional bulk with some large (millimetric) extra dimensions. It is worth mentioning that this scenario could in fact be realized from string theory. To simplify the discussion of this models we will follow Ref. [23], although we recommend the reader to see also the important early works, some of which are given in Refs. [24–27].

A Theoretical setup

Let us consider again five dimensions, with an orbifolded fifth dimension. Then, let us assume that the SM gauge fields live in the bulk, while fermions, ψ, and Higgs doublets H_i can either live in the bulk or on the $y = 0$ brane. The analysis will follow for two Higgs doublets to make a possible extension to supersymmetry obvious. The case with only one scalar doublet is straightforward. The lagrangian reads

$$\begin{aligned}\mathcal{L}_{5D} = &-\frac{1}{4}F^2_{MN} + \sum_i \left[(1-\varepsilon^{H_i})|D_M H_i|^2 + (1-\varepsilon^{\psi_i})i\bar{\psi}_i\Gamma^M D_M\psi_i\right] \\ &+\sum_i \left[\varepsilon^{H_i}|D_\mu H_i|^2 + \varepsilon^{\psi_i} i\bar{\psi}_i\sigma^\mu D_\mu\psi_i\right]\delta(y)\,, \end{aligned} \tag{5.1}$$

where $\varepsilon^F = 1\ (0)$ when the F-field lives on the boundary (bulk); $D_M = \partial_M + ig_5 V_M$; $V_M = V^a_M T^a$, with T^a the group generators and g_5 the 5D gauge coupling. Clearly a g'_5 should be introduced for $U(1)_Y$. Notice that the effective four dimensional couplings obeys $g = g_5/\sqrt{\pi R}$, thus the gauge sector should be strongly coupled on the bulk. Gauge and Higgs bosons living in the 5D bulk are assumed to be even under the $\mathcal{Z}_2$ symmetry. We will choose the even assignment for the ψ_L (ψ_R) components of fermions ψ which are doublets (singlets) under $\mathrm{SU}(2)_L$, this is in order to recover the low energy SM spectra.

At intermediate energies, below the compactification scale, M_c, the impact of this theory on standard (boson exchange) processes may be studied by integrating out the KK modes, and summing up the diagrams as we did for Eq. (3.10). Thus, let us introduce the useful small parameter

$$X = \sum_{n=1}^{\infty} \frac{2}{n^2}\frac{m_Z^2}{M_c^2} = \frac{\pi^2}{3}\frac{m_Z^2}{M_c^2}\,, \tag{5.2}$$

and do all the analysis at first order on X. Also, it is useful to introduce the effective mixing angle $s^2_\alpha = \sin^2\alpha = \varepsilon^{H_2}\, s^2_\beta + \varepsilon^{H_1}\, c^2_\beta$ where, as usual, $\tan\beta = \langle H_2\rangle/\langle H_1\rangle$, with $v^2 \equiv \langle H_1\rangle^2 + \langle H_2\rangle^2 \simeq (174\ GeV)^2$. In these terms the charged sector in the four dimensional effective theory and in the unitary gauge has the form [23]

$$\mathcal{L}^{cc}_{eff} = \frac{1}{2}M^2_W W\cdot W - gW\cdot\left[J - s^2_\alpha\, c^2_\theta\, X\, J^{KK}\right] - \frac{g^2}{2\,m_Z^2}X\, J^{KK}\cdot J^{KK}\,, \tag{5.3}$$

where $M^2_W = [1 - s^4_\alpha c^2_\theta X]\; m^2_W$; being $m^2_W = g^2v^2/2$ and θ the usual electroweak mixing angle. J_μ and J^{KK}_μ are the fermion currents of the zero and excited modes respectively. The last term in (5.3) is an effective four points interaction induced by the exchange of the heavy W^{KK}. Note that the tree level mass of the W receive a contribution from its mixing with the excited modes if the Higgs is a brane field ($\varepsilon^H = 1$). This lagrangian induce a tree level correction to the Fermi constant from the μ decay, that reads

$$\frac{G_F}{\sqrt{2}} = \frac{g^2}{8M_W^2}\left[1 + \varepsilon^{\ell_L} c_{2\alpha} c_\theta^2 X\right] . \tag{5.4}$$

On the other hand, for the neutral currents we get at the same limit

$$\begin{aligned}\mathcal{L}_{eff}^{nc} = &\frac{1}{2} M_Z^2\, Z\cdot Z - \frac{e}{s_\theta\, c_\theta} Z \cdot \left[J_Z - s_\alpha^2 X\, J_Z^{KK}\right] - eA\cdot J_{em} \\ &- \frac{1}{2\, M_Z^2}\frac{e^2}{s_\theta^2\, c_\theta^2} X J_Z^{KK} \cdot J_Z^{KK} - \frac{e^2}{2\, M_Z^2} X\, J_{em}^{KK}\cdot J_{em}^{KK} ,\end{aligned} \tag{5.5}$$

where $M_Z^2 = [1 - s_\alpha^4 X]\, m_Z^2$. As before the J's represent the usual four dimensional currents and J^{KK} that corresponding to the matter KK excitations, if they exist. Note that the zero mode of the photon (A) remain massless. From the W and Z masses combined with Eq. (5.4), one may relates the weak mixing angle to the G_F as

$$s_\theta^2\, c_\theta^2 = \frac{\pi\alpha}{\sqrt{2}\, G_F\, M_Z^2}\,(1+\Delta) , \tag{5.6}$$

where the parameter $\Delta = \left[\varepsilon^{\ell_L} c_{2\alpha} c_\theta^2 - s_\alpha^4 s_\theta^2\right] X$; and α is the fine structure constant.

Another ingredient that may be reinstalled on the theory is supersymmetry. Although it is not necessary to be considered, it is an interesting extension. After all, it seems plausible to exist if the high energy theory is string theory. If the theory is supersymmetric, the bulk fields come in $N = 2$ supermultiplets [25,24]. The on-shell field content of the gauge supermultiplet is $V = (V_\mu, V_5, \lambda^i, \Sigma)$ where λ^i ($i = 1, 2$) is a simplectic Majorana spinor and Σ a real scalar in the adjoint representation; (V_μ, λ^1) is even under $\mathcal{Z}_2$ and (V_5, Σ, λ^2) is odd. Matter and Higgs fields are arranged in $N = 2$ hypermultiplets that consist of chiral and antichiral $N = 1$ supermultiplets. The chiral $N = 1$ supermultiplets are even under $\mathcal{Z}_2$ and contain massless states. These will correspond to the SM fermions and Higgses.

Supersymmetry must be broken by some mechanisms that gives masses to all the superpartners which we may assume are of order M_c [24]. For some possible mechanism see Ref. [25]. In contrast with the case of four dimensional susy, where no extra effects appear at tree level after integrating out the superpartners, in the present case integrating out the scalar field Σ induces a tree-level contribution to M_W and Δ parameters. In Ref. [23] they were calculated to be given by $M_W^2 = \left[1 - \varepsilon^{H_1}\varepsilon^{H_2} s_{2\beta}^2 c_\theta^2 X\right] m_W^2$; and $\Delta = \left[\varepsilon^{\ell_L} c_{2\alpha}\, c_\theta^2 - s_\alpha^4 + \varepsilon^{H_1}\varepsilon^{H_2} s_{2\beta}^2\, c_\theta^2\right] X$ respectively. The form of the low energy lagrangian remains.

B Experimental constrains

There are two important effects of gauge KK boson states on collider experiments. (i) Mixing effects and (ii) real production of KK modes.

First, the mixings between the zero and the KK modes of gauge bosons modify the SM observables, affecting the Electroweak precision tests [7,23,49]. Lets consider for instance a specific non supersymmetric model that fixes fermions and one Higgs doublet to the boundary, while gauge bosons and another Higgs doublet propagates on the bulk. That fixes $s_\alpha = s_\beta$ in our expressions above. Then the model has two more parameters than the SM, given by s_β and X, or something equivalent. All observables will be expressed explicitly or implicitly in terms of these and the usual SM parameters. For example, Rizzo and Wells in Ref. [7], introduce the new parameter $V = \frac{M_W^2}{m_Z^2} X$ and the effective interaction couplings $g_W = g[1 - s_\beta^2 V]$ and $g_Z = g[1 - s_\beta^2 V/c_\theta^2]$ to get

$$
\begin{gathered}
G_F(\mu \text{ decay}) = \frac{\sqrt{2} g_W^2}{8 M_W^2}[1+V], \qquad \Gamma(Z \to f\bar{f}) = \frac{N_c M_Z}{12\pi}\left(\frac{g_Z}{2c_\theta}\right)^2 \left[v_f^2 + a_f^2\right], \\
Q_W = \frac{1}{M_Z^2}\left\{\frac{g^2(1 - s_\beta^2 V/c_\theta^2)^2}{c_\theta^2} + \frac{g^2 V}{c_\theta^4}\right\} a_e \left[v_u(2Z+N) + v_d(2N+Z)\right], \\
R = \frac{\sigma_{\rm NC}^\nu - \sigma_{\rm NC}^{\bar\nu}}{\sigma_{\rm CC}^\nu - \sigma_{\rm CC}^{\bar\nu}} = \left[\frac{g_Z^2}{c_\theta^2 M_Z^2} + \frac{g^2 V}{c_\theta^4 M_Z^2}\right]\left[\frac{g_W^2}{M_W^2} + \frac{g^2 V}{M_W^2}\right]^{-1}\left(\frac{1}{2} - s_\theta^2\right), \qquad (5.7) \\
A_f = \frac{2 v_f a_f}{v_f^2 + a_f^2}, \qquad \sin^2\theta_W^{\rm eff} = x + \frac{x(1-x)}{1-2x} V \left[c_\beta^4 - \frac{s_\beta^4}{1-x}\right], \\
M_W^2 = M_Z^2(1-x)\left\{1 + V\left[1 - 2s_\beta^2 - \frac{c_\beta^4(1-x) - s_\beta^4}{1-2x}\right]\right\},
\end{gathered}
$$

where Q_W is a measure of atomic parity violation, x is the solution to the equation $x(1-x) = \pi\alpha/\sqrt{2} G_F m_Z^2$; $v_f \equiv T_{3f} - 2Q_f s_\theta^2$ and $a_f \equiv T_{3f}$. Overall, the limit on M_c using the precision data measurements [52] is just about $M_c \gtrsim 3.3 - 3.8$ TeV [49].

Future colliders may be able to observe resonances due to KK modes if the compactification scale turns out to be on the TeV range. This needs a collider energy $\sqrt{s} \gtrsim M_c$. In hadron colliders (TEVATRON, LHC) the KK excitations might be directly produced in Drell-Yang processes $pp(p\bar{p}) \to \ell^-\ell^+X$ where the lepton pairs ($\ell = e, \mu, \tau$) are produced via the subprocess $q\bar{q} \to \ell^+\ell^+X$. This is the more useful mode to search for $Z^{(n)}/\gamma^{(n)}$ even $W^{(n)}$. Current search for Z' on this channels (CDF) impose $M_c > 510\ GeV$. Future bounds could be raised up to $650\ GeV$ in TEVATRON and $4.5\ TeV$ in LHC, which with $100\ fb^{-1}$ of luminosity can discover modes up to $M_c \approx 6\ TeV$.

Deviations on the cross sections due to virtual exchange of KK modes may be observed in both, hadron and lepton colliders. With a $20\ fb^{-1}$ of luminosity,

TEVATRONII may observe signals up to $M_c \approx 1.3\ TeV$. LEPII with a maximal luminosity of 200 fb^{-1} could impose the bound at 1.9 TeV, while NLC may go up to 13 TeV, which slightly improve the bounds coming from precision test.

VI GAUGE COUPLING UNIFICATION

Once we have assumed a low fundamental scale for quantum gravity, the natural question is whether the former picture of a Gran Unified Theory [53] should be abandoned and with it a possible gauge theory understanding of the quark lepton symmetry and gauge hierarchy. On the other hand, if string theory were the right theory above M an unique fundamental coupling constant would be expect, while the SM contains three gauge coupling constants. Then, it seems clear that, in any case, a sort of low energy gauge coupling unification is required. As pointed out in Ref. [26] and later explored in [27–31], if the SM particles live in higher dimensions, as in the model discused above, such a low GUT scale could be realized.

For comparison let us mention how one leds to gauge unification in four dimensions. Key ingredient in our discussion are the renormalization group equations (RGE) for the gauge coupling parameters that at one loop, in the $\overline{MS}$ scheme, read

$$\frac{d\alpha_i}{dt} = \frac{1}{2\pi} b_i \alpha_i^2 \tag{6.1}$$

where $t = ln\mu$. $\alpha_i = g_i^2/4\pi$; $i = 1, 2, 3$, are the coupling constants of the SM factor groups $U(1)_Y$, $SU(2)_L$ and $SU(3)_c$ respectively. The coefficient b_i receives contributions from the gauge part and the matter including Higgs field and its completely determinated by $4\pi b_i = \frac{11}{3} C_i(vectors) - \frac{2}{3} C_i(fermions) - \frac{1}{3} C_i(scalars)$, where $C_i(\cdots)$ is the index of the representation to which the $(\cdots)$ particles are assigned, and where we are considering Weyl fermion and complex scalar fields. Fixing the normalization of the $U(1)$ generator as in the $SU(5)$ model, we get for the SM $(b_1, b_2, b_3) = (41/10, -19/6, -7)$ and for the Minimal Supersymmetric SM (MSSM) $(33/5, 1, -3)$. Using Eq. (6.1) to extrapolate the values measured at the M_Z scale [52]: $\alpha_1^{-1}(M_Z) = 58.97 \pm .05$; $\alpha_2^{-1}(M_Z) = 29.61 \pm .05$; and $\alpha_3^{-1}(M_Z) = 8.47 \pm .22$, (where we have taken for the strong coupling constant the global average), one finds that only in the MSSM the three couplings merge together at the scale $M_{GUT} \sim 10^{16}$ GeV. This high scale naturally explains the long live of the proton and in the minimal $SO(10)$ framework one gets a very compelling and predictive scenario [54].

A different possibility for unification that does not involve supersymmetry is the existence of an intermediate left-right model [53] that breaks down to the SM symmetry at $10^{11-13}\ GeV$. It is worth mentioning that a non canonical normalization of the gauge coupling may, however, substantially change the above figures, predicting a different unification scale. Such a different normalization may arise either in some no minimal unified models, or in string theories where the SM group factors are realized on non trivial Kac-Moody levels [55,56]. Such scenarios are in general more complicated than the minimal $SU(5)$ or $SO(10)$ models since they introduce new exotic particles.

A Power law running

It is clear that the presence of KK excitations will affect the evolution of couplings in gauge theories and may alter the whole picture of unification of couplings. This question was first studied by Dienes, Dudas and Gherghetta (DDG) [26] on the base of the effective theory approach at one loop level. They found that above the compactification scale M_c one gets

$$\alpha_i^{-1}(M_c) = \alpha_i^{-1}(\Lambda) + \frac{b_i - \tilde{b}_i}{2\pi} \ln\left(\frac{\Lambda}{M_c}\right) + \frac{\tilde{b}_i}{4\pi} \int_{r\Lambda^{-2}}^{rM_c^{-2}} \frac{dt}{t} \left\{ \vartheta_3\left(\frac{it}{\pi R^2}\right) \right\}^{\delta}, \tag{6.2}$$

with Λ as the ultraviolet cut-off and δ the number of extra dimensions. The Jacobi theta function $\vartheta(\tau) = \sum_{-\infty}^{\infty} e^{i\pi\tau n^2}$ reflects the sum over the complete tower. Here b_i are the beta functions of the theory below M_c, and $\tilde{b}_i$ are the contribution to the beta functions of the KK states at each excitation level. The numerical factor r depends on the renormalization scheme. For practical purposes, we may approximate the above result by decoupling all the excited states with masses above Λ, and assuming that the number of KK states below certain energy μ between M_c and Λ is well approximated by the volume of a δ-dimensional sphere of radius $\frac{\mu}{M_c}$ given by $N(\mu, M_c) = X_\delta \left(\frac{\mu}{M_c}\right)^\delta$; with $X_\delta = \pi^{\delta/2}/\Gamma(1+\delta/2)$. The result is a power law behaviour of the gauge coupling constants [32]:

$$\alpha_i^{-1}(\mu) = \alpha_i^{-1}(M_c) - \frac{b_i - \tilde{b}_i}{2\pi} \ln\left(\frac{\mu}{M_c}\right) - \frac{\tilde{b}_i}{2\pi} \cdot \frac{X_\delta}{\delta} \left[\left(\frac{\mu}{M_c}\right)^\delta - 1\right], \tag{6.3}$$

which accelerates the meeting of the α_i's. In the MSSM the energy range between M_c and Λ –identified as the unification (string) scale M– is relatively small due to the steep behaviour in the evolution of the couplings [26,28]. For instance, for a single extra dimension the ratio Λ/M_c has an upper limit of the order of 30, and it substantially decreases for larger δ.

This same relation can be understood on the basis of a step by step approximation [29] as follows. We take the SM gauge couplings and extrapolate their values up to M_c then we add to the beta functions the contribution of the first KK levels, then we run the couplings upwards up to just below the next consecutive level where we stop and add the next KK contributions, and so on, until the energy μ. Despite the complexity of the spectra, the degeneracy of each level is always computable and performing a level by level approach of the gauge coupling running is possible. Above the N-th level the running receives contributions from b_i and of all the KK excited states in the levels below, in total $f_\delta(N) = \sum_{n=1}^{N} g_\delta(n)$, where

$g_\delta(n)$ represent the total degeneracy of the level n. Running for all the first N levels leads to

$$\alpha_i^{-1}(\mu) = \alpha_i^{-1}(M_c) - \frac{b_i}{2\pi}\ln\left(\frac{\mu}{M_c}\right) - \frac{\tilde{b}_i}{2\pi}\left[f_\delta(N)\ln\left(\frac{\mu}{M_c}\right) - \frac{1}{2}\sum_{n=1}^{N} g_\delta(n)\ln n\right]. \quad (6.4)$$

A numerical comparison of this expression with the power law running shows the accuracy of that approximation. Indeed, in the continuous limit the last relation reduces into Eq. (6.3). Thus, gauge coupling unification may now happen at TeV scales [26]. Next, we will discuss how accurated this unification is.

B One step unification

Many features of unification can be studied without bothering about the detailed subtleties of the running. Consider the generic form for the evolution equation

$$\alpha_i^{-1}(M_Z) = \alpha^{-1} + \frac{b_i}{2\pi}\ln\left(\frac{M}{M_Z}\right) + \frac{\tilde{b}_i}{2\pi}F_\delta\left(\frac{M}{M_c}\right), \quad (6.5)$$

where we have changed Λ to M to keep our former notation. Above, α is the unified coupling and F_δ is given by the expression between parenthesis in Eq. (6.4) or its correspondent limit in Eq. (6.3). Note that the information that comes from the bulk is being separated into two independent parts: all the structure of the KK spectra M_c and M are completely embedded into the F_δ function, and their contribution is actually model independent. The only (gauge) model dependence comes in the beta functions, $\tilde{b}_i$. Indeed, Eq. (6.5) is similar to that of the two step unification model where a new gauge symmetry appears at an intermediate energy scale. Such models are very constrained by the one step unification in the MSSM. The argument goes as follows: let us define the vectors: $\mathbf{b} = (b_1, b_2, b_3)$; $\tilde{\mathbf{b}} = (\tilde{b}_1, \tilde{b}_2, \tilde{b}_3)$; $\mathbf{a} = (\alpha_1^{-1}(M_Z), \alpha_2^{-1}(M_Z), \alpha_3^{-1}(M_Z))$ and $\mathbf{u} = (1,1,1)$, and construct the unification barometer [29] $\Delta\alpha \equiv (\mathbf{u}\times\mathbf{b})\cdot\mathbf{a}$. For single step unification models the unification condition amounts to the condition $\Delta\alpha = 0$. As a matter of fact, for the SM $\Delta\alpha = 41.13 \pm 0.655$, while for the MSSM $\Delta\alpha = 0.928 \pm 0.517$, leading to unification within two standard deviations. In this notation Eq. (6.5) leads to

$$\Delta\alpha = [(\mathbf{u}\times\mathbf{b})\cdot\tilde{\mathbf{b}}]\,\frac{1}{2\pi}F_\delta. \quad (6.6)$$

Therefore, for the MSSM, we get the constrain [57]

$$(7\tilde{b}_3 - 12\tilde{b}_2 + 5\tilde{b}_1)F_\delta = 0. \quad (6.7)$$

There are two solutions to the this equation: (a) $F_\delta(M/M_c) = 0$, which means $M = M_c$, bringing us back to the MSSM by pushing up the compactification scale to the unification scale. (b) Assume that the beta coefficients $\tilde{b}$ conspire to eliminate the term between brackets: $(7\tilde{b}_3 - 12\tilde{b}_2 + 5\tilde{b}_1) = 0$, or equivalently [26]

$$\frac{B_{12}}{B_{13}} = \frac{B_{13}}{B_{23}} = 1; \qquad \text{where} \qquad B_{ij} = \frac{\tilde{b}_i - \tilde{b}_j}{b_i - b_j}. \tag{6.8}$$

The immediate consequence of last possibility is the indeterminacy of F_δ, which means that we may put M_c as a free parameter in the theory. For instance we could choose $M_c \sim 10$ TeV to maximize the phenomenological impact of such models. It is compelling to stress that this conclusion is independent of the explicit form of F_δ. Nevertheless, the minimal model where all the MSSM particles propagate on the bulk does not satisfy that constrain [26,28]. Indeed, in this case we have $(7\tilde{b}_3 - 12\tilde{b}_2 + 5\tilde{b}_1) = -3$, which implies a higher prediction for α_s at low M_c. As lower the compactification scale, as higher the prediction for α_s. However, as discussed in Ref. [28] there are some scenarios where the MSSM fields are distributed in a nontrivial way among the bulk and the boundaries which lead to unification. There is also the obvious possibility of adding matter to the MSSM to correct the accuracy on α_s.

The SM case has similar complications. Now Eq. (6.5) turns out to be a system of three equation with three variables, then, within the experimental accuracy on α_i, specific predictions for M, M_c and α will arise. As $\Delta\alpha \neq 0$, the above constrain does not aply, instead the matter content should satisfy the consistency conditions [29]

$$Sign(\Delta\alpha) = Sign[(\mathbf{u} \times \mathbf{b}) \cdot \tilde{\mathbf{b}}] = -Sign(\tilde{\Delta}\alpha) \; ; \tag{6.9}$$

where $\tilde{\Delta}\alpha \equiv (\mathbf{u} \times \tilde{\mathbf{b}}) \cdot \mathbf{a}$. However, in the minimal model where all SM fields are assumed to have KK excitations one gets $\tilde{\Delta}\alpha = 38.973 \pm 0.625$; and $(\mathbf{u} \times \mathbf{b}) \cdot \tilde{\mathbf{b}}^{SM} = 1/15$. Hence, the constraint (6.9) is not fulfilled and unification does not occur. Extra matter could of course improve this situation [26,28,29]. Models with non canonical normalization may also modify this conclusion [29]. A particularly interesting outcome in this case is that there are some cases where, without introducing extra matter at the SM level, the unification scale comes out to be around 10^{11} GeV (for instance $SU(5) \times SU(5)$, $[SU(3)]^4$ and $[SU(6)]^4$). These models fit nicely into the new intermediate string scale models recently proposed in [58], and also with the expected scale in models with local $B - L$ symmetry. High order corrections has been considered in Ref. [30]. The analysis for the running of other coupling constants could be found in [26,31]. Two step models were also studied in [29].

Now well, with a TeV unification scale a extremely fast proton decay mediated by new gauge interactions may occur. There are two possible solutions to this problem. The obvious one is invoking an unified group that keeps the proton stable. A less trivial possibility was suggested in [26]. If the gauge bosons that mediate proton decay are odd under the $\mathcal{Z}_2$ symmetry of the orbifold, then their coupling to the quarks (fixed at the brane) is forbidden, and the proton remains stable. On the context of string theories it may also happen that gauge coupling unification occurs without the appearance of any extra gauge symmetry at the string scale. We should note, however, that this last mechanism does not remove the danger of having a fast proton decay induced by high order operators.

VII SPLITTING WAVE FUNCTIONS ON THICK WALLS

As we already mentioned some mechanism is needed in this theories to forbid dangerous higher dimension operators which lead to proton decay, large neutrino masses, etc., since they are now suppressed just by M. Without the knowledge of the theory above M it is difficult just to assume that such operators are not being induced. One might of course invoke global symmetries again. However, an interesting mechanism that explain how proton decay could get suppressed at the proper level appeared in [33]. It relays on the idea that the branes are being formed from an effective mechanism that traps the SM particles in it, resulting in a wall with thickness $L \sim M^{-1}$, where the fermions are stuck at different points. Then, fermion-fermion couplings get suppressed due to the exponentially small overlaps of their wave functions. This provides a framework for understanding both the fermion mass hierarchy and proton stability without imposing extra symmetries, but rather in terms of a higher dimensional geography [34]. Note that the dimension where the gauge fields propagate does not need to be orthogonal to the millimetric dimensions, but gauge fields may be restricted to live in a smaller part of that extra dimensions. Here we briefly summarize those ideas.

A Localizing wave functions on the brane

Let us start by assuming that the translational invariance along the fith dimension is being broken by a bulk scalar field Φ which develops a spatially varying expectation value $\langle\Phi\rangle(y)$. We assume that this expectation value have the shape of a domain wall transverse to the extra dimension and is centered at $y = 0$. With this background a bulk fermion will have a zero mode that is stuk at the zero of $\langle\Phi(y)\rangle$. To see this let us consider the action

$$S = \int d^4x\ \mathrm{d}y\, \overline{\Psi} \left[i\,\Gamma^M \partial_M + \langle\Phi\rangle(y) \right] \Psi, \tag{7.1}$$

in the chiral basis, as before. By introducing the expansions

$$\Psi_L(x,y) = \sum_n f_n(y)\psi_{nL}(x); \qquad \text{and} \qquad \Psi_R(x,y) = \sum_n g_n(y)\psi_{nR}(x); \tag{7.2}$$

where ψ_n are four dimensional spinors, we get for the y-dependent functions f_n and g_n the equations

$$(\partial_5 + \langle\Phi\rangle) f_n + \mu_n g_n = 0; \qquad \text{and} \qquad (-\partial_5 + \langle\Phi\rangle) g_n + \mu_n f_n = 0. \tag{7.3}$$

Therefore, the zero modes have the profiles [33]

$$f_0(y) \sim \exp\left[-\int_0^y ds\langle\Phi\rangle(s)\right] \qquad \text{and} \qquad g_0(y) \sim \exp\left[\int_0^y ds\langle\Phi\rangle(s)\right]; \tag{7.4}$$

up to normalization factors. Notice that when the extra space is supposed to be finite, both modes are normalizable. For the special choice $\langle\Phi\rangle(y) = 2\mu^2 y$, we get f_0 centered at $y = 0$ with the gaussian form

$$f_0(y) = \frac{\mu^{1/2}}{(\pi/2)^{1/4}} \exp\left[-\mu^2 y^2\right]. \tag{7.5}$$

The other mode has been projected out from our brane by being pushed away to the end of the space. Thus, our theory in the wall is a chiral theory. Notice that a negative coupling among Ψ and ϕ will instead project out the left handed part.

The generalization of this technique to the case of several fermions is straightforward. The action (7.1) is generalized to

$$S = \int d^5x \sum_{i,j} \bar{\Psi}_i [i\,\Gamma^M \partial_M + \lambda\langle\Phi\rangle - m]_{ij} \Psi_j\ , \tag{7.6}$$

where general Yukawa couplings λ and other possible five dimensional masses m_{ij} have been considered. For simplicity we will assume both terms diagonal. The effect of these new parameters is a shifting of the wave functions, which now are centered around the zeros of $\lambda_i\langle\Phi\rangle - m_i$. Taking $\lambda_i = 1$ with the same profile for the vacuum leads to gaussian distributions centered at $y_i = m_i/2\mu^2$. Thus, at low energies, the above action will describe a set of non interacting four dimensional chiral fermions localized at different positions in the fifth dimension.

Localization of gauge and Higgs bosons needs extra assumptions. The explanation of this phenomena is close related with the actual way the brane was formed. A field-theoretic mechanism for localizing gauge fields was proposed by Dvali and Shifman and was later extended and applied in [2]. There, the idea is to arrange for the gauge group to confine outside the wall; the flux lines of any electric sources turned on inside the wall will then be repelled by the confining regions outside and forced to propagate only inside the wall. This traps a massless gauge field on the wall. Since the gauge field is prevented to enter the confined region, the thickness L of the wall acts effectively as the size of the extra dimension in which the gauge fields can propagate. In this picture, the gauge couplings will exhibit power law running above the scale L^{-1}.

B Fermion mass hierarchies and proton decay

Let us consider the Yukawa coupling among the Higgs field and the leptons: $\kappa H L^T E^c$; where the massless zero mode l from L is localized at $y = 0$ while e from

E^c is localized at $y = r$. Let us also assume that the Higgs zero mode is delocalized inside the wall. Then the zero modes term of this coupling will generate the effective Yukawa action

$$S_{Yuk} = \int d^4x\, \kappa\, h(x) l(x) e^c(x) \int dy\ \phi_l(y)\ \phi_{e^c}(y)\ , \tag{7.7}$$

where ϕ_l and ϕ_{e^c} represent the gaussian profile of the fermionic modes. Last integral gives the overlap of the wave functions, which is exponentially suppressed [33] as

$$\int dy\ \phi_l(y)\ \phi_{e^c}(y)\ = e^{-\mu^2 r^2/2}. \tag{7.8}$$

This is a generic feature of this models. The effective coupling of any two fermion fields is exponentially suppressed in terms of their separation in the extra space. Thus, the explanation for the mass hierarchies becomes a problem of the cartography on the extra dimension. A more detailed analysis was presented in [34].

Let us now show how a fast proton decay is evaded in these models. Assume, for instance, that all quark fields are localized at $y = 0$ whereas the leptons are at $y = r$. Then, lets consider the following baryon and lepton number violating operator

$$S \sim \int d^5x\, \frac{(Q^T C_5 L)^\dagger (U^{cT} C_5 D^c)}{M^3}. \tag{7.9}$$

In the four dimensional effective theory, once we have introduced the zero mode wave functions, we get the suppressed action [33]

$$S \sim \int d^4x\ \lambda \times \frac{(ql)^\dagger (u^c d^c)}{M^2} \tag{7.10}$$

where $\lambda \sim \int dy\ \left[e^{-\mu^2 y^2}\right]^3 e^{-\mu^2 (y-r)^2} \sim e^{-3/4\mu^2 r^2}$. Then, for a separation of $\mu r = 10$ we obtain $\lambda \sim 10^{-33}$ which renders these operators completely safe even for $M \sim 1$ TeV. Therefore, we may imagine a picture where quarks and leptons are localized near opposite ends of the wall so that $r \sim L$. This mechanism, however, does not work for suppressing the other dangerous operator $(LH)^2/M$ responsible of a large neutrino masses.

VIII RANDALL SUNDRUM MODEL AND OTHER CURRENT TRENDS

To close our present discussion allow us to mention another important direction of research in this area. It was motivated by a seminal work of Randall and Sundrum [35], who proposed a drastic change on our present point of view of the way the bulk enters in the explanation of the hierarchy problem. Here we summarize some aspects of this model and some further trends [36–44] that may give a rough idea of the way this area is going.

A Mass Hierarchy from a Small Extra Dimension

Lets consider the following setup. A five dimensional space with an orbifolded fifth dimension. Consider two branes at the fixed points $y = 0, \pi$; with tensions σ and $-\sigma$ respectively. Assume that the bulk has a cosmological constant Λ. Contrary to our previous philosophy, here let us assume that all parameters are of the order of the Planck scale. Moreover, we will not longer assume that the bulk has a flat metric, instead we will consider a non factorizable geometry induced by the (explicitly) broken translational invariance. Thus, the more general metric that respects four dimensional Poincare invariance on the brane has the form:

$$ds^2 = G_{AB}dx^A dx^B = e^{-2\beta(y)} g_{\mu\nu}(x)dx^\mu dx^\nu - r^2 dy^2. \tag{8.1}$$

By solving the Einstein equations with this ansatz for the metric, it turns out that σ and Λ need to be related by the (fine tunning) condition

$$\Lambda = -\frac{\sigma^2}{6M^3}; \tag{8.2}$$

that is equivalent to the exact cancellation of the effective four dimensional cosmological constant. On the other hand, one gets $\beta(y) = kr|y|$, where $k^2 = \Lambda/6M^3$. The effective Planck scale is then given by

$$M_{P\ell}^2 = \frac{M^3}{k}\left(1 - e^{-2kr\pi}\right). \tag{8.3}$$

The effect of this metric on the brane fields parameters is non trivial. Lets consider for instance the Higgs action for the brane at the end of the space, where we assume all SM fields are fixed, it is given by

$$S_H = \int d^4x\sqrt{-g}\; e^{-4kr\pi}\left[e^{2kr\pi} g^{\mu\nu}\partial_\mu H \partial_\nu H - \lambda\left(H^2 - \hat{v}_0^2\right)^2\right]. \tag{8.4}$$

After introducing the normalization $H \to e^{kr\pi} H$ that recovers the canonical kinetic term, the above action becomes

$$S_H = \int d^4x\sqrt{-g}\left[g^{\mu\nu}\partial_\mu H \partial_\nu H - \lambda\left(H^2 - v^2\right)^2\right], \tag{8.5}$$

where the vacuum $v = e^{-kr\pi}\hat{v}_0$. Therefore, by choosing $kr \sim 12$, the physical mass of the Higgs field, its vacuum, and, thus all the SM masses appear at the TeV scale without needness or a large hierarchy on the radius, even though the original mass parameter $\hat{v}_0 \sim M \sim M_{P\ell}$. Notice that on the contrary, any field located on the other brane will get a mass of the order of M. Moreover, it also implies that no new particles exist in our world with masses larger than TeV.

B Some about the Phenomenology

Despite this impressive property of the model, which has attracted a lot of attention. It also turns out that the effective cut off of the theory is also at the TeV scale. Indeed, all high order operators get now suppressed just by $m_0 \sim e^{-kr\pi}M$, that is at the TeV range. The reason is as follows. Any operator of dimension n, Θ_n, is originally suppressed by the fundamental scale $M \sim M_{P\ell}$. However, under the change on the normalization of the fields, $\Theta_n \to e^{-nkr\pi}\Theta_n$. Therefore, on the effective theory we get the large enhancement $M \to m_0$. So far no definite solution to the dangerous presense of those operators exist, tough one might impose global symmetries again. On the other hand, this scenario has not yet been realized on the context of string theory.

Kaluza Klein modes in this model also get masses at the TeV scale, and the zero modes remain massless. They may get mass from brane contributions. The radion, however, will pick up a mass from the stabilization potential [36,37]. The intriguing possibility that the extra space could be actually infinite was presented in [38]. Some phenomenological bounds on the effect of KK gravitons are found in [39]. Neutrino masses were analyzed in [40]. Here, only the zero modes get tiny masses. Since the KK modes are heavy, they do not participate on the mixings. Thus, there is not a clear experimental signature for this models in the neutrino sector. A further application of this model, and its extensions, could be a possible resolution of the cosmological constant problem [41]

Cosmology on this models has received a lot of attention. It was early observed [42] that the Hubble parameter H is proportional to the density on the brane, ρ, instead of the usual $H \sim \sqrt{\rho}$ of the standard big bang cosmology. This could be disastrous for late cosmology and BBN. Nevertheless, at low energies the leading order on H has the right behaviour [37,43]. A big deal of work has been devoted to further study those ideas [44,45].

C Radion stabilization

Other interesting aspect of the RS framework is the recent considerations of the stabilization mechanism provided by a bulk scalar field. The idea is as follows. Consider the bulk scalar action

$$S_\phi = \frac{1}{2}\int d^5x\sqrt{-G}\left(G^{AB}\partial_A\phi\partial_B\phi - m^2\phi^2\right). \tag{8.6}$$

Let us assume that the scalar field satisfies certain boundary conditions associated to its couplings to the visible and hidden branes (at $y = 0$ and $y = \pi$ respectively). For instance

$$S_{h,v} = -\int d^4x\sqrt{-g_{h,v}}\,\lambda_{h,v}\left(\phi^2 - v_{h,v}^2\right)^2. \tag{8.7}$$

Those terms cause ϕ to develop a y-dependent vacuum which is determined classically by solving the equation of motion. Inserting this solution into S_ϕ and integrating over y yields an effective potential for the radius of the form [36]

$$V_{eff}(r) = 4ke^{-4kr\pi}\left(v_v - v_h e^{-\epsilon kr\pi}\right)^2 ; \tag{8.8}$$

where $\epsilon = m^2/4k^2 \ll 1$. This has a minimun at

$$kr = \left(\frac{4}{\pi}\right)\frac{k^2}{m^2}\ln\left[\frac{v_h}{v_v}\right] . \tag{8.9}$$

With the logarithmic part of the order of one, we just need k^2/m^2 of order 10 to get the right order in kr. Notice that despite the fact that we are only passing the small hierarchy on kr to the ratio among k and m, this model provides a stabilization potential that also gives a mass for the radion in the TeV range or so [36,37].

IX CONCLUDING REMARKS

The wave induced by the seminal works in [2,35] has generated a big industry that is still growing. Several aspects of this higher dimensional models have been investigated in the recent years and more is yet to come. As any new field, the study of physics on large extra dimensions is still confronting several criticism that eventually have possed serious challenges. So far the field have survived to many of these test, though several open questions remain. Several ingenious applications have attracted the attention of the community and created new directions of research. We mentioned along these notes what we believe are some of the best applications of the idea. However we must stress that some of them are still disconnected of the other parts. It is fair to say that yet a definite and comprehensive model of our world in this framework does not exist. So far, only separate pieces of the puzzle have been analyzed. There are many unsolved problems not just of technical nature but of fundamental nature. There is the problem of stabilization of the large extra dimensions; connecting all the phenomenology (neutrino masses, proton decay, etc.) in a single picture; and studying the impact of this scenario in other well established areas of physics. There is for instance the realization of Randall Sundrum and other models from string theory.

Besides the hope of observing deviations in the Newton law at small distances in the near future, neutrino physics on this models seems very promising too. On the other hand, it is possible that the collider experiments may only increase the lower bounds on the fundamental scale (as for other models of new physics), although we could also discover the first signals of extra dimensions on the next colliders. Besides, let us mention that while in these theories it seems possible to maintain gauge coupling unification, it is not yet clear whether it leads to a compelling scenario that may provide any light on the origin of other SM parameters. Certainly it seems clear that several other mechanisms must be invoked in contrast with the progress that it has been made on four dimensional theories [54]. There is, however, the hope that the years to come helps us either on conforming a more accurate picture or, perhaps on ruling out these ideas. Surely, the coming years will see a lot more on this topic, and perhaps new ideas and results could set it on the side of the well established world of physical theories. Meanwhile, most of the present results remain speculative, although well motivated. As always, Nature has the last word.

The present notes have been prepared intending to be a first introductive guide to the newborn field of models in large (and short) extra dimensions. Unable to make reference to all existing works in the area, we have tried to collect those we believed are relevant for our goal, although some important works could have been omitted. We advise the interested reader to consult the more extended literature that exist already.

Acknowledgments. I would like to thank the warm hospitality of the Particle Physics Group at the UMCP during the last two years. I have been beneficed by the collaboration of R.N. Mohapatra, S. Nandi, C.A. de S. Pires, and L. Díaz, and by the discussions on this topics with J.C. Pati, S. Nussinov, M. Luty, E. Pontón, C. Csaki, Z. Chacko, K. Dienes, to whom I am very thankful. This work was supported in part by CONACyT (México).

REFERENCES

1. Th. Kaluza, Sitzungober. Preuss. Akad. Wiss. Berlin, p. 966 (1921); O. Klein, Z. Phys. **37** (1926) 895.
2. N. Arkani-Hamed, S. Dimopoulos and G. Dvali, Phys. Lett.**B429** (1998) 263; Phys. Lett.**B436** (1999) 257; I. Antoniadis, *et al.*, Phys. Lett.**B436** (1999) 506; I. Antoniadis, S. Dimopoulos, G. Dvali, Nucl. Phys. **B516** (1998) 70.
3. J.C. Long, H.W. Chan and J.C. Price, Nucl. Phys. **B539** (1999) 23.
4. G. Giudice, R. Rattazzi and J. Wells, Nucl. Phys. **B544** (1999) 3; T. Han, J. Lykken and R. J. Zhang, Phys. Rev. D**59** (1999) 105006.
5. See for instance: E. Mirabelli, M. Perelstein and M. Peskin, Phys. Rev. Lett**82** (1999) 2236; S. Nussinov, R. Shrock, Phys. Rev. **D59** (1999) 105002; C. Balasz et. al, Phys. Rev. Lett**83** (1999) 2112; J. L. Hewett, Phys. Rev. Lett**82** (1999) 4765; P. Mathew, K. Sridhar and S. Raichoudhuri, Phys. Lett.**B450**S (1999) 343; T. G. Rizzo, Phys. Rev. D**59** (1999) 115010; K. Aghase and N. G. Deshpande, Phys. Lett. **B456** (1999) 60; K. Cheung and W. Y. Keung, Phys. Rev. D**60** (1999) 112003; L3 Coll. Phys. Lett.**B464** (1999) 135.
6. N. Arkani-Hamed, S. Dimopoulos and G. Dvali, Phys. Rev. D**59** (1999) 086004; V. Barger, *et al.*, Phys. Lett.**B461** (1999) 34; L.J. Hall and D. Smith. Phys. Rev. D**60** (1999) 085008; S. Cullen, M. Perelstein, Phys. Rev. Lett. **83** (1999) 268; C. Hanhart, *et al.*, nucl-th/0007016.
7. See for instance: P. Nath and M. Yamaguchi, Phys. Rev. **D60** (1999) 116004; *idem* 116006; P. Nath, Y. Yamada and M. Yamaguchi, Phys. Lett. **B466** (1999) 100; R. Casalbuoni, *et al.*, Phys. Lett. **B462** (1999) 48; M. Masip and A. Pomarol, Phys.

Rev. D **60** (1999) 096005; W. J. Marciano, Phys. Rev. D **60** (1999) 093006; I. Antoniadis, K. Benakli, M. Quiros, Phys. Lett. **B460** (1999) 176; M. L. Graesser, Phys. Rev. **D61** (2000) 074019; T. G. Rizzo, J. D. Wells, Phys. Rev. D **61** (2000) 016007.
8. V. A. Rubakov and M. E. Shaposhnikov, Phys. Lett. **B152** (1983) 136; K. Akama, in Lecture Notes in Physics, 176, Gauge Theory and Gravitation, Proceedings of the International Symposium on Gauge Theory and Gravitation, Nara, Japan, August 20-24, 1982, edited by K. Kikkawa, N. Nakanishi and H. Nariai, (Springer-Verlag, 1983), 267; M. Visser, Phys. Lett **B159** (1985) 22; E.J. Squires, Phys. Lett **B167** (1986) 286; G.W. Gibbons and D.L. Wiltshire, Nucl. Phys. **B287** (1987) 717.
9. I. Antoniadis, Phys. Lett. **B246** (1990) 377; I. Antoniadis, K. Benakli and M. Quirós, Phys. Lett. **B331** (1994) 313.
10. E. Witten, Nucl. Phys. **B 471** (1996) 135; P. Horava and E. Witten, Nucl. Phys. **B460** (1996) 506; *idem* **B475**(1996) 94.
11. J. Lykken, Phys. Rev. **D54** (1996) 3693.
12. For a recent review and references see R. N. Mohapatra, hep-ph/9910365; also see J.C. D'Olivo this same School.
13. K.R. Dienes, E. Dudas and T. Gherghetta, Nucl. Phys. **B557** (1999) 25; N. Arkani-Hamed, *et al.*, hep-ph/9811448.
14. G. Dvali and A.Yu. Smirnov, Nucl. Phys. **B563** (1999) 63.
15. R. N. Mohapatra, S. Nandi and A. Pérez-Lorenzana, Phys. Lett **B466** (1999) 115.
16. R. N. Mohapatra and A. Pérez-Lorenzana, Nucl. Phys. **B576** (2000) 466.
17. R. Barbieri, P. Creminelli and A. Strumia, hep-ph/0002199.
18. A. Faraggi and M. Pospelov, Phys. Lett.**B458** (1999) 237; G. C. McLaughlin, J. N. Ng, Phys. Lett. **B470** (1999) 157; nucl-th/0003023; A. Ioannisian, A. Pilaftsis, Phys.Rev. D**62** (2000) 066001.
19. R. N. Mohapatra and A. Pérez-Lorenzana, hep-ph/0006278.
20. A. Ioannisian and J. W. F Valle, hep-ph/9911349; E. Ma, M. Raidal and U. Sarkar, hep-ph/0006046.
21. R. N. Mohapatra, A. Pérez-Lorenzana and C. A. de S. Pires, To appear.
22. R. N. Mohapatra and A. Pérez-Lorenzana, Phys. Lett.**B468** (1999) 195.
23. A. Delgado, A. Pomarol and M. Quirós, hep-ph/9911252.
24. A. Pomarol and M. Quirós, Phys. Lett.**B438** (1998) 255; I. Antoniadis, *et al.*, Nucl. Phys **B544** (1999) 503; A. Delgado, A. Pomarol and M. Quirós, Phys. Rev. D**60** (1999) 095002; C. D. Carone, Phys.Rev. D61 (2000) 015008.
25. E. A. Mirabelli and M. E. Peskin, Phys. Rev. D **58** (1998) 065002.
26. K.R. Dienes, E. Dudas and T. Gherghetta, Phys. Lett.**B436** (1998) 55; Nucl. Phys. **B537** (1999) 47.
27. Z. Kakushadze, Nucl. Phys. **B548** (1999) 205;
28. D. Ghilencea and G.G. Ross, Phys. Lett.**B442** (1998) 165; C.D. Carone, Phys. Lett.**B454** (1999) 70; P. H. Frampton and A. Răsin, Phys. Lett. **B460** (1999) 313; A. Delgado and M. Quirós, Nucl. Phys. **B559** (1999) 235; D. Dumitru, S. Nandi, Phys.Rev. D**62** (2000) 046006.
29. A. Pérez-Lorenzana and R.N. Mohapatra, Nucl. Phys. **B559** (1999) 255;
30. H.-C. Cheng, B.A. Dobrescu and C.T. Hill, Nucl. Phys. **B573** (2000) 597; Z.

Kakushadze and T.R. Taylor, Nucl. Phys. **B562** (1999) 78; K. Huitu and T. Kobayashi, Phys. Lett. **B470** (1999) 90.
31. T. Kobayashi, *et al.*, Nucl.Phys. **B550** (1999) 99.
32. This power law running was early noted by T. Taylor and G. Veneziano, Phys. Lett. **B212** (1988) 147.
33. N. Arkani-Hamed, M. Schmaltz, Phys. Rev. D**61** (2000) 033005.
34. N. Arkani-Hamed, Y. Grossman and M. Schmaltz, hep-ph/9909411; E.A. Mirabelli and M. Schmaltz, Phys. Rev. D**61** (2000) 113011.
35. L. Randall and R. Sundrum, Phys. Rev. Lett**83** (1999) 3370.
36. W.D. Goldberger and M.B. Wise, Phys. Rev. Lett. **83** (1999) 4922.
37. C. Csaki, *et al.*, Phys. Rev. D**62** (2000) 045015.
38. L. Randall and R. Sundrum, Phys. Rev. Lett**83** (1999) 3370; N. Arkani-Hamed, *et al.*, Phys. Rev. Lett. **84** (2000) 586.
39. H. Davoudiasl, J.L. Hewett, T.G. Rizzo, hep-ph/0006041.
40. Y. Grossman and M. Neubert, Phys. Lett. **B474** (2000) 361;
41. For references see: N. Arkani-Hamed, *et al.*, Phys. Lett. **B480** (2000) 193; J.-W. Chen, M.A. Luty, E. Pontón, hep-th/0003067.
42. P. Binetruy, C. Deffayet and D. Langlois, Nucl. Phys. **B565** (2000) 269.
43. J.M. Cline, C. Grojean and G. Servant, Phys. Rev. Lett. **83** (1999) 4245; C. Csaki, *et al.*, Phys. Lett. **B462** (1999) 34; T. Shiromizu, K. Maeda and M. Sasaki, Phys. Rev. D**62** (2000) 024012.
44. For references see P. Kanti, *et al.*, Phys. Lett. **B468** (1999) 31; P. Kanti, K.A. Olive, M. Pospelov, Phys.Rev. D**61** (2000) 106004; Phys. Lett. **B481** (2000) 386; E.E. Flanagan, S.-H. H. Tye and I. Wasserman, Phys. Rev. D**62** (2000) 024011; R.N. Mohapatra, A. Pérez-Lorenzana, C.A. de S. Pires, hep-ph/0003328; C. Barcelo and M. Visser, Phys. Lett. **B482** (2000) 183.
45. For references on cosmology with large extra dimensions see D.H. Lyth, Phys. Lett. **B448** (1999) 191; N. Kaloper and A. Linde, Phys. Rev. D **59** (1999) 101303; N. Arkani-Hamed, *et al.*, Nucl.Phys. **B567** (2000) 189; G. Dvali, S.H.H. Tye, Phys. Lett. **B450** (1999) 72; R.N. Mohapatra, A. Pérez-Lorenzana, C.A. de S. Pires, hep-ph/0003089.
46. For a review see J. D. Lykken, hep-th/9612114; see also C. Ramírez this same School.
47. For a recent review see: K. Lane, hep-ph/0006143.
48. For a review see J. Polchinski, hep-th/9611050; see also H.H. García-Compeán this same School.
49. T.G. Rizzo, hep-ph/9910255; K. Cheung, hep-ph/0003306; I. Antoniadis and K. Benakli; hep-ph/0007226; and references therein.
50. For a discussion see for instance A. Dobado, A.L. Maroto, hep-ph/0007100.
51. N. Arkani-Hamed, S. Dimopoulos, hep-ph/9811353.
52. C. Caso *et al*, The Europhysics Journal **C3 Nos. 1-4** (1998) 1.
53. For a recent review see: R. N. Mohapatra, hep-ph/9911272.
54. See for instance: J.C. Pati, hep-ph/0005095; K.S. Babu, J.C. Pati and F. Wilczek, Nucl. Phys. **B566** (2000) 33.
55. For a recent discussions see: A. Pérez-Lorenzana, W.A. Ponce and A. Zepeda, Europhys. Lett. **39** (1997) 141; Mod. Phys. Lett. A **13** (1998) 2153; Phys.Rev. D**59**

(1999) 116004; A. Pérez-Lorenzana and W.A. Ponce, Europhys. Lett. **49** (2000) 296.
56. K.R. Dienes, Nucl. Phys. **B488** (1997) 141; K.R. Dienes, A.E. Faraggi and J. March-Russell, Nucl. Phys. **B467** (1996) 44.
57. B. Brahmachari and R.N. Mohapatra, Int. J. of Mod. Phys. A. **10** (1996) 1699.
58. K. Benakli,Phys. Lett.**B447** (1999) 51; C. Burgess, L. Ibañez and F. Quevedo, Phys. Lett.**B447** (1999) 257.

Topics on Strings, Branes and Calabi-Yau Compactifications[1]

Hugo García-Compeán[2] and Oscar Loaiza-Brito[3]

Departamento de Física
Centro de Investigación y de Estudios Avanzados del I.P.N.
Apdo. Postal 14-740, 07000, México D.F., México

Abstract. Basics of some topics on perturbative and non-perturbative string theory are reviewed. After a mathematical survey of the Standard Model of particle physics and GUTs, the bosonic string kinematics for the free case and with interaction is described. The effective action of the bosonic string and the spectrum is also discussed. Five perturbative superstring theories and their spectra is briefly outlined. Calabi-Yau three-fold compactifications of heterotic strings and their relation to some four-dimensional physics are given. T-duality in closed and open strings are surveyed. D-brane definition is provided and some of their properties and applications to brane boxes configurations, in particular to the cube model are discussed. Finally, non-perturbative issues like S-duality, M-theory, F-theory and basics of their non-perturbative Calabi-Yau compactifications are considered.

INTRODUCTION

String theory is by now, beyond the Standard Model (SM) of particle physics, the best and most sensible understanding of all of the basic components of matter and their interactions in an unified scheme. There are well known the 'aesthetic' problems arising in the standard model of particles, they include the hierarchy problem, the abundance of free parameters and the apparent arbitrariness of the flavor and gauge groups. The SM is for this reason commonly regarded as the low energy effective description of a more fundamental theory, which solves these problems (for a nice review see [1]). It is also widely recognized that Quantum Mechanics and General Relativity cannot be reconciliated in the context of a perturbative quantum field theory of point particles. Hence the nonrenormalizability of the general relativity can be regarded (similarly to the standard model case) as a genuine evidence that it is just an effective field theory and new physics associated

1) This survey is dedicated to the memory of the father of H. G.-C., Sr. Mariano García Salazar.
2) E-mail address: compean@fis.cinvestav.mx
3) E-mail address: oloaiza@fis.cinvestav.mx

CP562, *Particles and Fields: Ninth Mexican School,* edited by G. Herrera Corral and L. Nellen

to some fast degrees of freedom should exist at higher energies (for a review, see for instance [2]). String theory propose that these fast degrees of freedom are precisely the strings at the perturbative level and at the non-perturbative level the relevant degrees of freedom are higher-dimensional extended objects called D-branes (dual degrees of freedom).

At the perturbative level String Theory has intriguing generic predictions such as: (*i*) Spacetime supersymmetry, (*ii*) General Relativity and (*iii*) Yang-Mills fields. These subjects interesting by themselves are deeply interconnected in a rich way in string theory.

The study of theories involving D-branes is just in the starting stage and many surprises surely are coming up. Thus we are still at an exploratory stage of the whole structure of the string theory. Therefore the theory is far to be completed and we cannot give yet concrete physical predictions to take contact with collider experiments and/or astrophysical observations. However many aspects of theoretic character, necessary in order to make of string theory a physical theory, are quickly in progress. The purpose of these lectures are to overview the basic ideas to understand these progresses. This paper is an extended version of the lectures presented at the *Ninth Mexican School on Particles and Fields* held at Metepec Puebla. México. We don't pretend to be exhaustive and we will limit ourselves to describe the basics of string theory and some particular new developments like Calabi-Yau compactifications of the M and F theories. We apologize for omiting numerous original references and we prefer to cite review articles and some few seminal papers.

We first survey very briefly some basic concepts of the gravitational field and gauge theories pointing out the difficulties to put they together. After that we overview the string and the superstring theories, including Calabi-Yau compactifications and the relation of strings to physics in four dimensions from the perturbative point of view. T-duality, D-branes and brane boxes configurations is also considered. After that, we devote some time to describe the string dualities and the web of string theories connected by dualities. M and F theories are also briefly described. Finally we overview an approach to non-perturbative Calabi-Yau compactifications of M and F theories.

MOTIVATION FOR USING STRINGS

First we overview the basic structure of General Relativity (GR) and Yang-Mills (YM) theory in four dimensions. They are very different theories. GR for instance, is the dynamical theory of the spacetime metric while quantum YM theories and in general, Quantum Field Theory (QFT) describes the dynamical building blocks of matter in a fixed spacetime background. Here we survey basics aspects of GR and YM theory following closely Ref. [3].

The pure gravitational field is described by a pseudo-Riemannian metric $g_{\mu\nu}$ with $\mu, \nu = 0, 1, 2, 3$ (on a four-dimensional manifold M) satisfying the vacuum Einstein equations, $R_{\mu\nu} = 0$. Einstein equations can be derived from the Einstein-Hilbert action

$$S_{GR} = \frac{1}{16\pi G_N} \int_M d^4x \sqrt{-g} R, \tag{1}$$

where G_N is the Newton's constant. This constant together with $\hbar$ and c, deremines the Planck scale where gravitational effects in the quantum theory are relevant. The mass scale termed Planck mass is $M_{Pl} = \sqrt{\frac{\hbar c}{G_N}} = 1.2 \times 10^{-5} grams$ or equivalently the Planck length $L_{Pl} = \frac{\hbar}{M_{Pl}\, c} \approx 10^{-33} centimeters$.

On the other hand the SM of particle physics is described by the gauge field theory which is a quantum field theory provided with the gauge symmetry structure. If one wants to formulate the gauge theory on a pseudo-Riemmanian manifold spite of the metric $g_{\mu\nu}$ we require from an additional structure on the spacetime *i.e.* a connection A on a G-principal bundle on M: $G \to E \to M$, where G is the SM gauge group, $G = \mathrm{SU}_C(3) \times \mathrm{SU}_L(2) \times \mathrm{U}_Y(1)$. As usual, the gauge field $A_\mu(x)$ given by the connection one-form has associated the field strength $F_{\mu\nu} = \partial_\mu A^a_\nu - \partial_\nu A^a_\mu + i f^a_{bc} A^a_\mu A^b_\nu$ with f^a_{bc} being the structure constants of G. Given any representation $\mathcal{R}$ of G one can construct the *associated vector bundle* $V_{\mathcal{R}}$.

The Yang-Mills action is given by

$$S_{YM} = -\frac{1}{4g^2_{YM}} \int_M g^{\mu\mu'} g^{\nu\nu'} Tr_{\mathcal{R}} F_{\mu\nu} F_{\mu'\nu'}, \tag{2}$$

where $Tr_{\mathcal{R}}$ denotes the trace in the adjoint representation of G.

Now we would like to introduce fermions. The chiral fermions are sections of the the chiral spin bundles $\widehat{S}_\pm$ over spacetime manifold with *Spin* structure M, *i.e.* $\widehat{S} \to M$, where $\widehat{S} = \widehat{S}_+ \oplus \widehat{S}_-$. The fibers are the Clifford modules constructed with the Dirac matrices Γ^μ. Dirac operator is $\not{D} \equiv \Gamma^\mu D_\mu : \Gamma(\widehat{S}) \to \Gamma(\widehat{S})$ with D_μ being the spacetime covariant derivative. In even dimensions Dirac operator decomposes as: $\not{D} = \not{D}_+ \oplus \not{D}_-$ where $\not{D}_\pm : \Gamma(\widehat{S}_\pm) \to \Gamma(\widehat{S}_\mp)$ with

$$\not{D}_- \psi_+ = 0, \qquad \not{D}_+ \psi_- = 0, \tag{3}$$

and $\psi_\pm \in \Gamma(\widehat{S}_\pm)$.

The possibility to add a mass term in the above equation implies that that mass should be of order one in mass Planck unities M_{Pl}. But the mass of the low energies particle m should be much lower than the Planck mass M_{Pl}, *i.e.* $m << M_{Pl}$. A very nice solution can be given by introducing gauge fields in *complex* representations of the gauge group G. In that case the fermions should be sections of the original spin bundle $\widehat{S}$ coupled to the associated vector bundle $V_{\mathcal{R}}$. If $\mathcal{R} \ncong \widetilde{\mathcal{R}}$ then the corresponding bundles are not isomorphic $V_{\mathcal{R}} \ncong V_{\widetilde{\mathcal{R}}}$. So we have four possibilities

$$W_+ = \widehat{S}_+ \otimes V_{\mathcal{R}}, \qquad W_- = \widehat{S}_- \otimes V_{\mathcal{R}},$$

$$\widetilde{W}_+ = \widehat{S}_+ \otimes V_{\widetilde{\mathcal{R}}}, \qquad \widetilde{W}_- = \widehat{S}_- \otimes V_{\widetilde{\mathcal{R}}}.$$

CPT theorem implies that the fermions with different chirality are given by

$$\psi_+ \in \Gamma(\widehat{S}_+ \otimes V_{\mathcal{R}}), \qquad \widetilde{\psi}_- \in \Gamma(\widehat{S}_- \otimes V_{\widetilde{\mathcal{R}}}).$$

This explains why the mass term are not allowed in the right-hand of Eq. (3).

Now define $\mathcal{R} = \oplus_{i=1}^{15} \mathcal{R}_i$ and $\widetilde{\mathcal{R}} = \oplus_{i=1}^{15} \widetilde{\mathcal{R}}_i$ where $\mathcal{R}_i$ and $\widetilde{\mathcal{R}}_i$ are irreducible complex representations of the gauge group of the SM. Define the formal difference

$$\Delta \equiv U \ominus \widetilde{U} \tag{4}$$

between the general complex representation of a particle $U = U_0 \oplus \mathcal{R}$ and its corresponding complex conjugated $\widetilde{U} = U_0 \oplus \widetilde{\mathcal{R}}$ with U_0 being a *real* irreducible representation (irrep). Thus in the computation of Δ only the complex representations are relevant, *i.e.* $\Delta = \mathcal{R} \ominus \widetilde{\mathcal{R}}$.

On the other hand, spontaneously symmetry breaking is then the important mechanism to give mass to the fermions and gauge particles in the SM. The possibility to get lower symmetries through the breaking of the gauge group lead us to consider theories with higher symmetries than the SM and recuperate it by symmetry breaking. These are the Grand Unified Theories (GUTs). The extension of the gauge group G to another of higher dimensionality $\bar{G}$ was an exciting hope for understanding the 'aesthetic' problems of the SM mentioned in the introduction. One of the more successful GUT is the so called SU(5) GUT, where the gauge group $\bar{G}$ is SU(5) and it breaks to the SM group. Computation of Δ for this model consist in taking the formal difference between all irreducible representations of SU(5) and their complex conjugated ones, this gives

$$\Delta = 3\Big(\mathbf{5}^* \oplus \mathbf{10} \ominus \mathbf{5} \oplus \mathbf{10}^*\Big), \tag{5}$$

where $\mathbf{5}$ and $\mathbf{5}^*$ are the fundamental and anti-fundamental representations of SU(5), and $\mathbf{10}$ and $\mathbf{10}^*$ are the antisymmetric part of representation $\mathbf{5} \otimes \mathbf{5}$ of SU(5) and its complex conjugated. The '3' in the front part stands for the mysterious number of generations of quarks and leptons. We will come back later to comment about this mysterious number.

SU(5) is by itself a non-trivial maximal subgroup of SO(10). The GUT with gauge group SO(10) is another candidate for a unified model. The decomposition of irreps of SO(10) in terms of irreps of SU(5) is as follows: the fundamental representation of SO(10) $\mathbf{10}$ decomposes under SU(5) irreps as $\mathbf{10} = \mathbf{5} \oplus \mathbf{5}^*$. SO(10) has two complex conjugated spinor representations of 16 dimensions, they are: $\mathbf{16}$ and $\mathbf{16}^*$. They can be decomposed under SU(5) irreps as, $\mathbf{16} = \mathbf{1} \oplus \mathbf{5}^* \oplus \mathbf{10}$ and $\mathbf{16}^* = \mathbf{1} \oplus \mathbf{5} \oplus \mathbf{10}^*$. Then computation of Δ yields

$$\Delta = 3\Big(\mathbf{16} \ominus \mathbf{16}^*\Big). \tag{6}$$

Higher dimensionality group E_6 is the next candidate for a GUT. This group has complex representations which are: **27** and $\mathbf{27}^*$. Under SO(10) irreps, these representations decompose into the spinor, vector and identity irreps: *i.e.* $\mathbf{27} = \mathbf{16} \oplus \mathbf{10} \oplus \mathbf{1}$. Vector representation is real. Thus Δ is computed easily to get

$$\Delta = 3\Big(\mathbf{27} \ominus \mathbf{27}^*\Big). \tag{7}$$

Bigger exceptional groups like E_8 only has real representations and $\Delta = 0$.

The SM and GUTs are thus unable to answer the arbitrariness of the number of families of lepton and quarks (basically the '3' arising in Eqs. (5), (6) and (7)) as well as the arbitrariness of the gauge group. The hierarchy of lepton and quarks masses, the existence of the Higgs mechanism and the abundance of free parameters are 'aesthetic problems' as they don't contradict any experiment. However its is clear that the explanation of the origin has to come of somewhere beyond SM and GUTs. In the last 15 years we have learned that string theory has the necessary ingredients to solve these potential problems and it is a serious candidate to provide us with a complete unified theory of all known fundamental interactions of nature. In these lectures we attempt to give the very basic notions of some topics of perturbative and non-perturbative string theory.

PERTURBATIVE STRING AND SUPERSTRING THEORIES

In this section we overview some basic aspects of bosonic and fermionic strings. We focus mainly in the description of the spectrum of the theory in the light-cone gauge, the effective action, the description of spectra of the five consistent superstring theories and the perturbative Calabi-Yau compactifications (for details, precisions and further developments see for instance [4–9]).

First of all consider, as usual, the action of a relativistic point particle. It is given by $S = -m \int d\tau \sqrt{-\dot{X}^I \dot{X}_I}$, where X^I are D functions representing the coordinates of the $(D-1,1)$-dimensional Minkowski spacetime (the target space), $\dot{X}^I \equiv \frac{dX^I}{d\tau}$ and m can be identified with the mass of the point particle. This action is proportional to the length of the world-line of the relativistic particle.

In analogy with the relativistic point particle, the action describing the dynamics of a string (one-dimensional object) moving in a $(D-1,1)$-dimensional Minkowski spacetime (the target space) is proportional to the area $\mathbf{A}$ of the worldsheet Σ. We know from the theory of surfaces that such an area is given by $\mathbf{A} = \int \sqrt{det(-g)}$, where g is the induced metric (with signature $(-,+)$) on the worldsheet Σ. The background metric will be denoted by η_{IJ} and $\sigma^a = (\tau, \sigma)$ with $a = 0, 1$ are the local coordinates on the worldsheet. η_{IJ} and g_{ab} are related by $g_{ab} = \eta_{IJ}\partial_a X^I \partial_b X^J$ with $I, J = 0, 1, \ldots, D-1$. Thus the classical action of a relativistic string is given by the Nambu-Goto action

$$S_{NG}[X^I] = -T \int_\Sigma d\tau d\sigma \sqrt{-det(\partial_a X^I \partial_b X^J \eta_{IJ})}, \tag{8}$$

where $T = \frac{1}{2\pi\alpha'}$ is the string tension, X^I are D embedding functions of the worldsheet Σ into the target space X. Now introduce a metric h describing the intrinsic worldsheet geometry, we get a classically equivalent action to the Nambu-Goto action. This is the Polyakov action originally proposed by Brink, di Vecchia, Howe and Zumino

$$S_P[X^I, h_{ab}] = -\frac{1}{4\pi\alpha'} \int_\Sigma d^2\sigma \sqrt{-h} h^{ab} \partial_a X^I \partial_b X^J \eta_{IJ}, \tag{9}$$

where the X^I's are D scalar fields on the worldsheet. Such a fields can be interpreted as the coordinates of spacetime X (target space), $h = \det(h^{ab})$ and $h_{ab} = \partial_a X^I \partial_b X^J \eta_{IJ}$.

Polyakov action has the following symmetries: (*i*) Poincaré invariance, (*ii*) Worldsheet diffeomorphism invariance, and (*iii*) Weyl invariance (rescaling invariance). The energy-momentum tensor of the two-dimensional theory is given by

$$T^{ab} := \frac{1}{\sqrt{-h}} \frac{\delta S_P}{\delta h_{ab}} = \frac{1}{4\pi\alpha'} \Big(\partial^a X^I \partial^b X_I - \frac{1}{2} h^{ab} h^{cd} \partial_c X^I \partial_d X_I \Big). \tag{10}$$

Invariance under worldsheet diffeomorphisms implies that it should be conserved *i.e.* $\nabla_a T^{ab} = 0$, while the Weyl invariance gives the traceless condition, $T^a_a = 0$. The equation of motion associated with Polyakov action is given by

$$\partial_a \Big(\sqrt{-h} h^{ab} \partial_b X^I \Big) = 0. \tag{11}$$

Whose solutions should satisfy the boundary conditions for the open string: $\partial_\sigma X^I|_0^{\ell=\pi} = 0$ (Neumann) and for the closed string: $X^I(\tau,\sigma) = X^I(\tau, \sigma + 2\pi)$ (Dirichlet). Here $\ell = \pi$ is the characteristic length of the open string. The variation of S_P with respect to h^{ab} leads to the constraint equations: $T_{ab} = 0$. From now on we will work in the *conformal gauge*. In this gauge: $h_{ab} = \eta_{ab}$ and equations of motion (11) reduce to the Laplace equation in the flat worldsheet whose solutions can be written as linear superposition of plane waves.

The Closed String

For the closed string the boundary condition $X^I(\tau, \sigma) = X^I(\tau, \sigma + 2\pi)$, leads to the general solution of Eq. (11)

$$X^I(\tau,\sigma) = X_0^I + \frac{1}{\pi T}P^I\tau$$

$$+\frac{i}{2\sqrt{\pi T}}\sum_{n\neq 0}\frac{1}{n}\Big\{\alpha_n^I exp\Big(-i2n(\tau-\sigma)\Big) + \tilde{\alpha}_n^I exp\Big(-i2n(\tau+\sigma)\Big)\Big\} \qquad (12)$$

where X_0^I and P^I are the position and momentum of the center-of-mass of the string and α_n^I and $\tilde{\alpha}_n^I$ satisfy the conditions $\alpha_n^{I*} = \alpha_{-n}^I$ (left-movers) and $\tilde{\alpha}_n^{I*} = \tilde{\alpha}_{-n}^I$ (right-movers).

The Open String

For the open string the corresponding boundary condition is $\partial_\sigma X^I|_0^{\ell=\pi} = 0$ (this is the only boundary condition which is Lorentz invariant) and the solution is given by

$$X^I(\tau,\sigma) = X_0^I + \frac{1}{\pi T}P^I\tau + \frac{i}{\sqrt{\pi T}}\sum_{n\neq 0}\frac{1}{n}\alpha_n^I exp(-in\tau)\cos(n\sigma) \qquad (13)$$

with the matching condition $\alpha_n^I = \tilde{\alpha}_{-n}^I$.

Quantization

The quantization of the closed bosonic string can be carried over, as usual, by using the Dirac prescription to the center-of-mass and oscillator variables in the form

$$[X_0^I, P^J] = i\eta^{IJ},$$
$$[\alpha_m^I, \alpha_n^J] = [\tilde{\alpha}_m^I, \tilde{\alpha}_n^J] = m\delta_{m+n,0}\eta^{IJ},$$
$$[\alpha_m^I, \tilde{\alpha}_n^J] = 0. \qquad (14)$$

One can identify $(\alpha_n^I, \tilde{\alpha}_n^I)$ with the annihilation operators and the corresponding operators $(\alpha_{-n}^I, \tilde{\alpha}_{-n}^I)$ with the creation ones. In order to specify the physical states we first denote the center of mass state given by $|P^I\rangle$. The vacuum state is defined by $\alpha_m^I|0,P^I\rangle = 0$ with $m > 0$ and $P^I|0,P^I\rangle = p^I \mid 0,P^I\rangle$ and similar for the right movings (here $|0,P^I\rangle = |P^I\rangle \otimes |0\rangle$). For the zero modes these states have negative norm (ghosts). However one can choice a suitable gauge where ghosts decouple from the Hilbert space when $D = 26$.

Light-cone Quantization

Now we turn out to work in the so called *light-cone gauge*. In this gauge it is possible to solve explicitly the Virasoro constraints: $T_{ab} = 0$. This is done by removing the light-cone coordinates $X^{\pm} = \frac{1}{\sqrt{2}}(X^0 \pm X^D)$ leaving only the transverse coordinates X^i representing the physical degrees of freedom (with $i, j = 1, 2, \ldots, D-2$). In this gauge the Virasoro constraints are explicitly solved. Thus the independent variables are $(X_0^-, P^+, X_0^j, P^j, \alpha_n^j, \tilde{\alpha}_n^j)$. Operators α_n^- and $\tilde{\alpha}_n^-$ can be written in terms of α_n^j and $\tilde{\alpha}_n^j$ respectively as follows: $\alpha_n^- = \frac{1}{\sqrt{2\alpha' P^+}}(\sum_{m=-\infty}^{\infty} : \alpha_{n-m}^i \alpha_m^i : -2A\delta_n)$ and $\tilde{\alpha}_n^- = \frac{1}{\sqrt{2\alpha' P^+}}(\sum_{m=-\infty}^{\infty} : \tilde{\alpha}_{n-m}^i \tilde{\alpha}_m^i : -2A\delta_n)$. For the open string we get $\alpha_n^- = \frac{1}{2\sqrt{2\alpha' P^+}}(\sum_{m=-\infty}^{\infty} : \alpha_{n-m}^i \alpha_m^i : -2A\delta_n)$. Here $: \cdot :$ stands for the normal ordering and A is its associated constant.

In this gauge the Hamiltonian is given by

$$H = \frac{1}{2}(P^i)^2 + N - A \text{ (open string)}, \quad H = (P^i)^2 + N_L + N_R - 2A \text{ (closed string)} \tag{15}$$

where N is the operator number, $N_L = \sum_{m=-\infty}^{\infty} : \alpha_{-m}\alpha_m :$, and $N_R = \sum_{m=-\infty}^{\infty} : \tilde{\alpha}_{-m}\tilde{\alpha}_m :$. The *mass-shell condition* is given by $\alpha' M^2 = (N - A)$ (open string) and $\alpha' M^2 = 2(N_L + N_R - 2A)$ (closed string). For the open string, Lorentz invariance implies that the first excited state is massless and therefore $A = 1$. In the light-cone gauge A takes the form $A = -\frac{D-2}{2}\sum_{n=1}^{\infty} n$. From the fact $\sum_{n=1}^{\infty} n^{-s} = \zeta(s)$, where ζ is the Riemann's zeta function (which converges for $s > 1$ and has a unique analytic continuation at $s = -1$, where it takes the value $-\frac{1}{12}$) then $A = -\frac{D-2}{24}$ and therefore $D = 26$.

Spectrum of the Bosonic String

Closed Strings

The spectrum of the closed string can be obtained from the combination of the left-moving states and the right-moving ones. The ground state ($N_L = N_R = 0$) is given by $\alpha' M^2 = -4$. That means that the ground state includes a tachyon. The first excited state ($N_L = 1 = N_R$) is massless and it is given by $\alpha_{-1}^i \tilde{\alpha}_{-1}^j |0, P\rangle$. This state can be naturally decomposed into irreducible representations of the little group $SO(24)$ as follows

$$\alpha_{-1}^i \tilde{\alpha}_{-1}^j \mid 0, P\rangle = \alpha_{-1}^{[i} \tilde{\alpha}_{-1}^{j]} \mid 0, P\rangle + \left(\alpha_{-1}^{(i} \tilde{\alpha}_{-1}^{j)} - \frac{1}{D-2}\delta^{ij}\alpha_{-1}^k \tilde{\alpha}_{-1}^k\right) \mid 0, P\rangle$$
$$+ \frac{1}{D-2}\delta^{ij}\alpha_{-1}^k \tilde{\alpha}_{-1}^k \mid 0, P\rangle. \tag{16}$$

The first term of the rhs is interpreted as a spin 2 massless particle G_{ij} (*graviton*). The second term is a range 2 anti-symmetric tensor B_{ij}. While the last term is an scalar field Φ (*dilaton*). Higher excited massive states are combinations of irreducible representations of the corresponding little group SO(25).

Open Strings

For the open string, the ground state includes once again a tachyon since $\alpha' M^2 = -1$. The first exited state $N = 1$ is given by a massless vector field in 26 dimensions. The second excitation level is given by the massive states $\alpha^i_{-2} \mid 0, P\rangle$ and $\alpha^i_{-1}\alpha^j_{-1} \mid 0, P\rangle$ which are in irreducible representations of the little group SO(25).

Interacting Strings and the Effective Action

Interacting Strings

So far we have described the *free* propagation of a closed (or open) bosonic string. In what follows we consider the interaction of these strings. Here we focus in the closed string case, the open case requires from further definitions. The interaction of strings at the perturbative level is just the extension of the technique of Feynman diagrams for point particles to extended objets. The vacuum-vacuum amplitude $\mathcal{A}$ is given by

$$\mathcal{A} \sim \int \mathcal{D}h_{ab}\mathcal{D}X^I exp\Big(iS_P[X^I, h_{ab}]\Big). \tag{17}$$

The interacting case requires to sum over all loop diagrams. In the closed string case it means that we have to sum over all compact orientable surfaces with non-trivial boundary ($\partial\Sigma \neq 0$). In two dimensions these surfaces are completely characterized by their number of holes g (the genus) and boundaries b. The relevant topological invariant is the Euler number $\chi(\Sigma) = \frac{1}{4\pi}\int_\Sigma d^2\sigma\sqrt{-h}R^{(2)}$, where $R^{(2)}$ is the scalar curvature of the worldsheet Σ. In order to include the interaction of strings, the generalization of the Polyakov action consistent with its symmetries is given by

$$S = S_P[X^I, h_{ab}] + \frac{\Phi(X^I)}{4\pi}\int_\Sigma d^2\sigma\sqrt{-h}R^{(2)} + \frac{1}{2\pi}\int_{\partial\Sigma} dsK, \tag{18}$$

where $\Phi(X)$ is an scalar background field and represents the gravitational coupling constant of the two-dimensional Einstein-Hilbert Lagrangian. K in the above equation stands for the geodesic curvature of Σ. If we define the string coupling constant by $g_S \equiv e^{\Phi}$, then Eq. (17) for the case of closed strings generalizes to

$$\mathcal{A} \sim \sum_\chi g_S^{\chi(\Sigma)} \int \mathcal{D}h_{ab}\mathcal{D}X^I exp\Big(iS_P[X^I, h_{ab}]\Big). \tag{19}$$

The amplitude defined on-shell correspond to $g = 0$ and the rest ($g \geq 1$) corresponds to g-loop corrections.

The definition of correlation functions of operators requires of the idea of *vertex operators* $\mathcal{W}_\Lambda$. These operators are defined as

$$\mathcal{V}_\Lambda(k) = \int d^2\sigma\sqrt{-h}\mathcal{W}_\Lambda(\sigma,\tau)exp(ik\cdot X), \tag{20}$$

where $\mathcal{W}_\Lambda(\sigma,\tau)$ (with Λ being a generic massless field of the bosonic spectra) is a local operator assigned to some specific state of the spectrum of the theory. For instance for the tachyon it is given by $\mathcal{W}_T(\sigma,\tau) \sim \partial_a X_I \partial^a X^I$. While that for the graviton G with polarization ζ_{IJ} it is given by $W_G(\sigma,\tau) = \zeta_{IJ}\partial_a X^I \partial^a X^J$. Local operators $\mathcal{V}_\Lambda$ are diffeomorphism and conformal invariant and therefore more convenient to define scattering amplitudes.

Thus one can define the scattering amplitude of the vertex field operators by their corresponding invariant operators $\mathcal{V}_\Lambda$. In perturbation theory the scattering amplitude is given by

$$\mathcal{A}(\Lambda_1, k_1; \dots \Lambda_N, k_N) \sim \sum_\chi g_S^{\chi(\Sigma)} \int \mathcal{D}h\mathcal{D}X exp\Big(iS_P[X^I, h_{ab}]\Big) \prod_{i=1}^N \mathcal{V}_{\Lambda_i}(k_i). \tag{21}$$

This scattering amplitude is, of course, proportional to the correlation function of the product of N invariant operators $\mathcal{V}_{\Lambda_i}(k_i)$ as follows

$$\mathcal{A}(\Lambda_1, k_1; \dots \Lambda_N, k_N) \propto \langle \prod_{i=1}^N \mathcal{V}_{\Lambda_i}(k_i) \rangle. \tag{22}$$

Effective String Actions

In order to make contact with the spacetime physics we now describe how the spacetime equations of motion come from conformal invariance conditions for the non-linear sigma model in curved spaces. The immediate generalization of the Polyakov action is

$$S = -\frac{1}{4\pi\alpha'}\int_\Sigma d^2\sigma\sqrt{-h}h^{ab}\partial_a X^I \partial_b X^J G_{IJ}(X), \tag{23}$$

where $G_{IJ}(X)$ is an arbitrary background metric of the curved target space X. The perturbation of this metric $G_{IJ}(X) = \eta_{IJ} + h_{IJ}(X)$ in the partition function $Z \sim exp(-S[X^I, \eta_{IJ} + h_{IJ}])$, leads to an expansion in powers of h_{IJ}. This partition function can be easily interpreted as containing the information of the interaction of the string with a coherent state of gravitons with invariant operator $\mathcal{V}_G(k) = \int d^2\sigma\sqrt{-h}\mathcal{W}_G(\sigma,\tau)exp(ik\cdot X)$ with $\mathcal{W}_G = h^{ab}\partial_a X^I \partial_b X^J h_{IJ}(X)$.

On the other hand, the Polyakov action can be generalized to be consistent with all symmetries and with the massless spectrum of the bosonic closed strings in the form of a non-linear sigma model

$$\widehat{S} = \frac{1}{4\pi\alpha'}\int_\Sigma d^2\sigma\sqrt{-h}\bigg[\Big(h^{ab}G_{IJ}(X) + i\varepsilon^{ab}B_{IJ}(X)\Big)\partial_a X^I \partial_b X^J + \alpha'\Phi(X)R^{(2)}\bigg], \tag{24}$$

where $G_{IJ}(X)$ is the target space curved *metric*, $B_{IJ}(X)$ is an anti-symmetric field, also called the *Kalb-Ramond* field, and $\Phi(X)$ is the scalar field called the *dilaton* field. From the viewpoint of the two-dimensional non-linear sigma model these *background* fields can be regarded as *coupling constants* and the renormalization group techniques become applied. The computation of the quantum *conformal anomaly* by using the dimensional regularization method, leads to express the energy-momentum trace as a linear combination

$$T_a^a = -\frac{1}{2\alpha'}\beta^G_{IJ}h^{ab}\partial_a X^I \partial_b X^J - \frac{i}{2\alpha'}\beta^B_{IJ}\varepsilon^{ab}\partial_a X^I \partial_b X^J - \frac{1}{2}\beta^\Phi R^{(2)}, \tag{25}$$

where β are the one-loop beta functions associated with each coupling constant or background field. They are explicitly computed and give

$$\beta^G_{IJ} = \alpha'\Big(R_{IJ} + 2\nabla_I\nabla_J\Phi - \frac{1}{4}H_{IKL}H_J^{\ KL}\Big) + O(\alpha'^2), \tag{26}$$

$$\beta^B_{IJ} = \alpha'\Big(-\frac{1}{2}\nabla^K H_{KIJ} + \nabla^K\Phi H_{KIJ}\Big) + O(\alpha'^2), \tag{27}$$

$$\beta^\Phi = \alpha'\Big(\frac{D-26}{6\alpha'} - \frac{1}{2}\nabla^2\Phi + \nabla_K\Phi\nabla^K\Phi - \frac{1}{24}H_{IJK}H^{IJK}\Big) + O(\alpha'^2), \tag{28}$$

where $H_{IJK} = \partial_I B_{JK} + \partial_J B_{KI} + \partial_K B_{IJ}$. Weyl invariance at the quantum level implies the vanishing of the conformal anomaly and therefore the vanishing of each beta function. This leads to three coupled field equations for the background fields. These conditions for these fields can been regarded as equations of motion derivable from the spacetime action in D dimensions

$$S = \frac{1}{2\kappa_0^2}\int_X d^D x\sqrt{-G}e^{-2\Phi}\Big(R + 4\nabla_I\Phi\nabla^I\Phi - \frac{1}{12}H_{IJK}H^{IJK} - \frac{2(D-26)}{3\alpha'} + O(\alpha')\Big), \tag{29}$$

where κ_0 is a normalization constant.

It is interesting to see that a redefinition of background metric under the transformation in D dimensions $\widetilde{G}_{IJ}(X) = exp(2\varpi(X))G_{IJ}$ with $\varpi = \frac{2}{D-2}(\Phi_0 - \Phi)$ leads to the background action in the 'Einstein frame'

$$\widehat{S} = \frac{1}{2\kappa^2}\int d^D x\sqrt{-\widetilde{G}}\Big(\widetilde{R} - \frac{4}{D-2}\nabla_I\widetilde{\Phi}\nabla^I\widetilde{\Phi} - \frac{1}{12}e^{-8\widetilde{\Phi}/(D-2)}H_{IJK}H^{IJK} - \frac{2(D-26)}{3\alpha'}e^{4\widetilde{\Phi}/(D-2)} + O(\alpha')\Big), \tag{30}$$

where $\widetilde{R} = e^{-2\varpi}[R - 2(D-1)\nabla^2\varpi - (D-2)(D-1)\partial_I\varpi\partial^I\varpi]$ and $\widetilde{\Phi} = \Phi - \Phi_0$. The form of this action will of extreme importance later when we describe the strong/weak coupling duality in effective supergravity actions of the different superstring theories types. In the above action $\kappa \equiv \kappa_0 e^{\Phi_0} = \kappa_0 \cdot g_S$ is the gravitational coupling constant in D dimensions, *i.e.* $\kappa = \sqrt{8\pi G_N}$.

A very close procedure can be performed for the open string and compute its effective low energy action. It was done about 15 years ago and for gauge fields with constant field strength F_{IJ} it is given by the Dirac-Born-Infeld action

$$S_O = -T \int d^D x e^{-\Phi} \sqrt{-det(G_{IJ} + B_{IJ} + 2\pi\alpha' F_{IJ})}. \tag{31}$$

Later we shall talk about some applications of this effective action.

Superstrings

In bosonic string theory there are two bold problems. The first one is the presence of tachyons in the spectrum. The second one is that there are no spacetime fermions. Here is where superstrings come to the rescue. A superstring is described, despite of the usual bosonic fields X^I, by fermionic fields $\psi^I_{L,R}$ on the worldsheet Σ. Which satisfy anticommutation rules and where the L and R denote the left and right worldsheet chirality respectively. The action for the superstring is given by

$$L_{SS} = -\frac{1}{8\pi} \int d^2\sigma \sqrt{-h} \Big(h^{ab} \partial_a X^I \partial_b X_I + 2i\bar{\psi}^I \gamma^a \partial_a \psi_I - i\bar{\chi}_a \gamma^b \gamma^a \psi^I (\partial_b X_I - \frac{i}{4} \bar{\chi}_b \psi_I) \Big), \tag{32}$$

where ψ^I and χ_a are the superpartners of X^I and the tetrad field e^a respectively. In the superconformal gauge ($h_{ab} = \eta_{ab}$ and $\chi_a = 0$) and in light-cone coordinates it can be reduced to

$$L_{SS} = \frac{1}{2\pi} \int \Big(\partial_L X^I \partial_R X_I + i\psi^I_R \partial_L \psi_{I\ R} + i\psi^I_L \partial_R \psi_{I\ L} \Big). \tag{33}$$

In analogy to the bosonic case, the local dynamics of the worldsheet metric is manifestly conformal anomaly free at the quantum level if the critical spacetime dimension D is 10. Thus the string oscillates in the 8 transverse dimensions. The action (32) is invariant under: (i) worldsheet supersymmetry, (ii) Weyl transformations, (iii) super-Weyl transformations, (iv) Poincaré transformations and (v) Worldsheet reparametrizations. The equation of motion for the $X's$ fields is the same that in the bosonic case (Laplace equation) and whose general solution is given by Eqs. (12) or (13). Equation of motion for the fermionic field is the Dirac equation in two dimensions. Constraints here are more involved and they are called the *super-Virasoro constraints.* However in the light-cone gauge, everything simplifies and the transverse coordinates (eight coordinates) become the bosonic physical degrees of freedom together with their corresponding supersymmetric partners. Analogously to the bosonic case, massless states of the spectrum come into representations of the little group SO(8) which is a subgroup of SO(9, 1), while that the massive states lie into representations of the little group SO(9).

For the closed string there are two possibilities for the boundary conditions of fermions: (*i*) periodic boundary conditions (Ramond (**R**) sector) $\psi^I_{L,R}(\sigma) = +\psi^I_{L,R}(\sigma + 2\pi)$ and (*ii*) anti-periodic boundary conditions (Neveu-Schwarz (**NS**) sector) $\psi^I_{L,R}(\sigma) = -\psi^I_{L,R}(\sigma + 2\pi)$. Solutions of Dirac equation satisfying these boundary conditions are

$$\psi^I_L(\sigma,\tau) = \sum_n \bar{\psi}^I_{-n} exp\Big(- in(\tau + \sigma)\Big), \qquad \psi^I_R(\sigma,\tau) = \sum_n \psi^I_n exp\Big(- in(\tau - \sigma)\Big), \tag{34}$$

where $\bar{\psi}^I_{-n}$ and ψ^I_n are fermionic modes of left and right movers respectively.

In the case of the fermions in the **R** sector n is integer and it is semi-integer in the **NS** sector.

The quantization of the superstring come from the promotion of the fields X^I and ψ^I to operators whose oscillator variables are operators satisfying the relations $[\alpha^I_n, \alpha^J_m]_- = n\delta_{m+n,0}\eta^{IJ}$ and $[\psi^I_n, \psi^J_m]_+ = \eta^{IJ}\delta_{m+n,0}$, where $[,]_-$ and $[,]_+$ stand for commutator and anti-commutator respectively.

The zero modes of α are diagonal in the Fock space and its eigenvalue can be identified with its momentum. For the **NS** sector there is no fermionic zero modes but they can exist for the **R** sector and they satisfy a Clifford algebra $[\psi^I_0, \psi^J_0]_+ = \eta^{IJ}$. The Hamiltonian for the closed superstring is given by $H_{L,R} = N_{L,R} + \frac{1}{2}P^2_{L,R} - A_{L,R}$. For the **NS** sector $A = \frac{1}{2}$, while for the **R** sector $A = 0$. The mass is given by $M^2 = M^2_L + M^2_R$ with $\frac{1}{2}M^2_{L,R} = N_{L,R} - A_{L,R}$.

There are five consistent superstring theories: Type IIA, IIB, Type I, SO(32) and $E_8 \times E_8$ heterotic strings, represented by HO and HE respectively. In what follows of this section we briefly describe the spectrum in each one of them.

Type II Superstring Theories

In this case the theory consist of closed strings only. They are theories with $\mathcal{N} = 2$ spacetime supersymmetry. There are 8 scalar fields (representing the 8 transverse coordinates to the string) and one Weyl-Majorana spinor. There are 8 left-moving and 8 right-moving fermions.

In the **NS** sector there is still a tachyon in the ground state. But in the supersymmetric case this problem can be solved through the introduction of the called **GSO** projection. This projection eliminates the tachyon in the **NS** sector and it acts in the **R** sector as a ten-dimensional spacetime chirality operator. That means that the application of the **GSO** projection operator defines the chirality of a massless spinor in the **R** sector. Thus from the left and right moving sectors, one can construct states in four different sectors: (*i*) **NS-NS**, (*ii*) **NS-R**, (*iii*) **R-NS** and (*iv*) **R-R**. Taking into account the two types of chirality L and R one has two possibilities:

a)– The **GSO** projections on the left and right fermions produce different chirality in the ground state of the **R** sector (*Type IIA*).
b)–**GSO** projection are equal in left and right sectors and the ground states in the **R** sector, have the same chirality (*Type IIB*). Thus the spectrum for the Type IIA and IIB superstring theories is:

- *Type IIA*

  The **NS-NS** sector has a symmetric tensor field G_{IJ} (spacetime metric), an antisymmetric tensor field B_{IJ} and a scalar field Φ (dilaton). In the **R-R** sector there is a vector field A_I associated with a 1-form $A_{(1)}$ ($A_I \Leftrightarrow A_{(1)}$) and a rank 3 totally antisymmetric tensor $A_{IJK} \Leftrightarrow A_{(3)}$ and by Hodge duality in ten dimensions also we have $A_{(5)}$, $A_{(7)}$ and $A_{(9)}$. In general the **R-R** sector consist of p-forms $F_{(p)} = dA_{(p-1)}$ (where $A_{(p)}$ are called RR fields) on the ten-dimensional spacetime X with p *even i.e.* $F_{(2)}, F_{(4)}, \ldots, F_{(10)}$. In the **NS-R** and **R-NS** sectors we have two gravitinos with opposite chirality and the supersymmetric partners of the mentioned bosonic fields.

- *Type IIB*

  In the **NS-NS** sector Type IIB theory has exactly the same spectrum that of Type IIA theory. On the **R-R** sector it has a scalar field $a \Leftrightarrow A_{(0)}$ (the axion field), an antisymmetric tensor field $B'_{IJ} \Leftrightarrow A_{(2)}$ and a rank 4 totally antisymmetric tensor $D_{IJKL} \Leftrightarrow A_{(4)}$ whose field strength is self-dual *i.e.*, $F_{(5)} = dA_{(4)}$ with $*F_{(5)} = +F_{(5)}$. Similar than for the case of Type IIA theory one has also the Hodge dual fields $A_{(6)}$, $A_{(8)}$, $A_{(10)}$. In general, RR fields in Type IIB theory are given by p-forms $F_{(p)} = dA_{(p-1)}$ on the spacetime X with p *odd i.e.* $F_{(1)}, F_{(3)}, \ldots, F_{(11)}$. The **NS-R** and **R-NS** sectors do contain two gravitinos with the same chirality and the corresponding fermionic matter.

Type I Superstrings

In this case the L and R degrees of freedom are identified. Type I and Type IIB theories have the same spectrum, except that in the former one the states which are not invariant under the change of orientation of the worldsheet, are projected out. This worldsheet parity Ω interchanges the left and right modes. Type I superstring theory is a theory of breakable closed strings, thus it incorporates also open strings. The Ω operation leave invariant only one half of the spacetime supersymmetry, thus the theory is $\mathcal{N} = 1$.

The spectrum of bosonic massless states in the **NS-NS** sector is: G_{IJ} (spacetime metric) and Φ (dilaton) from the closed sector and B_{IJ} is projected out. On the **R-R** sector there is an antisymmetric field B_{IJ} of the closed sector. The open string sector is necessary in order to cancel tadpole diagrams. A contribution to the spectrum come from this sector. Chan-Paton factors can be added at the boundaries of open strings. Hence the cancellation of the tadpole are needed 32 labels at each end. Therefore in the **NS-NS** sector there are 496 gauge fields in the adjoint representation of SO(32).

Heterotic Superstrings

This kind of theory involves only closed strings. Thus there are left and right sectors. The left-moving sector contains a bosonic string theory and the right-moving sector contains superstrings. This theory is supersymmetric on the right sector only, thus the theory contains $\mathcal{N} = 1$ spacetime supersymmetry. The momentum at the left sector P_L lives in 26 dimensions, while P_R lives in 10 dimensions. It is natural to identify the first ten components of P_L with P_R. Consistency of the theory tell us that the extra 16 dimensions should belong to the root lattice $E_8 \times E_8$ or a $\mathbf{Z}_2$-sublattice of the SO(32) weight lattice.

The spectrum consists of a tachyon in the ground state of the left-moving sector. In both sectors we have the spacetime metric G_{IJ}, the antisymmetric tensor B_{IJ}, the dilaton ϕ and finally there are 496 gauge fields A_I in the adjoint representation of the gauge group $E_8 \times E_8$ or SO(32).

All these Types of superstring theories do admit a low energy effective description in terms of a supergravity theory. These theories involves the corresponding background fields of their spectra. Supergravity actions of these diverse types will be constructed later to study strong/weak coupling duality in string theory.

Calabi-Yau Compactifications in Perturbative String Theory

In order to connect superstring theories to the observed 4-dimensional spacetime physics, we have to reduce the critical dimension $D = 10$ to four dimensions. To preserve certain supersymmetry consistent with chirality in four dimensions it is necessary to require some properties to the ten dimensional spacetime X. Perhaps the simplest ansatz is to assume that the four-dimensional Minkowski spacetime M and a six-dimensional internal space $\mathcal{K}$ *factorizes* as $X \cong M \times \mathcal{K}$, where $\mathcal{K}$ has tiny dimensions and unobservable in our present experiments. It is worth to say that this factorization ansatz is not unique and other possibility is the *warped compactification* of the celebrated Randall-Sundrum scenarios, which are nicely reviewed in Ref. [10].

It is useful to classify the compactifications according to how much supersymmetries is broken, because this number is related with the quantum corrections that we shall consider. We choose $\mathcal{K}$ to be a manifold with the property that a certain number of supersymmetries are preserved[4].

---

4) Cosmological constant is generated by perturbation theory. Strings propagating in a $\mathcal{K}$ mani-

We are now looking for conditions in the background which leave some supersymmetry unbroken. These conditions are given by null variations of the Fermi fields.

Consider the *diagonal* metric for ten-dimensional spacetime X given by $G_{IJ} = f(y)\eta_{\mu\nu} + G_{mn}(y)$ where y denotes the compactified coordinates and $I, J = 0, ..., 9$, $\mu, \nu = 0, ..., 3$, $m, n = 4, ..., 9$. For $D = 10$, $\mathcal{N} = 1$ heterotic string theory the Fermi fields variations are:

- *gravitino*: $\nabla\psi_\mu = \Delta_\mu\varepsilon$, $\delta\psi_m = (\partial_m + \frac{1}{4}\Omega^-_{mnp}\Gamma^{np})\varepsilon$,
- *dilatino*: $\delta\xi = (\Gamma^m\partial_m\phi - \frac{1}{12}\Gamma^{mnp}H_{mnp})\varepsilon$,
- *gaugino*: $\delta\lambda = F_{mn}\Gamma^{mn}\varepsilon$,

where ε is a Weyl-Majorana spinor in ten dimensions, Ω^-_{mnp} is the internal component of $\Omega^-_{MNP} = \omega_{MNP} - \frac{1}{2}H_{MNP}$, Γ are the Dirac matrices.

The compactification ansatz $X = M \times \mathcal{K}$ breaks the Lorentz group SO(9,1) into $\mathrm{SO}(3,1) \times \mathrm{SO}(6)$. In the spinor representation **16** the Weyl-Majorana supersymmetry parameter $\varepsilon_{\alpha\beta}$ decomposes as $\varepsilon(y) \to \varepsilon_{\alpha\beta}(y) + \varepsilon^*_{\alpha\beta}(y)$ under $\mathbf{16} \to (\mathbf{2}, \mathbf{4}) \oplus (\mathbf{2}^*, \mathbf{4}^*)$. The general form of $\varepsilon_{\alpha\beta}$ is $\varepsilon_{\alpha\beta} = u_\alpha\zeta_\beta(y)$ with u_α an arbitrary Weyl spinor. When we put the condition that Fermi fields variations vanish, then each internal spinor $\zeta_\beta(y)$ gives the minimal ($\mathcal{N} = 1$) $D = 4$ supersymmetric algebra.

Now, by the null fermi fields variations we can find conditions in the background fields assuming that $H_{mnp} = 0$. These are:

- $\delta\zeta = 0 \Rightarrow \partial_m\Phi = 0$,
- $\delta\psi = 0 \Longrightarrow G_{\mu\nu} = \eta_{\mu\nu}$,
- $\delta\psi_m = 0 \Rightarrow \nabla_m\zeta = 0$.

The last equation tell us that ζ_β is covariantly constant on the internal space $\mathcal{K}$, and implies that $\mathcal{K}$ is Ricci-flat. This is because $[\nabla_m, \nabla_n]\zeta = \frac{1}{4}R_{mnpq}\Gamma^{pq}\zeta = 0$. For this reason, in general Γ^{pq} do not belong to SO(6) but to SU(3), which is a subgroup that leaves one component of the spinor ζ invariant. Thus the compact manifold $\mathcal{K}$ must have SU(3) holonomy. The second unbroken susy condition implies that the warped factor $f(y)$ in metric is 1 and the metric G_{IJ} is unwarped. Finally, the first condition implies that the dilation is constant. This is a *Calabi-Yau* three-fold. A Calabi-Yau three-fold is also a Kähler manifold in which the first Chern class zero *i.e.* $c_1(T\mathcal{K}) = 0$. Any Calabi-Yau manifold possesses a unique Ricci-flat metric. When we consider $\mathcal{N} = 1$ heterotic string theory on Calabi-Yau three-fold we obtain a four-dimensional chiral theory with spacetime supersymmetry $\mathcal{N} = 1$. In fact, compactification on manifolds of SU(3) holonomy preserves 1/4 of supersymmetry. If we consider $\mathcal{N} = 2$ theories (for example, type II superstrings)

---

fold in which all supersymmetries are broken distablizes the Minkowski vacuum.

in $D = 10$ dimensions, after compactification on a Calabi-Yau three-fold we obtain $\mathcal{N} = 2$ theories in $D = 4$.

In addition to the CY-threefold structure for $\mathcal{K}$ the unbroken susy condition $\delta\lambda^a = 0 = F^a_{mn}\Gamma^{mn}\varepsilon$, leads to the equations in complex coordinates

$$F_{IJ} = F_{\bar{I},\bar{J}} = 0, \qquad G^{I\bar{J}}F_{I\bar{J}} = 0. \tag{35}$$

These equations require to specify a gauge subbundle V of a $E_8 \times E_8$ gauge bundle over $\mathcal{K}$ and a gauge connection A on V with curvature F. The condition $F_{IJ} = F_{\bar{I},\bar{J}} = 0$ tell us that the subbundle V as well as the corresponding connection should be holomorphic. The second condition $G^{I\bar{J}}F_{I\bar{J}} = 0$ is the celebrated Donaldson-Uhlenbeck-Yau equation for A. This equation has a unique solution if the bundle V is stable and if it is satisfied the integrability condition $\int_{\mathcal{K}} \Omega^{n-1} \wedge c_1(V) = 0$, where Ω is the Kähler form of $\mathcal{K}$. There is a further condition to be satisfied by the connection A, the Bianchi identity for H and F, it is given by

$$dH = trR \wedge R - \frac{1}{30} trF \wedge F. \tag{36}$$

The only solution is $trR \wedge R \propto trF \wedge F$ which implies that $c_2(T\mathcal{K}) = c_2(V)$. This situation is usually known as the *standard embedding* of the spin connection in the gauge connection ant it is a method to determine the connection A on V.

Thus in the compactification of phenomenological interest of the heterotic theory with the ansatz $X = M \times \mathcal{K}$, the internal space has to be a Calabi-Yau three-fold and one has to specify a stable, holomorphic vector bundle V over X (or $\mathcal{K}$) satisfying $c_1(V) = 0$ and $c_2(V) = c_2(TX)$.

If V is a SU(n) vector bundle over X the subgroups of $E_8 \times E_8$ that commutes are E_6, SO(10) and SU(5) for $n = 3, 4, 5$ respectively. This leads to GUTs in four dimensions justly with the gauge groups E_6, SO(10) or SU(5).

Massless Spectrum

In order to describe the impact of the characteristics of $\mathcal{K}$ and V on the properties of the spectrum of the four dimensional theory we start by decomposing the ten-dimensional Dirac operator under $M \times \mathcal{K}$ into

$$\not{D}^{(10)} = \sum_{I=0}^{9} \Gamma^I D_I = \not{D}^{(4)} + \not{D}_{\mathcal{K}}, \tag{37}$$

where $\not{D}^{(4)} = \sum_{I=0}^{3} \Gamma^I D_I$ and $\not{D}_{\mathcal{K}} = \sum_{J=4}^{9} \Gamma^J D_J$. Dirac equation in ten dimensions is

$$\not{D}^{(10)}\Psi(x^I, y^J) = (\not{D}^{(4)} + \not{D}_{\mathcal{K}})\Psi(x^I, y^J). \tag{38}$$

Thus the spectrum of the Dirac operator $\not{D}_{\mathcal{K}}$ on $\mathcal{K}$ determines the massive spectrum of fermions in four dimensions.

In ten dimensions the Lorentz group only has *real* spinor representations and the Clifford modules decomposes as: $S^{(10)} = S_+^{(10)} \oplus S_-^{(10)}$. Positive and negative chirality are distinguised by $\Gamma^{(10)} = \Gamma^0\Gamma^1 \ldots \Gamma^9$. CPT theorem implies that we must take only one chirality

$$\Gamma^{(10)}\Psi = +\Psi. \tag{39}$$

Decompose the spinor representation of SO(1,9) under SO(1,3) × SO(6) with $\Gamma^{(10)} = \Gamma^{(4)} \cdot \Gamma^{(6)}$ where $\Gamma^{(4)} = i\Gamma^0\Gamma^1\Gamma^2\Gamma^3$ and $\Gamma^{(6)} = -i\Gamma^4\Gamma^5 \ldots \Gamma^9$. One solution with $\Gamma^{(10)} = +1$ is given by $\Gamma^{(4)} = \Gamma^{(6)}$ and then the spin bundle decomposes under $M \times \mathcal{K}$ as

$$\widehat{S}^{(10)} = \left(\widehat{S}_+^{(4)} \otimes \widehat{S}_+^{\mathcal{K}}\right) \oplus \left(\widehat{S}_-^{(4)} \otimes \widehat{S}_-^{\mathcal{K}}\right). \tag{40}$$

Now solve the Dirac equation with the ansatz $\Psi(x^I, y^J) = \sum_m \phi_m(x^I) \otimes \chi_m(y^J) = \sum_m \psi_m$ and $\not{D}'_{\mathcal{K}}\chi_m = \lambda_m \chi_m$. It leads to

$$(\not{D}'^{(4)} + \lambda_m)\psi_m = 0 \tag{41}$$

where $\not{D}'^{(4)} = \Gamma^{(4)} \not{D}^{(4)}$ and $\not{D}'_{\mathcal{K}} = \Gamma^{(4)} \not{D}_{\mathcal{K}}$.

$\not{D}_{\mathcal{K}}$ is an elliptic operator on the compact manifold $\mathcal{K}$, this implies that that operator has a *finite* number of fermion zero modes. Massless fermions in four dimensions originate as zero modes of the Dirac operators $\not{D}^{\mathcal{K}}$ of the internal manifold $\mathcal{K}$. By the Atiyah-Singer theorem, a topological invariant of $\mathcal{K}$ containing the information of the chiral fermions on $\mathcal{K}$ is given by the index of the Dirac operator

$$Index(\not{D}_{\mathcal{K}}) = N_+^{\lambda=0} - N_-^{\lambda=0}, \tag{42}$$

for chiral fermions on $\mathcal{K}$ with $\Gamma^{(6)} = \pm 1$. Here $N_\pm^{\lambda=0}$ are the number of positive and negative chiral zero modes. In $2k+2$ dimensions this index is vanishing. We need to couple gauge fields coming from the heterotic string theory. Recall that they are $E_8 \times E_8$ valued gauge fields.

The *standard embedding* of the spin connection in the gauge connection leads to the chain of maximal subgroups: $SO(6) \times SO(10) \subset SO(16) \subset E_8$. This breaks SO(16) to SO(10). The computation of the Δ for this case yields $\Delta = \oplus_i L_i \otimes \mathcal{R}_i$ where L_i are irreps of SO(6) and $\mathcal{R}_i$ are complex irreps of SO(10). These latter determine the irreps where are distributed the massless fermions of the four-dimensional theory. The former irreps L_i determine the number of fermionic chiral zero modes described by the topological index $\delta = Index(D_{\mathcal{K}})$. This is given by

$$\delta = N_{\Gamma^{(6)}=+1}^{\lambda=0} - N_{\Gamma^{(6)}=-1}^{\lambda=0} = \int_{\mathcal{K}} ch(V) td(\mathcal{K}) = \frac{1}{2}\int_{\mathcal{K}} c_3(V)$$

and from the solution $\Gamma^{(6)} = \Gamma^{(4)}$ it determines the chiral fermion families in four dimensions

$$\delta = N^{\lambda=0}_{\Gamma^{(4)}=+1} - N^{\lambda=0}_{\Gamma^{(4)}=-1}. \tag{43}$$

Thus the theory in four dimensions has $\Delta = \oplus_i \delta_i \mathcal{R}_i$ where

$$\Delta = \delta\Big(\mathbf{16} \ominus \mathbf{16}^*\Big), \tag{44}$$

where $\delta = \chi(\mathcal{K})/2$ with $\chi(\mathcal{K})$ is the Euler number of $\mathcal{K}$.

Some Physics in Four Dimensions

In this section we intend to make contact with some four-dimensional physics. The development of this line of work is known as *string phenomenology*. Recent reviews of this topic at the light of string dualities is given in Ref. [11]. In the present short review we follows Ref. [5,12].

Continuous and Discrete Symmetries

In building models coming from string theory, *there are no global internal symmetries in spacetime* (there are no continuous global symmetries in all string theories). This is because if there is an internal symmetry, there should be a vector field in the spectrum because the properties of SCFT and it has the same properties of the gauge field of that symmetry.

Take for instance Type I or Type II superstring theory. We know from Noether's theorem of the two-dimensional theory that associated with the Type I or II superconformal symmetry there is a worldsheet conserved supercharge, $Q = \frac{1}{2\pi i}\oint(dzd\theta J - d\bar{z}d\bar{\theta}\bar{J})$, where, by uses of this symmetry, J should be a $(\frac{1}{2},0)$ tensor superfield and $\bar{J}$ is a $(0,\frac{1}{2})$ tensor superfield. The associated bosonic vertex operators (when we combine these tensors with the fermionic fields $\widetilde{\psi}^I$ and ψ^I respectively) have the property to couple with left and (or) right-moving parts of Q, giving rise to a spacetime gauge symmetry.

The absence of internal global symmetries in spacetime physics coming from string theory may help to understand some exciting problems of particle physics like the existence of the non-zero neutrino masses, which lie in the violation of the leptonic number. This was argued recently by Witten in [13].

However there are generically some discrete symmetries in string models. For example, T-duality which is an infinite dimensional one, or some models inherited from the point group of orbifold constructions which are finite dimensional, are in fact regarded as discrete symmetries. The importance of this kind of symmetries relay in the fact that they are useful for model building, hierarchy of masses and other related problems.

P, C, T Symmetries

We will see how discrete spacetime symmetries P, C and T are broken in string theory. If string theory is correct, when we compactify (for example on Calabi-Yau manifolds, orbifolds, tori , etc.) one must obtain the same symmetries (or broken symmetries) as these of the SM.

- *P-symmetry*

  Parity symmetry is violated by gauge interactions in SM. In string theory there are an analogous situation. Take for instance the heterotic string. The massless states in ten dimension are labeled by irreps of the little group SO(8). The action of parity symmetry reverses the spinor representations $\mathbf{8_s}$ and $\mathbf{8_s}'^*$ of the left and right-moving sectors. The symmetry is realized if the corresponding gauge representations $\mathcal{R}$ and $\mathcal{R}'^*$ are equal.

  However, $\mathcal{R}$ is the adjoint representation while that $\mathcal{R}'^*$ is empty. This tell us that parity symmetry is broken and the gauge couplings are chiral. But although in ten dimensions the spectrum is chiral, when we compactify to four dimensions, the spectrum could be turned out into a non-chiral one.

  For example, for toroidal compactifications, the spectrum is no-chiral, but for Z_3 orbifold (compactification) the spectrum it is. Other kinds of compactifications produce chiral gauge couplings.

  The chirality of the spectrum can be expressed by a topological quantity called *Index* as we saw in the last subsection. Since the index is a topological invariant quantity, it does not suffers any change under continuous transformations of the CFT.

- *C and CP Symmetry*

  Charge conjugation symmetry is also broken in SM. This is because C leaves spacetime invariant, but conjugates the gauge generators. As in SM, in string theories we require that conjugate representations (for example in SM, the fermions) satisfy $\mathcal{R}_\pm = \mathcal{R}^*_\pm$. From this we can see that chiral gauge couplings do not satisfy C symmetry. For the orbifold example this is also true.

  Consider now the CP symmetry. This symmetry takes $\mathcal{R}_+ \to \mathcal{R}^*_-$. Thus, any gauge coupling satisfies it as a consequence of CPT invariance. In the case of the orbifold there is a symmetry of the action which reverses X^k into ψ^k with $k = 3, 5, 7, 9$, and all the λ^I (I odd). This is a CP symmetry in 4-dimensions. So, Z_3 orbifold is CP symmetric.

- *CPT Symmetry*

  In string theory, as in local Lorentz-invariant quantum field theory, CPT symmetry is preserved. In string perturbation theory we use the θ-operation[5]. This is defined as $\theta(X^{0,3}) = \psi^{0,3}$ and vice versa. In Euclidean time this can be represented by a π-rotation in the plane (iX^0, X^3). In this context is clear that

[5] This is basically, the same argument used in field theory to prove CPT symmetry.

the action of θ is a symmetry, that reverses time and includes parity (in X^3). In order to show that this action also includes charge conjugation consider the S-matrix,

$$\langle \alpha, out|\beta, in\rangle = \langle \bar{\mathcal{V}}_\alpha \mathcal{V}_\beta\rangle, \tag{45}$$

where $\mathcal{V}$ is the vertex operator and we are only considering vertex operators to the initial and final states. The action of θ is

$$\langle \alpha, out|\beta, in\rangle = \langle \theta \cdot \bar{\mathcal{V}}_\alpha \theta \cdot \mathcal{V}_\beta\rangle = \langle \theta\bar{\beta}, out|\theta\bar{\alpha}, in\rangle. \tag{46}$$

When we apply CPT operation, we can see that it is antiunitary and it is θ combined with the conjugation of the vertex operator

$$\langle CPT \cdot \beta, out|CPT \cdot \alpha, in\rangle = \langle \alpha, out|\beta, in\rangle. \tag{47}$$

The manner in what we saw that CPT is an exact symmetry in string theory is only applicable to the perturbative sector. However for the non-perturbative sector, we can argue that SM, or field theory is the low energy limit of the string theory, so we take CPT symmetry for this low energy limit, and then we put it in 10 dimensions.

Effective Actions in Four Dimensions

First of all, it is important to emphasize that consistent four-dimensional superstring models which are chiral lead to $\mathcal{N} = 1$ supersymmetric theories. At $\mathcal{N} \geq 2$ supersymmetry spoils chirality. Thus in order to consider phenomenologically string models in four dimensions we restrict ourselves to construct $\mathcal{N} = 1$ supersymmetric actions. Mostly of the material we describe here is at the review by F. Quevedo [12] and we recommend it for checking details.

The corresponding spectrum of massless particles are composed by graviton-gravitino multiplet (G_{IJ}, ψ^I) and the gauge-gaugino multiplet $(A_I^\alpha, \lambda^\alpha)$. Also there are matter and moduli fields, in the form of chiral multiplets (Z, χ). In the case of the dilaton field Φ, it couples to the antisymmetric tensor B_{IJ} to form the linear $\mathcal{N} = 1$ multiplet (Φ, B_{IJ}, ρ). We can construct a $\mathcal{N} = 1$ chiral multiplet (S, χ_S) from the linear multiplet, where S is obtained by a duality transformation of the dilaton field. This transformation is given by $S = a + ie^{\Phi}$ and $\nabla_I a \equiv \varepsilon_{IJKL}\nabla^J B^{KL}$.

Thus the whole theory can be described as a $\mathcal{N} = 1$ supergravity theory coupled only to gauge and chiral multiplets. The most general Lagrangian which describes these fields depends on three arbitrary functions of the chiral multiplets. These fields are: (*i*) $K(Z, Z^*)$, the Kähler potential which is a real function which determines the kinetic terms of the chiral fields. The corresponding Lagrangian is given by $L_{kin} = K_{ZZ^*}\partial_I Z \partial^I Z^*$. (*ii*) $W(Z)$, the superpotential, which is a holomorphic function of the chiral multiplets. (*iii*) $f_{ab}(Z)$, the gauge kinetic function (holomorphic), and determines the gauge kinetic couplings in the corresponding Lagrangian $L_{gauge} = \text{Re}f_{ab}(Z)F^a_{IJ}F^{IJb} + \text{Im}f_{ab}(Z)F^a_{IJ}\widetilde{F}^{IJb}$. This function contributes to gaugino masses.

There are another quantity called the scalar potential $V = V_F + V_D$, where $V_F(Z, Z^*) = exp(\frac{K}{M_{Pl}^2})\{D_Z W K^{-1}_{ZZ^*} D_Z W^* - 3\frac{|W|^2}{M_{Pl}^2}\}$, $D_Z W = W_Z + W\frac{K_Z}{M_{Pl}^2}$ and $V_D = (\text{Re}f^{-1})_{ab}(K_Z, T^a Z)(K_{Z^*}, T^b Z^*)$.

Thus, the problem to find an effective four-dimensional action, is to calculate the functions K, W and f_{ab} when we are giving a specific string model. To do this, we have to use all the symmetries we have and taking into account that four dimensional string models are governed by two perturbation expansions. That is, an expansion in the sigma model (controlled by the size of the extra dimensions) and the proper string perturbation (in terms of the dilaton field) of the string coupling constant g_S.

First of all we consider only couplings generated at string tree-level. For the sigma model, we also take only the tree-level expansion. Using symmetries as the four dimensional Poincaré symmetry, supersymmetry, gauge symmetries and the axionic symmetry, we can extract the dependence of the effective action on the dilaton field S. Then, at tree-level, the functions K, W, f are given by $K = -log(S + S^*) + \widehat{K}(T, U, Q)$, $W = Y_{IJK}Q^I Q^J Q^K$ and $f_{ab} = S\delta_{ab}$ with $\widehat{K}$ an undetermined function. Our purpose is to find approximated expressions to this functions in the tree-level of the string perturbation theory, but otherwise exact in the CFT.

It can be showed that by using the axionic symmetries that at all orders in sigma-model expansion, superpotential W does not depend on T and U, so it is just a function of the matter fields Q^I. Thus W does not admit any kind of corrections in the sigma model[6]. The superpotential W does not depends on S as well. We know that S is the string loop-counting parameter and this implies that W is also an exact expression at tree-level string perturbation theory, *i.e.*, does not admit any radiative corrections.

Now, we are interested in finding a useful expression for K. This is more difficult, because we only can calculate it for some simple cases. Take for example a Calabi-Yau compactification whit $h_{1,1} = 1$ and $h_{2,1} = 0$. This give us that $K = -log(S + S^*) - 3log(T + T^* + QQ^*)$. When we write the Kähler potential as an expansion in matter fields, it is possible to extract an exact tree-level expression. The expansion is given by

$$K = -log(S + S^*) + K^M(T, T^*, U, U^*) + K^Q(T, T^*, U, U^*)QQ^*$$
$$+\widehat{Z}(T, T^*, U, U^*)(QQ + Q^*Q^*) + \mathcal{O}(Q^3).$$

For some $(2, 2)$ orbifold and Calabi-Yau models, it has been computed the quantities K^M, K^Q and $\widehat{Z}$.

---

[6] The reason for this, is that the field T, related with the size of the extra dimensions, comes from the internal components of the metric and determines the form that the loop expansion of the worldsheet action takes.

Consider now the loop corrections. We have seen that the superpotential (which is an holomorphic function) does not admit radiative corrections. However, for the Kähler potential this is different. We have just to calculate order by order in the loop expansions the corresponding expression for K_{loop}. On the other hand, the gauge kinetic function f_{ab} is also holomorphic and we know the expression in an exact manner for the tree level. Loop corrections to this function have a great importance due to this function determines the gauge coupling. Here we do not get expressions for this corrections, but it is important to say that there are no further corrections to f_{ab} beyond one loop, as in the standard supersymmetric theories.

In general, we have problems to determine how supersymmetry is broken at low energies. We can not solve this problem within perturbative string theory. We need work in the non-perturbative sector of the theory. But this sector, despite of many efforts and excellent results, we do not yet have a complete non-perturbative version of string theory. However there are some interesting non-perturbative mechanisms to break supersymmetry as gaugino condensation, composite goldstinos and instantons. The reader interested in these and another issues of non-perturbative string phenomenology is encouraged to consult [11,14].

T-DUALITY, D-BRANES AND BRANE CONFIGURATIONS

This section has the purpose of introducing basic ideas about T-duality in closed and open string theory. The open string case leads in a natural way to the definition of D-branes (for reviews of D-branes see [5,15,16]). These objects are of extreme importance since they are precisely the solitonic degrees of freedom which realize the strong-weak coupling duality in superstring theory. This duality is also known as string S-duality. T and S dualities relate the five perturbative superstring theories discussed previously and their compactifications in diverse dimensions. Moreover, the strong coupling limit of HE and Type II string theory (and their compactifications) suggest that there is an eleven-dimensional theory which has the eleven-dimensional supergravity as low energy limit. This prospect of theory is widely known as M-theory. The name come from the words: mystery, magic, mother, etc. Compactifications to diverse lower dimensions than ten gives more evidence of the existence of this theory. The fundamental degrees of freedom of this unified theory are unknown, but macroscopically they include membranes and fivebranes. 'Matrix Theory' is an attempt to give the dof's of M-theory. The proposal is that these degrees of freedom are the D0-branes. The worldvolume effective theory of a gas of N D0-branes is a SU(N) quantum mechanics. Large N-limit reproduces the description of membranes and fivebranes and some other results of eleven dimensional supergravity (for some reviews the reader can consult [17,18]).

D-branes also, are very important tools to study the strong coupling of supersymmetric theories in various dimensions. Different properties as chirality, dualities etc. are encoded in the engineering of brane configurations. The moduli space of these susy gauge theories is described by the Higgs and the Coulomb branches of the corresponding brane configuration. Many field theory results are understanding in terms of a geometrical language and many generalizations have been established motivated by the brane engineering (more about this topic can be found in Ref. [19]).

Finally, the presence of branes leads to modify the prescription of Calabi-Yau or orbifold compactifications and new non-perturbative are possible. In these sections we will discuss some of these interesting topics.

Toroidal Compactification, T-duality and D-branes

D-branes are, despite of the dual fundamental degrees of freedom in string theory, extremely interesting and useful tools to study nonperturbative properties of string and field theories (for some reviews see [15,16]). Non-perturbative properties of supersymmetric gauge theories can be better understanding as the world-volume effective theory of some configurations of intersecting D-branes (for a review see [19]). D-branes also are very important to connect gauge theories with gravity. This is the starting point of the AdS/CFT correspondence or Maldacena's conjecture. We don't review this interesting subject in this paper, however the reader can consult the excellent review [20]. Roughly speaking D-branes are static solutions of string equations which satisfy Dirichlet boundary conditions. That means that open strings can end on them. To explain these objects we follow the traditional way, by using T-duality on open strings we will see that Neumann conditions are turned out into the Dirichlet ones. To motivate the subject we first consider T-duality in closed bosonic string theory.

T-duality in Closed Strings

The general solution of Eq. (12) in the conformal gauge can be written as $X^I(\sigma,\tau) = X^I_R(\sigma^-) + X^I_L(\sigma^+)$, where $\sigma^{\pm} = \sigma \pm \tau$. Now, take one coordinate, say X^{25} and compactify it on a circle of radius R. Thus we have that X^{25} can be identified with $X^{25} + 2\pi Rm$ where m is called the *winding number*. The general solution for X^{25} with the above compactification condition is

$$X^{25}_R(\sigma^-) = X^{25}_{0R} + \sqrt{\frac{\alpha'}{2}} P^{25}_R(\tau - \sigma) + i\sqrt{\frac{\alpha'}{2}} \sum_{l\neq 0} \frac{1}{l} \alpha^{25}_{R,l} exp\Big(- il(\tau - \sigma)\Big)$$

$$X^{25}_L(\sigma^+) = X^{25}_{0L} + \sqrt{\frac{\alpha'}{2}} P^{25}_L(\tau + \sigma) + i\sqrt{\frac{\alpha'}{2}} \sum_{n\neq 0} \frac{1}{l} \alpha^{25}_{L,l} exp\Big(- il(\tau + \sigma)\Big), \qquad (48)$$

where

$$P^{25}_{R,L} = \frac{1}{\sqrt{2}}\Big(\frac{\sqrt{\alpha'}}{R}n \mp \frac{R}{\sqrt{\alpha'}}m\Big). \tag{49}$$

Here n and m are integers representing the discrete momentum and the winding number, respectively. The latter has not analogous in field theory. While the canonical momentum is given by $P^{25} = \frac{1}{\sqrt{2\alpha'}}(P^{25}_L + P^{25}_R)$. Now, by the mass shell condition, the mass of the perturbative states is given by $M^2 = M^2_L + M^2_R$, with

$$M^2_{L,R} = -\frac{1}{2}P^I P_I = \frac{1}{2}(P^{25}_{L,R})^2 + \frac{2}{\alpha'}(N_{L,R} - 1). \tag{50}$$

We can see that for all states with $m \neq 0$, as $R \to \infty$ the mass become infinity, while $m = 0$ implies that the states take all values for n and form a continuum. At the case when $R \to 0$, for states with $n \neq 0$, mass become infinity. However in the limit $R \to 0$ for $n = 0$ states with all m values produce a continuum in the spectrum. So, in this limit the compactified dimension disappears. For this reason, we can say that the mass spectrum of the theories at radius R and $\frac{\alpha'}{R}$ are identical when we interchange $n \Leftrightarrow m$. This symmetry is known as *T-duality.*

The importance of T-duality lies in the fact that the T-duality transformation is a parity transformation acting on the left and right moving degrees of freedom. It leaves invariant the left movers and changes the sign of the right movers (see Eq. (50))

$$P^{25}_L \to P^{25}_L, \qquad P^{25}_R \to -P^{25}_R. \tag{51}$$

The action of T-duality transformation must leave invariant the whole theory (at all order in perturbation theory). Thus, all kind of interacting states in certain theory should correspond to those states belonging to the dual theory. In this context, also the vertex operators are invariant. For instance the tachyonic vertex operators are

$$\mathcal{V}_L = exp(iP^{25}_L X^{25}_L), \qquad \mathcal{V}_R = exp(iP^{25}_R X^{25}_R). \tag{52}$$

Under T-duality, $X^{25}_L \to X^{25}_L$ and $X^{25}_R \to -X^{25}_R$; and from the general solution Eq. (13), $\alpha^{25}_{R,i} \to -\alpha^{25}_{R,i}$, $X^{25}_{0R} \to -X^{25}_{0R}$. Thus, T-duality interchanges $n \Leftrightarrow m$ (Kaluza-Klein modes $\Leftrightarrow$ winding number) and $R \Leftrightarrow \frac{\alpha'}{R}$ in closed string theory.

T-duality in Open Strings

Now, consider *open strings* with Neumann boundary conditions. Take again the 25th coordinate and compactify it on a circle of radius R, but keeping Neumann conditions. As in the case of closed string, center of mass momentum takes only discrete values $P^{25} = \frac{n}{R}$. While there is not analogous for the winding number.

So, when $R \to 0$ all states with nonzero momentum go to infinity mass, and do not form a continuum. This behavior is similar as in field theory, but now there is something new. The general solutions are

$$X_R^{25} = \frac{X_0^{25}}{2} - \frac{a}{2} + \alpha' P^{25}(\tau - \sigma) + i\sqrt{\frac{\alpha'}{2}} \sum_{l\neq 0} \frac{1}{l} \alpha_l^{25} exp\Big(- i2l(\tau - \sigma)\Big),$$

$$X_L^{25} = \frac{X_0^{25}}{2} + \frac{a}{2} + \alpha' P^{25}(\tau + \sigma) + i\sqrt{\frac{\alpha'}{2}} \sum_{l\neq 0} \frac{1}{l} \alpha_l^{25} exp\Big(- i2l(\tau + \sigma)\Big) \qquad (53)$$

where a is a constant. Thus, $X^{25}(\sigma,\tau) = X_R^{25}(\sigma^-) + X_L^{25}(\sigma^+) = X_0^{25} + \frac{2\alpha' n}{R}\tau + oscillator\ terms$. Taking the limit $R \to 0$, only the $n = 0$ mode survives. Because of this, the string seems to move in 25 spacetime dimensions. In other words, the strings vibrate in 24 transversal directions. T-duality provides a new T-dual coordinate defined by $\widetilde{X}^{25}(\sigma,\tau) = X_L^{25}(\sigma^+) - X_R^{25}(\sigma^-)$. Now, taking $\widetilde{R} = \frac{\alpha'}{R}$ we have $\widetilde{X}^{25}(\sigma,\tau) = a + 2\widetilde{R}\sigma n + oscillator\ terms$. Using the boundary conditions at $\sigma = 0, \pi$ one has $\widetilde{X}^{25}(\sigma,\tau)\mid_{\sigma=0} = a$ and $\widetilde{X}^{25}(\sigma,\tau)\mid_{\sigma=\pi} = a + 2\pi\widetilde{R}n$. Thus, we started with an open bosonic string theory with Neumann boundary conditions, and T-duality and a compactification on a circle in the 25th dimension, give us Dirichlet boundary conditions in such a coordinate. We can visualize this saying that an open string has its endpoints fixed at a hyperplane with 24 dimensions.

Strings with $n = 0$ lie on a 24 dimensional plane space (D24-brane). Strings with $n = 1$ has one endpoint at a hyperplane and the other at a different hyperplane which is separated from the first one by a factor equal to $2\pi\widetilde{R}$, and so on. But if we compactify p of the X^i directions over a T^p torus ($i = 1, ..., p$). Thus, after T-dualizing them we have strings with endpoints fixed at hyperplane with $25 - p$ dimensions, the D$(25 - p)$-brane.

Summarizing: the system of open strings moving freely in spacetime with p compactified dimensions on T^p is equivalent, under T-duality, to strings whose enpoints are fixed at a D$(25-p)$-brane *i.e.* obeying Neumann boundary conditions in the X^i longitudinal directions ($i = 1, \ldots, p$) and Dirichlet ones in the transverse coordinates X^m ($m = p + 1, ..., 25$).

The effect of T-dualizing a coordinate is to change the nature of the boundary conditions, from Neumann to Dirichlet and vice versa. If one dualize a longitudinal coordinate this coordinate will satisfies the Dirichlet condition and a Dp-brane becomes a D$(p - 1)$-brane. But if the dualized coordinate is one of the transverse coordinates the Dp-brane becomes a D$(p + 1)$-brane.

T-duality also acts conversely. We can think to begin with a closed string theory, and compactify it on to a circle in the 25th coordinate, and then by imposing Dirichlet conditions, obtain a D-brane. This is precisely what occurs in Type II theory, a theory of closed strings.

Spectrum and Wilson Lines

Now, we will see how does emerges a gauge field on the Dp-brane world-volume. Again, for the mass shell condition for open bosonic strings and because T-duality $M^2 = (\frac{n}{\alpha'}\widetilde{R})^2 + \frac{1}{\alpha'}(N-1)$. The massless state ($N=1$, $n=0$) implies that the gauge boson $\alpha_{-1}^I \mid 0\rangle$ ($U(1)$ gauge boson) lies on the D24-brane world-volume. On the other hand, $\alpha_{-1}^{25} \mid 0\rangle$ has a *vev* (vacuum expectation value) which describes the position $\widetilde{X}^{25}$ of the D-brane after T-dualizing. Thus, we can say in general, there is a gauge theory $U(1)$ over the world volume of the Dp-brane.

Consider now an *orientable open string.* The endpoints of the string carry charge under a non-Abelian gauge group. For Type II theories the gauge group is $U(N)$. One endpoint transforms under the fundamental representation $\mathbf{N}$ of $U(N)$ and the other one, under its complex conjugate representation (the anti-fundamental one) $\mathbf{N}^*$.

The ground state wave function is specified by the center of mass momentum and by the charges of the endpoints. Thus implies the existence of a basis $\mid k; ij\rangle$ called *Chan-Paton basis.* States $\mid k; ij\rangle$ of the Chan-Paton basis are those states which carry charge 1 under the ith $U(1)$ generator and -1 under the jth $U(1)$ generator. So, we can decompose the wave function for ground state as $\mid k; a\rangle = \sum_{i,j=1}^{N} \mid k; ij\rangle \lambda_{ij}^a$ where λ_{ij}^a are called *Chan-Paton factors.* From this, we see that it is possible to add degrees of freedom to endpoints of the string, that are precisely the Chan-Paton factors.

This is consistent with the theory, because the Chan-Paton factors have a Hamiltonian which do not posses dynamical structure. So, if one endpoint to the string is prepared in a certain state, it always will remains the same. It can be deduced from this, that $\lambda^a \longrightarrow U\lambda^a U^{-1}$ with $U \in U(N)$. Thus, the worldsheet theory is symmetric under $U(N)$, and this global symmetry is a gauge symmetry in spacetime. So the vector state at massless level $\alpha_{-1}^I \mid k, a\rangle$ is a $U(N)$ gauge boson.

When we have a gauge configuration with non trivial line integral around a compactified dimension (i.e a circle), we said there is a Wilson line. In case of open strings with gauge group $U(N)$, a toroidal compactification of the 25th dimension on a circle of radius R. If we choice a background field A^{25} given by $A^{25} = \frac{1}{2\pi R} diag(\theta_1, ..., \theta_N)$ a Wilson line appears. Moreover, if $\theta_i = 0$, $i = 1, ..., l$ and $\theta_j \neq 0$, $j = l+1, ..., N$ then gauge group is broken: $U(N) \longrightarrow U(l) \times U(1)^{N-l}$. It is possible to deduce that θ_i plays the role of a Higgs field. Because string states with Chan-Paton quantum numbers $\mid ij\rangle$ have charges 1 under i^{th} $U(1)$ factor (and -1 under j^{th} $U(1)$ factor) and neutral with all others; canonical momentum is given now by $P_{(ij)}^{25} \Longrightarrow \frac{n}{R} + \frac{(\theta_j - \theta_i)}{2\pi R}$. Returning to the mass shell condition it results,

$$M_{ij}^2 = \left(\frac{n}{R} + \frac{\theta_j - \theta_i}{2\pi R}\right)^2 + \frac{1}{\alpha'}(N-1). \tag{54}$$

Massless states ($N = 1, n = 0$) are those in where $i = j$ (diagonal terms) or for which $\theta_j = \theta_i$ ($i \neq j$). Now, T-dualizing we have $\widetilde{X}_{ij}^{25}(\sigma, \tau) = a + (2n + \frac{\theta_j - \theta_i}{\pi})\widetilde{R}\sigma +$ *oscillator terms*. Taking $a = \theta_i \widetilde{R}$, $\widetilde{X}_{ij}^{25}(0, \tau) = \theta_i \widetilde{R}$ and $\widetilde{X}_{ij}^{25}(\pi, \tau) = 2\pi n\widetilde{R} + \theta_j \widetilde{R}$.

This give us a set of N D-branes whose positions are given by $\theta_j \widetilde{R}$, and each set is separated from its initial positions ($\theta_j = 0$) by a factor equal to $2\pi\widetilde{R}$. Open strings with both endpoints on the same D-brane gives massless gauge bosons. The set of N D-branes give us $U(1)^N$ gauge group. An open string with one endpoint in one D-brane, and the other endpoint in a different D-brane, yields a massive state with $M \sim (\theta_j - \theta_i)\widetilde{R}$. Mass decreases when two different D-branes approximate to each other, and are null when become the same. When all D-branes take up the same position, the gauge group is enhanced from $U(1)^N$ to $U(N)$. On the D-brane world-volume there are also scalar fields in the adjoint representation of the gauge group $U(N)$. The scalars parametrize the transverse positions of the D-brane in the target space X.

D-Brane Action

With the massless spectrum on the D-brane world-volume it is possible to construct a low energy effective action. Open strings massless fields are interacting with the closed strings massless spectrum from the **NS-NS** sector. Let ξ^a (with $a = 0, \ldots, p$) be the world-volume coordinates on W. The effective action is the gauge invariant action well known as the Dirac-Born-Infeld (DBI)-action

$$S_D = -T_p \int_W d^{p+1}\xi e^{-\Phi} \sqrt{det(G_{ab} + B_{ab} + 2\pi\alpha' F_{ab})}, \tag{55}$$

where T_p is the tension of the D-brane, G_{ab} is the world-volume induced metric, B_{ab} is the induced antisymmetric field, F_{ab} is the Abelian field strength on W and Φ is the dilaton field.

For N D-branes the massless fields turns out to be $N \times N$ matrices and the action turns out to be non-Abelian DBI-action (for a nice review about the Born-Infeld action in string theory see [21])

$$S_D = -T_p \int_W d^{p+1}\xi e^{-\Phi} Tr\Big(\sqrt{det(G_{ab} + B_{ab} + 2\pi\alpha' F_{ab})} + O([X^m, X^n]^2)\Big) \tag{56}$$

where $m, n = p+1, \ldots, 9$. The scalar fields X^m representing the transverse positions become $N \times N$ matrices and so, the spacetime become a noncommutative spacetime. We will come back later to this interesting point.

Ramond-Ramond Charges

D-branes are coupled to Ramond-Ramond (RR) fields G_p [15]. The complete effective action on the D-brane world-volume W which taking into account this coupling is

$$S_D = -T_p \int_W d^{p+1}\xi \Big\{ e^{-\Phi} \sqrt{det(G_{ab} + B_{ab} + 2\pi\alpha' F_{ab})}$$

$$+i\mu_p \int_W \sum_p C_{(p+1)} Tr\left(e^{2\pi\alpha'(F+B)}\right)\Big\} \tag{57}$$

where μ_p is the RR charge. RR charges can be computed by considering the anomalous behavior of the action at intersections of D-branes. Thus RR charge is given by

$$Q_{RR} = ch(j!E)\sqrt{\widehat{A}(TX)}, \tag{58}$$

where $j : W \hookrightarrow X$. Here E is the Chan-Paton bundle over X, $\widehat{A}(TX)$ is the genus of the spacetime manifold X. This gives an ample evidence that the RR charges take values not in a cohomology theory, but in fact, in a K-Theory. This result was developed by Witten in Ref. [22] in the context of non-BPS brane configurations worked out by Sen [23].

Finally, RR charges and RR fields do admit a classification in terms of topological K-theory. The inclusion of a B-field turns out the effective theory non-commutative and a suitable generalization of the topological K-theory is needed. The *right* generalization seems to be the K-Homology and the K-theory of C^* algebras [24]. This subject is right now under intensive investigation.

D-brane Configurations and Susy Gauge Theories

As an application of the D-brane theory we consider in this section the brane box models, in particular we focus on the *cube* model discussed in [25].

The dynamics of D-branes in certain configurations of intersecting branes encodes many field-theoretical facts about supersymmetric theories in several dimensions (for a review, see [19]). Gauge theories in $(p+1)$ dimensions with sixteen supercharges can be obtained as the world-volume theories of flat infinite Dp-branes. In the context of theories with eight supersymmetries in p dimensions, it was shown in [26] that such theories can be realized by considering Dp-branes with a world-volume which is finite in one direction, in which the D-brane ends on NS fivebranes. The brane is suspended between NS fivebranes spanning 012345. The low energy theory in the non-compact dimensions of the D-brane is p-dimensional. It is still a gauge theory, but the presence of the NS branes breaks half of the supersymmetries, so eight supercharges remain. This construction has been generalized in several directions, and has yielded the realization of a large family of models in several dimensions. This setup has also been exploited to compute different exact quantum results in these theories. For a review of such achievements, see [19].

A nice property of the interplay of field theories and configurations of branes is that the intersections of branes can sometimes support chiral zero modes. This opens the possibility of studying chiral gauge theories using branes. The simplest such example is provided by the realization of six-dimensional theories with eight supersymmetries, which are chiral. These can be realized in the setup described above by taking $p = 6$, *i.e.* one considers D6 branes extending along 0123456, and which are bounded in 6 by NS branes with world-volume along 012345.

Chirality if a fragile property, in the sense that toroidal compactifications or too much supersymmetry spoil it. Thus, in order to obtain chiral theories in four dimensions one has to consider theories with only four supercharges. Their realization in terms of branes requires new ingredients. A fairly general family of brane configurations realizing generically chiral gauge theories in four dimensions was introduced in Ref. [27].

The idea is a clever extension of the philosophy in [26]. It consists in realizing first a five-dimensional theory with eight supercharges, by using D5 branes along 012346, suspended between NS branes with world-volume along 012345. Then, the D5 brane is bounded in the direction 4, by using a new set of NS branes oriented along 012367 (denoted NS$'$ branes). The low energy theory is four-dimensional, since the world-volume of the D5 brane along 46 is a finite rectangle. Such configurations are known as brane box models. The presence of the new kind of branes breaks a further half of the supersymmetries, and so the theory has only four supercharges. Furthermore, the intersections of NS, NS$'$ and D5 branes introduce chirality in the four dimensional theory. There is no complete understanding of the quantum effects of these gauge theories in terms of branes.

The Cube Brane Box Configurations and Susy Theories in Two Dimensions

Here we introduce certain supersymmetric configurations of NS, NS$'$, and NS$''$ branes, and D4 branes in Type IIA superstring theory. They give rise to two-dimensional $(0, 2)$ field theories. These configurations are obtained in the spirit of the brane box configurations in [27], by considering D-branes which are finite in several directions. As explained before, they belong to a natural sequence of brane box models yielding chiral theories in six, four and two dimensions (taking D branes compact in one, two and three directions, respectively).

Let us consider the ingredients of the brane configurations which we will use in this paper. Brane configurations consist of:

- NS fivebranes located along (012345).
- NS' fivebranes located along (012367).
- NS" fivebranes located along (014567).
- D4 branes located along (01246).

In this configuration the D4 branes are finite in the directions 246. They are bounded in the direction 2 by the NS$''$ branes, in the direction 4 by the NS$'$ branes, and in the direction 6 by the NS branes. For the D4 branes to be suspended in this way, it is necessary that the coordinates of all branes in 89 should be equal. It is also required that two NS branes joined by a D4 brane should have the same position in 7, and analogously that two NS$'$ branes joined by a D4 brane should have the same position in 5, and that two NS$''$ branes should have the same position in 3.

The low-energy effective field theory on the D4 branes is two-dimensional, since 01 are the only non-compact directions in their world-volume. The presence of each kind of NS fivebrane breaks one half of the supersymmetries, and altogether they break to 1/8 of the original supersymmetry. A further half is broken by the D4 branes, and the world-volume theory has $(0, 2)$ supersymmetry in two dimensions. Since the D brane is bounded by NS fivebranes, the world-volume gauge bosons will not be projected out and there will be a gauge group for each box in the model. The $U(1)_R$ R-symmetry of the field theory is manifest as the rotational symmetry in the directions 89.

We note that there are a variety of other objects that can be introduced in the configuration without breaking the supersymmetry. For instance, there are three kinds of D6 branes that can be introduced, namely D6 branes along 0124789, D6$'$ branes along 0125689, and D6$''$ branes along 0134689. They provide vector-like flavors for the gauge groups. These extensions are quite well-known from other contexts, and we will not study them in the present paper.

There is a first rough classification we can make in these brane configurations, according to whether the directions 246 are taken compact or not. If some of these directions are non-compact, then there will be some semi-infinite box, which will represent some global symmetry. For definiteness we will center on the case in which all three directions are compact, with lengths R_2, R_4 and R_6. Extension of our results to other cases is straightforward.

A generic configuration consists of a three-dimensional grid of k NS branes, k' NS' branes and k'' NS" branes dividing the 246 torus into a set of $kk'k''$ boxes. We will often think about these configurations as infinite periodic arrays of boxes in $\mathbf{R}^3$, quotiented by an infinite discrete group of translations in a three-dimensional lattice Λ. This point of view is particularly useful to define models in which the unit cell has non-trivial identifications of sides [28].

(0,2) Effective Theory on the D4 Brane

The effective field theory on the only non-compact directions 01 of the D4 branes world-volume is a (0,2) gauge theory in two dimensions. These theories are described in the (0,2) superspace $(y^\alpha, \theta^+, \bar{\theta}^+)$. There are three basic kinds of multiplets which we will use.

- The $(0, 2)$ *gauge multiplet* V', which contains gauge bosons v_α, $\alpha = 0, 1$, and one fermion χ_-.
- The $(0, 2)$ *chiral multiplet* Φ', contains one complex scalar ϕ and one chiral fermion ψ_+.
- The $(0, 2)$ *Fermi multiplet*, Λ, is described by an anticommuting superfield. Its complete θ expansion contains a chiral spinor λ_-, an auxiliary field G, and

a holomorphic function E depending on the chiral (0,2) superfields Φ'_i. The Fermi multiplet Λ satisfies the constraint $\bar{\mathcal{D}}_+\Lambda = \sqrt{2}E(\Phi')$, with $\bar{\mathcal{D}}_+ E = 0$. Here $\bar{\mathcal{D}}_+$ represents the supersymmetric covariant derivative. The expansion in components for the Fermi superfield is

$$\Lambda = \lambda_- - \sqrt{2}\theta^+ G - i\theta^+\bar{\theta}^+(D_0 + D_1)\lambda_- - \sqrt{2}\bar{\theta}^+ E(\Phi') \tag{59}$$

with D_α denoting the usual supersymmetric derivative.

Gauge theories involving these fields are described by a Lagrangian with the following structure,

$$L = L_{gauge} + L_{ch} + L_F + L_{D,\theta} + L_J. \tag{60}$$

As usual, L_{gauge} is the kinetic term of the gauge multiplet given by

$$L_{gauge} = \frac{1}{8g^2}\int d^2y d\theta^+ d\bar{\theta}^+ \mathrm{Tr}(\bar{\Upsilon}\Upsilon) \tag{61}$$

where Υ is the field strength of V'.

The term L_{ch} contains the kinetic energy and gauge couplings of the (0,2) chiral superfields Φ'_i. It is given by

$$L_{ch} = -\frac{i}{2}\int d^2y d^2\theta \sum_i \left(\bar{\Phi}'_i(\mathcal{D}_0 - \mathcal{D}_1)\Phi'_i\right), \tag{62}$$

where $\mathcal{D}_0$ and $\mathcal{D}_1$ are the (0,2) gauge covariant derivatives with respect to V'.

The term L_F describes the dynamics of the Fermi multiplets Λ, and certain interactions. It is given by

$$L_F = -\frac{1}{2}\int d^2y d^2\theta \sum_a (\bar{\Lambda}_a\Lambda_a). \tag{63}$$

Substitution of (69) into (63) leads to

$$\begin{aligned} L_F = \int d^2y \sum_a \Big\{ & i\bar{\lambda}_{-,a}(D_0 + D_1)\lambda_{-,a} + |G_a|^2 - |E_a|^2 \\ & - \sum_j (\bar{\lambda}_{-,a}\frac{\partial E_a}{\partial \phi_j}\psi_{+,j} + \frac{\partial \bar{E}_a}{\partial \bar{\phi}_j}\bar{\psi}_{+,j}\lambda_{-,a})\Big\}. \end{aligned} \tag{64}$$

The Fayet-Iliopoulos and theta angle terms are encoded in the (0,2) Lagrangian $L_{D,\theta}$ which is written as

$$L_{D,\theta} = \frac{t}{4}\int d^2y d\theta^+ \mathrm{Tr}(\Upsilon|_{\bar{\theta}^+=0}) + h.c. \tag{65}$$

where $t = \frac{\theta}{2\pi} + ir$.

Finally (0,2) models do admit an additional interaction term L_J which depends on a set of holomorphic functions $J^a(\Phi')$ of the chiral superfields. There is one such function for each Fermi superfield. They satisfy the relation $\sum_a E_a J^a = 0$. This interaction is the (0,2) analog of the superpotential, and its Lagrangian L_J is given by

$$L_J = -\frac{1}{\sqrt{2}} \int d^2y d\theta^+ \sum_a \left(\Lambda_a J^a|_{\bar{\theta}^+=0} \right) - h.c. \ . \tag{66}$$

The expansion of this term in components is

$$L_J = -\int d^2y \sum_a \left(G_a J^a + \sum_j \lambda_{-,a} \psi_{+,j} \frac{\partial J^a}{\partial \phi_j} \right) - h.c. \ . \tag{67}$$

After combining the Lagrangians L_F and L_J and solving for the equations of motion for the auxiliary fields G, the relevant interaction terms in the Lagrangian (we are not listing the gauge interactions and D-terms here) are

$$\sum_a (|J^a(\phi)|^2 + |E^a(\phi)|^2) - \sum_{a,j} (\bar{\lambda}_{-,a} \frac{\partial E_a}{\partial \phi_j} \psi_{+,j} + \lambda_a \frac{\partial J^a}{\partial \phi_j} \psi_{+,j} + h.c.). \tag{68}$$

The first term contains the scalar potential, and the second the Yukawa couplings. Notice that the choice of the functions E and J completely defines the interactions of the theory.

For (0,2) theories in two dimensions we have just one U(1) R-symmetry group acting on the superspace coordinates $(\theta^+, \bar{\theta}^+)$. This is right-moving R-symmetry and it acts as $\theta^+ \to e^{i\beta}\theta^+$, $\bar{\theta}^+ \to e^{-i\beta}\bar{\theta}^+$, leaving θ^- , $\bar{\theta}^-$ invariant.

Interpretation of the Linear Sigma Model

Up to here we have introduced a large family of two-dimensional $(0,2)$ gauge theories. Since $(2,2)$ and $(0,2)$ theories have been traditionally used as world-sheet descriptions of string theories propagating on some target space, it is a natural question whether the (classical) Higgs branch or our models has any geometrical interpretation of the kind. In this section we are to show that it describes the dynamics of a type IIB D1 brane on a $\mathbf{C}^4/\Gamma$ singularity, with Γ an abelian subgroup of $SU(4)$. The main tool for reaching this conclusion will be a T-duality performed on the brane box model along the directions 246.

In this section we perform a T-duality on the brane box models along the directions 246. The main tool will be the well known T-duality relation between a set of n parallel NS fivebranes and n Kaluza-Klein monopoles. The discussion in this subsection parallels that in [28].

Let us start with the simplest case of a brane box model formed by a unit cell of $k \times k' \times k''$ boxes, with trivial identifications of faces. In this case the T-duality along the directions 246 is particularly easy. We start with k NS branes along

012345, k' NS$'$ branes along 012367, and k'' NS$''$ branes along 014567. The T-duality along 2 transforms the NS$''$ branes into k'' Kaluza-Klein monopoles. These will be described by a multi-center Taub-NUT metric, with non-trivial geometry on the directions 2$'$,3,8,9, with 2$'$ denoting the coordinate dual to 2. Notice that, since the 3,8,9 coordinates of the initial NS$''$ branes coincided, so do the coordinates of the corresponding k'' centers of the Taub-NUT metric, so that it contains singularities of type $A_{k''-1}$.

Similarly, the T-duality along 4 transforms the k' NS$'$ branes into k' Kaluza-Klein monopoles with world-volume along 012367, and represented by a nontrivial geometry on 4$'$,5,8,9. Again, since the centers of the Kaluza-Klein monopoles coincide, such geometry will contain singularities of type $A_{k'-1}$. Finally, the T-duality along 6 turns the k NS branes into k Kaluza-Klein monopoles. Their world-volume spans 012345, and they are represented by a non-trivial geometry along 6$'$,7,8,9. Since again all the centers coincide, there will be A_{k-1} singularities.

Thus, the final T-dual of the grid of NS, NS$'$ and NS$''$ branes is type IIB string theory with a complicated geometry in the directions 2$'$,3,4$'$,5,6$'$,7,8,9. One can think of it roughly as some 'superposition' of the Kaluza-Klein monopoles we have described. Even without a quantitative knowledge of such metric, we can describe the relevant features for our purposes. One such feature is that the number of unbroken supersymmetries constrains the manifold to be a Calabi-Yau four-fold. Also, from our remarks above we know the existence of certain (complex) surfaces of singularities of type A_{k-1}, $A_{k'-1}$ and $A_{k''-1}$ singularities. If we introduce complex coordinates $w_1 = \exp(x^7 + ix^{6'})$,$w_2 = \exp(x^5 + ix^{4'})$, $w_3 = \exp(x^3 + ix^{2'})$, and $w_4 = x^9 + ix^8$, the surface of A_{k-1} singularities is defined roughly by $w_1 = w_4 = 0$, the surface of $A_{k'-1}$ singularities is defined by $w_2 = w_4 = 0$, and the surface of $A_{k''-1}$ singularities is given by $w_3 = w_4 = 0$. At the origin $w_1 = w_2 = w_3 = w_4 = 0$ all surfaces meet and the singularity is worse. It can be described as a quotient singularity of type $\mathbf{C}^4/\Gamma$, with $\Gamma = \mathbf{Z}_k \times \mathbf{Z}_{k'} \times \mathbf{Z}_{k''}$. This discrete group is generated by elements θ, ω, η, whose action on $(z_1, z_2, z_3, z_4) \in \mathbf{C}^4$ is as follows:

$$\begin{array}{llll} \theta: & (z_1, z_2, z_3, z_4) & \to & (e^{2\pi i/k} z_1, z_2, z_3, e^{-2\pi i/k} z_4) \\ \omega: & (z_1, z_2, z_3, z_4) & \to & (z_1, e^{2\pi i/k'} z_2, z_3, e^{-2\pi i/k'} z_4) \\ \eta: & (z_1, z_2, z_3, z_4) & \to & (z_1, z_2, e^{2\pi i/k''} z_3, e^{-2\pi i/k''} z_4). \end{array} \tag{69}$$

In this description it becomes clear that there may be further surfaces of singularities when the greatest common divisor of any two of k, k', k'' is not 1, in analogy with the discussion in [28]. This will not be relevant for our purposes and we do not develop the issue further.

After the T-duality, the initial D4 branes become D1 branes located at a point in the four-fold. When the initial D4 branes are bounded by the grid of NS, NS$'$ and NS$''$ branes, the T-dual D1 branes will be located precisely at the $\mathbf{C}^4/\Gamma$ singular point. The field theories introduced previously correspond to the field theories appearing in the world-volume of such D1 brane probes. In addition, the structure of the singularity controls the spectrum and dynamics of the field theory (for a recent review see [29].

Brane boxes models generating gauge theories in two dimensions with enchanced chiral (0,4), (0,6) and (0,8) supersymmetry can be constructed and are also described at [25].

NON-PERTURBATIVE STRING THEORY

Strong-Weak Coupling String Duality

We have described the massless spectrum of the five consistent superstring theories in ten dimensions. Additional theories can be constructed in lower dimensions by compactification of some of the ten dimensions. Thus the ten-dimensional spacetime X looks like the product $X = \mathcal{K}^d \times \mathbf{R}^{1,9-d}$, with $\mathcal{K}$ a suitable compact manifold or orbifold. Depending on which compact space is taken, it will be the quantity of preserved supersymmetry.

All five theories and their compactifications are parametrized by: the string coupling constant g_S, the geometry of the compact manifold $\mathcal{K}$, the topology of $\mathcal{K}$ and the spectrum of bosonic fields in the **NS-NS** and the **R-R** sectors. Thus one can define the *string moduli space* $\mathcal{M}$ of each one of the theories as the space of all associated parameters. Moreover, it can be defined a map between two of these moduli spaces. The dual map is defined as the map $\mathcal{S} : \mathcal{M} \to \mathcal{M}'$ between the moduli spaces $\mathcal{M}$ and $\mathcal{M}'$ such that the strong/weak region of $\mathcal{M}$ is interchanged with the weak/strong region of $\mathcal{M}'$. One can define another map $\mathcal{T} : \mathcal{M} \to \mathcal{M}'$ which interchanges the volume V of $\mathcal{K}$ for $\frac{1}{V}$. One example of the map $\mathcal{T}$ is the equivalence, by T-duality, between the theories Type IIA compactified on $\mathbf{S}^1$ at radius R and the Type IIB theory on $\mathbf{S}^1$ at raduis $\frac{1}{R}$. The theories HE and HO constitutes another example of the $\mathcal{T}$ map. In this section we will follows the Sen's review [30]. Another useful references are [31–35]. Type IIB theory is self-dual with respect the $\mathcal{S}$ map.

Type IIB-IIB Duality

The Type IIB theory is self-dual. In order to see that write the bosonic part of the action of Type IIB supersting theory

$$S_{\mathbf{IIB}} = \frac{1}{2\kappa^2}\int_X d^{10}x\sqrt{-G_{\mathbf{IIB}}}e^{-2\Phi}\Big(R + 4\partial_I\Phi\partial^I\Phi - \frac{1}{2}H_{IJK}H^{IJK}\Big)$$

$$-\frac{1}{4\kappa^2}\int_X d^{10}x\sqrt{-G_{IIB}}\Big(|F_{(1)}|^2 + |\widetilde{F}_{(3)}|^2 + \frac{1}{2}|\widetilde{F}_{(5)}|^2\Big) - \frac{1}{4\kappa^2}\int_X A_{(4)}\wedge H_{(3)}\wedge F_{(3)}, \tag{70}$$

where $\widetilde{F}_{(3)} = dA_{(2)} - a \wedge H_{(3)}$ and $\widetilde{F}_{(5)} = dA_{(4)} - \frac{1}{2}A_{(2)} \wedge H_{(3)} + \frac{1}{2}B_{(2)} \wedge F_{(3)}$. This action is clearly invariant under

$$\Phi' = -\Phi, \qquad G'_{IJ} = e^{-\Phi}G_{IJ},$$
$$B_{(2)} = A_{(2)} \qquad A'_{(2)} = -B_{(2)}, \qquad A'_{(4)} = A_{(4)}. \tag{71}$$

This symmetry leads to an identification of a fundamental string F1 with a D1-brane ($B_{(2)} = A_{(2)}$) and the interchanging of a pair of D3-branes.

Type I-SO(32)-Heterotic Duality

In order to analyze the duality between Type I and SO(32) heterotic string theories we recall the spectrum of both theories. These fields are the dynamical fields of a supergravity Lagrangian in ten dimensions. Type I string theory has in the **NS-NS** sector the following fields: the metric $G^{\mathbf{I}}_{IJ}$, the dilaton $\Phi^{\mathbf{I}}$ and in the **R-R** sector: the antisymmetric tensor $A^{\mathbf{I}}_{IJ}$. Also there are 496 gauge bosons $A^{a\mathbf{I}}_I$ in the adjoint representation of the gauge group SO(32). For the SO(32) heterotic string theory the spectrum consist of: the spacetime metric $G^{\mathbf{H}}_{IJ}$, the dilaton field $\Phi^{\mathbf{H}}$, the antisymmetric tensor $B^{\mathbf{H}}_{IJ}$ and 496 gauge fields $A^{a\mathbf{H}}_I$ in the adjoint representation of SO(32). Both theories have $\mathcal{N} = 1$ spacetime supersymmetry. The effective action for the massless fields of the Type I supergravity effective action $S_{\mathbf{I}}$ is defined as

$$S_{\mathbf{I}} = \frac{1}{2\kappa^2}\int_X d^{10}x\sqrt{-G^{\mathbf{I}}}e^{-2\Phi^{\mathbf{I}}}\Big(R + 4(\nabla\Phi^{\mathbf{I}})^2 - \frac{1}{12}|\widetilde{F}_{(3)}|^2\Big)$$
$$-\frac{1}{g^2}\int_X d^{10}x\sqrt{-G^{\mathbf{I}}}e^{-\Phi^{\mathbf{I}}}Tr(F^{\mathbf{I}}_{IJ}F^{\mathbf{I}IJ}) \tag{72}$$

where $\widetilde{F}_{(3)} = F_{(3)} - \frac{\alpha'}{4}[\omega_{3Y}(A) - \omega_{3L}(\omega)]$.

While the heterotic action $S_{\mathbf{H}}$ is defined as

$$S_{\mathbf{H}} = \frac{1}{2\kappa^2}\int_X d^{10}x\sqrt{-G^{\mathbf{H}}}e^{-2\Phi^{\mathbf{H}}}\Big[R + 4(\nabla\Phi^{\mathbf{H}})^2 - \frac{1}{12}|\widetilde{H}_{(3)}|^2 - \frac{\alpha'}{8}Tr(F^{\mathbf{H}}_{IJ}F^{\mathbf{H}IJ})\Big] \tag{73}$$

where $\widetilde{H}_{(3)} = dB_{(2)} - \frac{\alpha'}{4}[\omega_{3Y}(A) - \omega_{3L}(\omega)]$.

The comparison of these two actions leads to the following identification of the fields

$$G^{\mathbf{I}}_{IJ} = e^{-\Phi^{\mathbf{H}}}G^{\mathbf{H}}_{IJ}, \qquad \Phi^{\mathbf{I}} = -\Phi^{\mathbf{H}},$$
$$A^{a\mathbf{I}}_I = A^{a\mathbf{H}}_I, \qquad \widetilde{F}^{\mathbf{I}}_{(3)} = \widetilde{H}^{\mathbf{H}}_{(3)}. \tag{74}$$

This give us many information, the first relation tell us that the metrics of both theories are the same. The second relation interchanges the $B_{(2)}$ field in the **NS-NS** sector and the $A_{(2)}$ field in the **R-R** sector. That implies the interchanging of heterotic strings and Type I D1-branes. The third relation identifies the gauge fields coming from the Chan-Paton factors from the Type I side with the gauge fields coming from the 16 compactified internal dimensions of the heterotic string. Finally, the opposite sign for the dilaton relation means that the string coupling constant $g_S^{\mathbf{I}}$ is inverted $g_S^{\mathbf{H}} = 1/g_S^{\mathbf{I}}$ within this identification, and interchanges the strong and weak couplings of both theories leading to the explicit realization of the $\mathcal{S}$ map.

M-Theory

We have described how to construct dual pairs of string theories. By the uses of the $\mathcal{S}$ and the $\mathcal{T}$ maps a network of theories can be constructed in various dimensions all of them related by dualities. However new theories can emerge from this picture, this is the case of M-theory. M-theory (the name come from 'mystery', 'magic', 'matrix', 'membrane', etc.) was originally defined as the strong coupling limit for Type IIA string theory [31].

It is known that Type IIA theory can be obtained from the dimensional reduction of the eleven dimensional supergravity theory (a theory known from the 70's years) and given by the Cremmer-Julia-Scherk Lagrangian

$$I_{11} = \frac{1}{2\kappa_{11}^2} \int_Y d^{11}x \sqrt{-G_{11}} \Big(R - |dA_3|^2 \Big) - \frac{1}{6} \int_Y A_{(3)} \wedge F_{(4)} \wedge F_{(4)}, \tag{75}$$

where Y is the eleven dimensional manifold. If we assume that the eleven-dimensional spacetime factorizes as $Y = X \times \mathbf{S}^1_R$, where the compact dimension has radius R. Usual Kaluza-Klein dimensional reduction leads to get the ten-dimensional metric, an scalar field and a vector field. $A_{(3)}$ from the eleven dimensional theory leads to $A_{(3)}$ and $A_{(2)}$ in the ten-dimensional theory. The scalar field turn out too be proportional to the dilaton field of the **NS-NS** sector of the Type IIA theory. The vector field from the KK compactification can be identified with the $A^{\mathbf{IIA}}$ field of the **R-R** sector. From the three-form in eleven dimensions are obtained the RR field $A_{(3)}$ of the Type IIA theory. Fin ally, the $A_{(2)}$ field is identified with the NS-NS B-field of field strength $H_{(3)} = dB_{(2)}$. Thus the eleven-dimensional Lagrangian leads to the Type IIA supergravity in the weak coupling limit ($\Phi \to 0$ or $R \to 0$). The ten-dimensional IIA supergravity describing the bosonic part of the low energy limit of the Type IIA superstring theory is

$$S_{\mathbf{IIA}} = \frac{1}{2\kappa^2} \int_X d^{10}x \sqrt{-G^{\mathbf{IIA}}} e^{-2\Phi^{\mathbf{IIA}}} \Big(R + 4(\nabla\Phi^{\mathbf{IIA}})^2 - \frac{1}{12}|H_{(3)}|^2 \Big)$$

$$-\frac{1}{4\kappa^2} \int_X d^{10}x \sqrt{-G^{\mathbf{IIA}}} \Big(|F_{(2)}|^2 + |\widetilde{F}_{(4)}|^2 \Big) - \frac{1}{4\kappa^2} \int_X B_{(2)} \wedge dA_{(3)} \wedge dA_{(3)} \tag{76}$$

where $H_{(3)} = dB_{(2)}$, $F_{(2)} = dA_{(1)}$ and $\widetilde{F}_{(4)} = dA_{(3)} - A_{(1)} \wedge H_{(3)}$.

It is conjectured that there exist an eleven dimensional fundamental theory whose low energy limit is the 11 dimensional supergravity theory and that it is the strong coupling limit of the Type IIA superstring theory. At the present time the degrees of freedom (dof's) are still unknown, through at the macroscopic level they should be membranes and fivebranes (also called M2-branes and M5-branes).

Horava-Witten Theory

Just as the M-theory compactification on $\mathbf{S}^1_R$ leads to the Type IIA theory, Horava and Witten realized that orbifold compactifications leads to the $E_8 \times E_8$ heterotic theory in ten dimensions HE (see for instance [33]). More precisely

$$\mathrm{M}/(\mathbf{S}^1/\mathbf{Z}_2) \Longleftrightarrow HE \tag{77}$$

where $\mathbf{S}^1/\mathbf{Z}_2$ is homeomorphic to the finite interval I and the M-theory is thus defined on $Y = X \times I$. From the ten-dimensional point of view, this configuration is seen as two parallel planes placed at the two boundaries ∂I of I. Dimensional reduction and anomalies cancellation conditions imply that the gauge degrees of freedom should be trapped on the ten-dimensional planes X with the gauge group being E_8 in each plane. While that the gravity is propagating in the bulk and thus both copies of X's are only connected gravitationally.

F-Theory

F-Theory was formulated by C. Vafa, looking for an analog theory to M-Theory for describing non-perturbative compactifications of Type IIB theory (for a review see [34,30]). Usually in perturbative compactifications the parameter $\lambda = a + iexp(-\Phi/2)$ is taken to be constant. F-theory generalizes this fact by considering variable λ. Thus F-theory is defined as a twelve-dimensional theory whose compactification on the elliptic fibration $T^2 - \mathcal{K} \to B$, gives the Type IIB theory compactified on B (for a suitable space B) with the identification of $\lambda(\vec{z})$ with the modulus $\tau(\vec{z})$ of the torus T^2. These compactifications can be related to the M-theory compactifications through the known S mapping $\mathcal{S} : IIA \to M/\mathbf{S}^1$ and the $\mathcal{T}$ map between Type IIA and IIB theories. This gives

$$F/\mathcal{K} \times \mathbf{S}^1 \Longleftrightarrow M/\mathcal{K}. \tag{78}$$

Thus the spectrum of massless states of F-theory compactifications can be described in terms of M-theory. Other interesting F-theory compactifications are the Calabi-Yau compactifications

$$F/CY \Leftrightarrow H/K3. \tag{79}$$

NON-PERTURBATIVE CALABI-YAU COMPACTIFICATIONS

M-theory Vacua

In this section we review some Calabi-Yau compactifications of M and F-Theories. In the first part of these lecture we described the perturbative CY compactifications, the purpose of the present section is see how these compactifications behaves in the light of duality and D-brane theory (for excellent reviews see [36,37]). The presence of D-branes or M-branes, in the case of M theory, modifies the perturbative CY compactifications, here we briefly describe these modifications.

Assume that the eleven-dimensional spacetime is $Y = M \times \mathbf{S}^1/\mathbf{Z}_2 \times \mathcal{K}$, with $\mathcal{K}$ being a Calabi-Yau three-fold. Here we consider that $\mathcal{K}$ is a elliptic fibration, since they are favored by CY compactifications of M and F theories. These spacetime corresponds of having two copies (planes) of $X = M \times \mathcal{K}$ at the two boundaries of the orbifold. According to the Horava-Witten theory, anomalies cancellation involves that one $\mathcal{N} = 1$ vector supermultiplet of the E_8 super Yang-Mills theory has to be captured in each orbifold fixed plane X_i, $i = 1, 2$.

According to the perturbative description it is necessary to specify now two stable or semi-stable holomorphic vector bundles V_i on $\mathcal{K}$ with arbitrary group structure. For the heterotic-M theory compactifications the structure group has to be a subgroup of E_8. For simplicity we restrict ourselves to $\mathrm{SU}(n_i)$ vector bundles V_i over $\mathcal{K}$. The presence of fivebranes is of extreme importance here, since it allows more flexibility to construct such vector bundles V_i which leads to more realistic particle physics models. From the modified Bianchi identity and the anomaly cancellation condition of the orbifold system and the fivebranes wrapped on holomorphic two-cycles of $\mathcal{K}$, leads that these bundles are subject to the cohomological constraint of the second Chern classes $c_2(V_1) + c_2(V_2) + [W] = c_2(T\mathcal{K})$, where $[W]$ is the topological class associated to the fivebranes.

The description of the low-energy physics requires of the computation of the first three Chern classes of the holomorphic bundles V_i over $\mathcal{K}$ and thus determine completely a *non-perturbative vacuum*. M and F theories compactifications require that $\mathcal{K}$ must be a holomorphic elliptic fibration. Thus the construction of these bundles are nontrivial.

Construction of the Gauge Bundles over Elliptic Fibrations

An holomorphic elliptically fibered Calabi-Yau three-fold is a fibration

$$\mathcal{K} \xrightarrow{\pi} B$$

where B is an auxiliary complex two-dimensional manifold, π is an holomorphic mapping, and for each $b \in B$, $\pi^{-1}(\{b\})$ is isomorphic to an elliptic curve E_b. In addition we require the existence of a global section $\sigma : B \to \mathcal{K}$ of this fibration.

The elliptic fibration can be characterized by a single line bundle $\mathcal{L}$ over B, $\mathcal{L} \to B$, whose fiber is the cotangent space to the elliptic curve, T^*E_b. This bundle satisfies the condition: $\mathcal{L} = K_B^{-1}$ with K_B being the canonical bundle of B, under the usual condition that the canonical bundle $K_{\mathcal{K}}$ has vanishing first Chern class $c_1(K_{\mathcal{K}}) = 0$. While the global section is specified giving the bundles $K_B^{-\otimes 4}$ and $K_B^{-\otimes 6}$.

These conditions are known to be satisfied by base spaces B corresponding to *del Pezzo, Hirzebruch* and *Enriques* surfaces.

For elliptic fibrations, Friedman, Morgan and Witten [38] found that the second Chern class of the holomorphic tangent bundle $T\mathcal{K}$ can be written in terms of the Chern classes of B as follows

$$c_2(T\mathcal{K}) = c_2(B) + 11c_1^2(B) + 12\sigma c_1(B), \tag{80}$$

where $c_1(B)$ and $c_2(B)$ are the first and the second class of B and σ is a two-form and represents the Poincaré dual of mentioned global section of the elliptic fibration.

One can construct the semi-stable $\mathrm{SU}(n_i)$ holomorphic bundles V_i on $\mathcal{K}$ through the specification of two line bundles $\widehat{\mathcal{L}}$ with first Chern class $\eta \equiv c_1(\widehat{\mathcal{L}})$ and $\widehat{\mathcal{W}}$ with corresponding first Chern class $c_1(\widehat{\mathcal{W}})$ depending on some parameters $n, \sigma, c_1(B), \eta$ and λ. Thus the bundle $\widehat{\mathcal{W}}$ is completely specified by the elliptic fibration and the line bundle $\widehat{\mathcal{L}}$. The condition that $c_1(\widehat{\mathcal{W}}) \in \mathbf{Z}$ leads to the relation $\lambda = m + \frac{1}{2}$ for n odd and $\lambda = m$ and $\eta = c_1(B)$ mod 2, for n even, $m \in \mathbf{Z}$. Thus the Chern classes of the $\mathrm{SU}(n)$ gauge bundle V are

- $c_1(V) = 0$
- $c_2(V) = \eta\sigma - \frac{1}{24}c_1^2(B)(n^3 - n) + \frac{1}{2}(\lambda^2 - \frac{1}{4})n\eta(\eta - nc_1(B))$
- $c_3(V) = 2\lambda\sigma\eta(\eta - nc_1(B))$.

In order to construct realistic particle physics models we take a given base space B and compute its corresponding Chern classes $c_1(B)$ and $c_2(B)$. Compute the relevant Chern classes of the $\mathrm{SU}(n)$ gauge bundles V. The constraints above reduce the number of consistent physical non-perturbative vacua. Given appropriate η and λ one can determine completely the physical Chern classes.

Counting the Number of Families

The number of families of leptons and quarks of the four-dimensional theory is related to the number of zero modes of the Dirac operator coupled to gauge field, which is a connection on the $\mathrm{SU}(n)$ bundle V on the elliptic fibered Calabi-Yau three-fold $\mathcal{K}$. In order to count the number of the chiral fermionic zero modes, one can consider the following cases:

- $SU(3) \times E_6 \subset E_8 : \mathbf{248} = (\mathbf{8}, \mathbf{1}) \oplus (\mathbf{1}, \mathbf{78}) \oplus (\mathbf{3}, \mathbf{27}) \oplus (\mathbf{3}^*, \mathbf{27}^*)$.

- $SU(4) \times SO(10) \subset E_8$: $\mathbf{248} = (\mathbf{15}, \mathbf{1}) \oplus (\mathbf{1}, \mathbf{45}) \oplus (\mathbf{4}, \mathbf{16}) \oplus (\mathbf{4^*}, \mathbf{16^*})$.
- $SU(5) \times SU(5) \subset E_8$: $\mathbf{248} = (\mathbf{24}, \mathbf{1}) \oplus (\mathbf{1}, \mathbf{24}) \oplus (\mathbf{10}, \mathbf{5}) \oplus (\mathbf{10^*}, \mathbf{5^*}) \oplus (\mathbf{5}, \mathbf{10^*}) \oplus (\mathbf{5^*}, \mathbf{10})$.

The matter representations appear in the fundamental representation of the gauge group SU(n). The index of the Dirac operator gives

$$\delta = index(\not{D}_{\mathcal{K}}) = \int_{\mathcal{K}} ch(V) td(\mathcal{K}) = \frac{1}{2} \int_{\mathcal{K}} c_3(V) \tag{81}$$

where $td(\mathcal{K})$ is the Todd class of $\mathcal{K}$. From explitit formula for $c_3(V)$ we get that the number of generations is given by $\delta = \lambda\eta(\eta - nc_1(B))$.

Acknowledgements

We are very grateful to the organizers of the *Ninth Mexican School on Particles and Fields* for the opportunity to give these lectures. One of us O. L-B. is supported by a CONACyT graduate fellowship. It is a pleasure to thank A. Güijosa, A. Pérez-Lorenzana, F. Quevedo, N. Quiroz, M. Ruiz-Altaba, and A.M. Uranga for very useful discussions and enjoyable collaboration.

REFERENCES

1. R.N. Mohapatra, *Unification and Supersymmetry* second edition, Springer-Verlag (1992).
2. E. Alvarez, "Quantum Gravity: an Introduction to Some Recent Results", Rev. Mod. Phys. **61** (1989) 561.
3. E. Witten, "Physics and Geometry", Proc. Int. Congress Math. (1986) pp. 267.
4. M. Green, J.H. Schwarz and E. Witten, *Superstring Theory*, Two volumes, Cambridge Univerity Press, Cambridge (1987).
5. J. Polchinski, *String Theory*, Two volumes, Cambridge University Press, Cambridge (1998).
6. D. Lüst and S. Theisen, *Lectures on String Theory*, Lectures Notes in Physics, Springer-Verlag, Berlin (1989).
7. B. Hatfield, *Quantum Field Theory of Point Particles and Strings,* Addison-Wesley Publishing, Redhood City, (1992).
8. H. Ooguri and Z. Yin, "Lectures on Perturbative String Theory", hep-th/9612254.
9. E. Kiritsis, "Introduction to Superstring Theory", hep-th/9709062.
10. S. Kachru, "Lectures on Warped Compactifications and Stringy Brane Constructions", hep-th/0009247.
11. F. Quevedo, "Superstring Phenomenology: An Overview", Nucl. Phys. Proc. Suppl. **62** (1998) 134, hep-ph/9707434; L.E. Ibañez, "The Second String (Phenomenology)

Revolution", Class. Quantum Grav. **17** (2000) 1117; G. Aldazabal, L.E. Ibañez and F. Quevedo, "On Realistic Brane Worlds From Type I Strings", hep-th/0005033.
12. F. Quevedo, "Lectures on Superstring Phenomenology", in *Workshop on Particles and Fields and Phenomenology of Fundamental Interactions* Eds. J.C. D'Olivo, A. Fernández and M.A. Pérez, AIP (1996).
13. E. Witten, "Lepton Number and Neutrino Masses", hep-ph/0006332.
14. M. Dine, "TASI Lectures on M Theory Phenomenology", hep-th/0003175.
15. J. Polchinski, "TASI Lectures on D-Branes", hep-th/9611050.
16. C.V. Johnson, "D-brane Premier", hep-th/0007170.
17. T. Banks, "Matrix Theory", Nucl. Phys. Proc. Suppl. **67** (1998) 180, hep-th/9710231; "TASI Lectures on Matrix Theory", hep-th/9911068.
18. W. Taylor IV, "The M(atrix) Model of M-Theory", hep-th/0002016.
19. A. Giveon and D. Kutasov, "Brane Dynamics and Gauge Theory", Rev. Mod. Phys. **71** (1999) 983, hep-th/9802067.
20. S.S. Gubser, J.M. Maldacena, H. Ooguri and Y. Oz, "Large N Field Theories, String Theory and Gravity", Phys. Rept. **323** (2000) 183, hep-th/9905111.
21. A.A. Tseytlin, "Born-Infeld Action, Supersymmetry and String Theory", hep-th/9908105.
22. E. Witten, "D-branes and K-theory", JHEP **9812** (1998) 019, hep-th/9810188.
23. A. Sen, "Non-BPS States and Branes in String Theory", hep-th/9904207.
24. E. Witten, "Overview of K-theory Applied to Strings", hep-th/0007175.
25. H. García-Compeán and A.M. Uranga, "Brane Box Realization of Chiral Gauge Theories in Two Dimensions", Nucl. Phys. B **539** (1999) 329, hep-th/9806177.
26. A. Hanany and E. Witten, "Type IIB Superstrings, BPS Monopoles, and Three-dimensional Gauge Dynamics" Nucl. Phys. B **492** (1997) 152, hep-th/9611230.
27. A. Hanany and A. Zaffaroni, "On the Realization of Chiral Four-dimensional Gauge Theories Using Branes", JHEP **05** (1998) 001, hep-th/9801134.
28. A. Hanany and A. Uranga, "Brane Boxes and Branes on Singularities", JHEP **9805** (1998) 013, hep-th/9805139.
29. A.M. Uranga, "From Quiver Diagrams to Particle Physics", hep-th/0007173.
30. A. Sen, "An Introduction to Non-perturbative String Theory", hep-th/9802051.
31. E. Witten, "String Theory Dynamics in Various Dimensions", Nucl.Phys. B443 (1995) 85, hep-th/9503124.
32. J.H. Schwarz, "Lectures on Superstring and M Theory Dualities", Nucl. Phys. Proc. Suppl. **55B** (1997) 1, hep-th/9607201.
33. P.K. Townsend, "Four Lectures on M-Theory, hep-th/9612121.
34. C. Vafa, Lectures on Strings and Dualities, hep-th/9702201.
35. E. Kiritsis, "Introduction to Non-perturbative String Theory", hep-th/9708130.
36. B. Andreas, "$\mathcal{N} = 1$ Heterotic/F-Theory Duality", Fortsch. Phys. **47** (1999) 587, hep-th/9808159.
37. B.A. Ovrut, "$\mathcal{N} = 1$ Supersymmetric Vacua in Heterotic M-Theory", hep-th/9905115.
38. S. Sethi, C. Vafa and E. Witten, "Constraints on Low Dimensional String Compactifications", Nucl. Phys. B **480** (1996) 213, hep-th/9606122; R. Friedman, J. Morgan and E. Witten, "Vector Bundles and F-theory", Commun. Math. Phys. **187** (1997) 679, hep-th/9701162.

Dark Matter and Energy of Universe

Luis Masperi[1]
Centro Latinoamericano de Fí sica
Av. Venceslau Bráz 71 Fundos, 22290-140 Rio de Janeiro, Brazil

Abstract. We describe the different components of the present energy of the universe starting from the well established radiation and luminous matter, following with the dark baryonic matter determined by primordial nucleosynthesis, the likely cold dark matter with its theoretical candidates, the sure but not yet defined hot dark matter represented by neutrinos and the evidence of dark energy given by cosmological constant, vacuum energy or quintessence.

I EXPANDING UNIVERSE

We wish to discuss the inventory of the mass (energy) of the universe.

Assuming that for large scale the universe is homogeneous, we may use the Robertson-Walker metric [1]

$$ds^2 = dt^2 - R^2(t)\left(\frac{dr^2}{1-kr^2} + r^2 d\theta^2 + r^2 \sin^2\theta \, d\varphi^2\right) , \tag{1}$$

where k is a constant. It may be seen that the 3-dimensional space is closed, open or flat for $k > 0$, $k < 0$ or $k = 0$ respectively.

The Einstein equation relates the curvature of the metric to the energy-momentum tensor through the Newton constant. For a perfect fluid characterized by energy density ρ and pressure p the energy-momentum tensor is

$$T^{\mu\nu} = (\rho + p)\, u^\mu u^\nu - g^{\mu\nu} p , \tag{2}$$

where u^μ is the covariant velocity and $g^{\mu\nu} = (+1, -1, -1, -1)$ diagonal tensor.

The Einstein equation takes therefore the form of the Friedmann ones

$$\frac{\dot{R}^2}{R^2} + \frac{k}{R^2} = \frac{8\pi}{3} G_N \, \rho , \tag{3}$$

$$\frac{\ddot{R}}{R} = -\frac{4\pi}{3} G_N \, (\rho + 3p) . \tag{4}$$

---

[1] E-mail: masperi@cbpf.br. On leave of absence from Centro Atómico Bariloche, Argentina.

CP562, *Particles and Fields: Ninth Mexican School,* edited by G. Herrera Corral and L. Nellen

The Hubble parameter is $H = \frac{\dot{R}}{R}$ · Its value today is expressed in terms of a constant h as

$$H_0 = 100h \; km \; \sec^{-1} Mpc^{-1} \; , \tag{5}$$

and the critical density is that which comes from Eq.(3) for $k = 0$ i.e.

$$\rho_c = \frac{3H_0^2}{8\pi G_N} \; . \tag{6}$$

Since $G_N = \frac{1}{m_{pl}^2}$ with the Planck mass $m_{pl} = 10^{19} GeV$, using the equalities of the Appendix

$$\rho_c = 1.88 \; h^2 10^{-29} gr \; cm^{-3}. \tag{7}$$

One important experimental result is the determination of the present value of the Hubble constant which confirms that the universe is still expanding. All current data are consistent with [2]

$$h = 0.65 \pm 0.05 \; . \tag{8}$$

According to Eq.(3) if the actual universe density is larger, equal or smaller than the critical value of Eqs.(7, 8), the universe is closed, flat or open respectively.

II RADIATION AND LUMINOUS MATTER

One contribution that can be easily estimated is that of the cosmic background radiation (CBR) discovered in 1964 by Penzias and Wilson which has been extremely well evaluated by the COBE satellite [3] as a black body radiation of $T = 2.728 \pm 0.002 \; ^\circ K$. From the Stefan-Boltzmann law and using the Appendix

$$\rho_R \sim T^4 \sim \left(2 \; 10^{-4} \; eV\right)^4 \simeq 10^{-33} \; gr \; cm^{-3} \; , \tag{9}$$

which corresponds to 400 photons per cm^3 and is much smaller than the critical density. Of course, since the decrease of ρ_R is faster with the universe expansion than that of nonrelativistic matter

$$\rho_R \propto \frac{1}{R^4} \; , \; \rho_M \propto \frac{1}{R^3} , \tag{10}$$

radiation must have dominated in the past.

It is usual to refer the different contributions of density to the critical value so that at present from Eqs.(9) and (7)

$$\Omega_R = \frac{\rho_R}{\rho_C} \sim 10^{-4} \; . \tag{11}$$

The next obvious contribution to density is that of luminous matter. This is evaluated summing over all the masses M in a volume V using a typical mass to luminosity ratio $\frac{M}{L} \simeq \frac{5M_o}{L_o}$ related to that of sun

$$\rho_L = \frac{M}{L}\sum_i \frac{L_i}{V} \simeq 5\frac{M_\odot}{L_\odot}\lambda \; . \tag{12}$$

From the measures of fluxes coming from bright regions of galaxies

$$\lambda = (2 \pm 0.6)\, 10^8 L_\odot \; h \; Mpc^{-3} \; , \tag{13}$$

where the factor h comes because the distances are calculated through the Hubble law which gives the redshift

$$z = H_o d \; . \tag{14}$$

Inserting the mass of the sun M_o into Eq.(12) one obtains

$$\rho_L = (0.7 \pm 0.2)\, 10^{-31} h \; gr \; cm^{-3} \; , \tag{15}$$

which again gives a small fraction of the critical density

$$\Omega_L \simeq (0.003 \pm 0.001)\, h^{-1} \sim 0.005 \; . \tag{16}$$

III DARK BARYONIC MATTER

It is astonishing that it is possible to infer the total density of ordinary baryonic matter from the primordial synthesis of light nuclei D, ^3He, ^4He and ^7Li because this is based on well known nuclear physics. The deuterium abundance is particularly relevant because it cannot be produced in stellar processes. The recent determination [4] of the ratio of abundances $\frac{D}{H} = (3.4 \pm 0.3)\, 10^{-5}$ in uncontaminated distant clouds through absorption of quasar radiation is explained by a ratio of baryon to photon number densities.

$$\eta = \frac{n_B}{n_\gamma} = (5 - 7)\, 10^{-10} \; . \tag{17}$$

Though this number of baryons seems very small, since its mass is around $1GeV$ i.e. almost 13 orders of magnitude larger than the energy of photons, it leads to a mass density

$$\rho_B = (3.8 \pm 0.4)\, 10^{-31} gr \; cm^{-3} \; , \tag{18}$$

and a fraction of critical density

$$\Omega_B = (0.020 \pm 0.002)\, h^{-2} \sim 0.05 \ , \tag{19}$$

ten times larger than the luminous matter.

This is the first evidence of dark matter, the baryonic one which could reside in the MACHOs detected by microlensing that might account for up to one third of the galactic halo, and probably more abundantly in diffuse gas very difficult to detect except in the x-ray emitting intracluster gas.

Why must there be a galactic halo? It happens that according to Kepler law, if the mass of the galaxy is M the orbital velocity v of an object at a distance r from its centre is given by

$$\frac{v^2}{r} = G_N \frac{M}{r^2} \ , \tag{20}$$

so that one would expect v^2 to decrease as r^{-1} for large distances. But astronomers have observed that for increasing distances, through rare stars and atomic emission of 21 cm line, v remains constant indicating that M is not concentrated in the luminous part of the galaxy but increases as r. This proves the existence of a dark halo which on the whole gives to the galaxy a mass ten times larger than its luminous part. Note that if all the halo is baryonic this would appear to cover the ratio between Eqs.(19) and (16), leaving no room for diffuse gas.

But a very important observation has been done for clusters of galaxies where most of baryons reside in the hot x-ray emitting intracluster gas. It has been determined [5] that the ratio of baryonic and total cluster mass, to explain its motion, is

$$f_{gas} = (0.070 \pm 0.002)\ h^{-\frac{3}{2}} \ , \tag{21}$$

so that assuming that clusters having a size of around $10 Mpc$ are good samples for the composition of matter in universe, the total matter fraction of critical density is

$$\Omega_M = \frac{\Omega_B}{f_{gas}} \sim 0.4 \tag{22}$$

This result, apart from giving evidence that most of baryonic dark matter must be in gaseous form, indicates that there is much more matter of non-baryonic type.

IV HOT AND COLD DARK MATTER

Which is the nature of this non-baryonic dark matter (DM) is still a mystery.

Among the known weak interacting particles, the obvious candidates are neutrinos. They are a form of hot dark matter (HDM) because they are light particles that were certainly relativistic when their thermal equilibrium was frozen, i.e. when the rate of their reactions became smaller than the Hubble parameter which occurred at $T \sim 1MeV$. Therefore, except for the fact that now the temperature of neutrinos is slightly smaller than that of photons of CBR, the density of neutrinos in the universe is certainly not much lower than that of photons i.e. around 100 per cm^3 for each species. Since it has been determined by the accelerator LEP that there are 3 kinds of light neutrinos, if their masses were of a few eV, neutrinos might explain all DM. However this seems to be unlikely because if DM was relativistic when the structures started to be formed, the small scale inhomogeneities would be washed out and the large scale ones would be the first to appear, contrary to the observations that have shown that galaxies appeared at $z \sim 5$ i.e. earlier than clusters.

The question if neutrinos have mass is one of the most important issues of present elementary particle physics. Of great importance has been the observation of atmospheric neutrino oscillation at Superkamiokande [6] indicating that there is a difference of mass between two kinds of neutrinos of around $0.1eV$. This leads to a lower bound of one type of neutrino $m_\nu \geq 0.1eV$ and therefore, even though neutrinos cannot explain all DM, it must be

$$\Omega_\nu \geq \Omega_L \ . \tag{23}$$

Therefore one must look for a cold dark matter (CDM) candidate i.e. nonrelativistic particle when structures began.

One possibility is that when thermal equilibrium froze, its mass was larger than temperature so that the density was suppressed in comparison with that of photons. Such particles are thought to emerge from supersymmetric (SUSY) extensions of the standard model which assume that for each fermion there is a boson and viceversa (i) giving stability to quantum corrections of Higgs mass, (ii) predicting high energy unification of coupling constants for electromagnetic, weak and strong interactions and (iii) being an ingredient of string theories which include gravity. The lightest supersymmetric particle would be the neutralino, fermion mixture of photino, zino and higgsino. Since the scale for breaking of SUSY should be ≤ 1 TeV to avoid excessive quantum corrections to the Higgs mass, it is reasonable to expect a neutralino mass $m_\chi \sim 100\ GeV$.

It is possible to estimate the density of neutralinos compared to that of photons from the freeze-out condition at temperature $T_\chi < m_\chi$

$$n_\chi \left(\sigma v\right) = H\left(T_\chi\right) \simeq \frac{T_\chi^2}{m_{pl}} \simeq \frac{n_\gamma}{m_{pl} T_\chi} \ . \tag{24}$$

Considering that the thermal averaged cross-section (σv) at freeze-out is due to annihilation with a coupling $\alpha \sim 10^{-2}$, the dimensional assumption

$$(\sigma v) \sim \frac{\alpha^2}{m_\chi^2} \sim 10^{-36} cm^2 \tag{25}$$

corresponds to a weak interaction. Taking $m_\chi \sim 10T_\chi$ from Eqs.(24) and (25) it turns out that

$$n_\chi \sim 10^{-12} n_\gamma \ , \tag{26}$$

and being at present m_χ around 15 orders of magnitude larger than the energy of a photon, recalling Eq.(11) it is possible to fit

$$\Omega_\chi \sim 0.35 \ . \tag{27}$$

explaining all DM, as it was done many years ago [7] .

Accelerators have so far excluded the lightest supersymmetric particle with $m_\chi <$ $100\ GeV$. A very intense search of these weak interacting massive particles (WIMP) is pursued both through direct detection by the recoil of Ge nucleus or indirect detection by high energy neutrinos coming from the annihilation $\chi\chi \to \nu\bar{\nu}$ possible if χ are gravitationally captured by sun, which is one of the aims of the South Pole Amanda neutrino telescope, or also by annihilation in the galactic halo with production of antiprotons that might be revealed by the AMS project.

Another candidate for CDM which was never in thermal equilibrium is the axion, a theoretical particle introduced [8] to avoid the CP violation in strong interactions. It is a neutral pseudoscalar particle which, when the temperature of universe falls below the confinement scale $\wedge_{QCD} \sim 200\ MeV$ acquires a mass because of its mixture with the state of pion. The mass of the axion turns out to be

$$m_a \sim \frac{\wedge_{QCD}^2}{f_a} \ , \tag{28}$$

where f_a is of the order of a superheavy quark mass to which it is coupled. Because of this interaction, the coupling of axion with electromagnetic fields is of the form

$$\mathfrak{L}_{a\gamma\gamma} \sim \frac{\alpha_{em}}{f_a} a \ \mathbf{E}.\mathbf{B} \tag{29}$$

which gives an extremely large lifetime for the decay $a \to 2\gamma$. With the choice $f_a \sim 10^{12}\ GeV$, from Eq.(28) $m_a \sim 10^{-5}\ eV$ which is a crucial value as it will be seen below.

How is it possible that with such a tiny mass the axion is a nonrelativistic particle? This would come from the fact that the equation of motion of the axion field in the expanding universe is

$$\ddot{a} + 3H(t)\,\dot{a} + m_a^2(t)\,a = 0 \ . \tag{30}$$

When the Hubble parameter is larger than the axion mass the friction term of Eq.(30) dominates, and the solution is $a =$ constant everywhere. Afterwards when $m_a > H$ a starts the oscillation giving rise to the particles which, due to the uniformity of the field in space, have almost vanishing momenta so that they are nonrelativistic.

It is a delicate matter to evaluate the contribution to universe energy density of an initial non-alignment of the field a when the QCD mass effects were very small. But due to the fact that m_a decreases with f_a, the potential is flatter and misalignment larger turning out that

$$\Omega_a \sim \left(\frac{10^{-5}eV}{m_a}\right)^{1.1} . \tag{31}$$

Therefore axions may close the universe if they have the correct mass. Experiments are underway [9] using the interaction Eq.(29) to transform the axion with an intense magnetic field into a wave in a resonant cavity and finally decide whether this particle exists.

On the other extreme of mass scale it is also possible that superheavy relics of the age of the grand unification theories (GUT) $m_{GUT} \geq 10^{15}GeV$ form a part, not necessarily very large, of the galactic halo. The interest of these objects is that they might be the origin, through a very slow decay, of the ultra-high energy cosmic rays (UHECR) [10] . Of course the question is how could they survive till the present epoch. One alternative is that they are particles [11] which interact with the known ones only through forces of gravitational order. Another is that they are cosmic strings either with attached monopoles [12] or stabilized by superconducting currents [13] . This possible origin of UHECR may be elucidated by the Auger observatory which is beginning to be built.

V COSMOLOGICAL CONSTANT AND QUINTESSENCE

But DM does not seem the end of the story regarding the energy of the universe. There is a growing evidence that the energy is the critical one

$$\Omega_o = \Omega_M + \Omega_\Lambda = 1 , \tag{32}$$

with $\Omega_\Lambda \simeq 0.6$ corresponding to a cosmological constant.

This constant Λ had been introduced by Einstein in his equations in a way equivalent to add to the energy-momentum tensor Eq.(2) a term

$$T^{\mu\nu}_{vac} = \rho_{vac}\, g^{\mu\nu} , \tag{33}$$

with the definition $\rho_{vac} = \frac{\Lambda}{(8\pi G_N)}$. Note that Eq.(33) gives a negative pressure $p = -\rho_{vac}$. This justifies the equivalence of ρ_{vac} with a vacuum energy density because thermodynamically this system increases its energy with volume in agreement with a negative pressure.

Now, with Eqs.(3), (4) and (32), the deceleration parameter is

$$q_o = -\frac{\left(\frac{\ddot{R}}{R}\right)_o}{H_o^2} = \frac{1}{2} + \frac{3}{2}\sum_i \Omega_i w_i \ , \tag{34}$$

where $p_i = w_i \rho_i$ for each component, so that $w_i = \frac{1}{3}, 0, -1$ for radiation, nonrelativistic matter and vacuum respectively. For the case of Eq.(32) $q_o = \frac{1}{2} - \frac{3}{2}\Omega_\Lambda = -0.4$ and the universe now would be actually accelerating! This is precisely what is emerging from the observation of distant supernovae [14] .

The existence of a cosmological constant leads to an agreement between the relative large age of the universe and large present Hubble parameter H_o because it acts as opposing the effect of gravitational attraction. Another impressive evidence [15] is the fit of the so-called acoustic peak of the anisotropy of CBR which favours $\Omega_o = 1$ and $\Omega_\Lambda = 0.6$, issue that will be completely settled by the future satellites MAP and Planck.

It is interesting [16] that also the reionization of intergalactic matter corresponding to the evolution of universe since a redshift $z \sim 5$ is in favour of CDM plus a cosmological constant contribution $\Omega_\Lambda \sim 0.6$.

From the point of view of quantum field theory it is very hard to explain why the vacuum energy should have a value of the order of Eq.(7) which, using the Appendix, can be rewritten as

$$\rho_c \sim (0.001\ eV)^4 \ . \tag{35}$$

In fact, since if SUSY were exact the vacuum energy should be zero, one might expect that ρ_c would depend on the scale of SUSY breaking i.e. $\rho_c \sim (1\ TeV)^4$ which is an estimation wrong by 60 orders of magnitude!

One may prefer to think that for some not yet explained reason [17] based on SUSY and extra dimensions, Einstein cosmological constant and vacuum energy exactly compensates and that what one observes today is a sort of dynamical cosmological constant due to a uniform field [18] φ called "quintessence". Things would go in the following way.

Since the energy-momentum tensor for a scalar field is

$$T^{\mu\nu} = \frac{\partial \mathfrak{L}}{\partial \partial_\mu \varphi}\partial^\nu \varphi - g^{\mu\nu}\mathfrak{L} \ , \tag{36}$$

for the spatially homogeneous φ with a potential V, comparing with Eq.(2)

$$\rho = \frac{1}{2}\varphi^2 + V \ , p = \frac{1}{2}\varphi^2 - V \ . \tag{37}$$

On the other hand the equation of motion, similar to Eq.(30), is

$$\ddot{\varphi} + 3H(t)\,\dot{\varphi} + V'(\varphi) = 0 \ . \tag{38}$$

Therefore, as discussed for the axion, when $H > m_\varphi$ friction dominates, φ is constant and from Eq.(37) $p = -\rho$ as for a cosmological constant. The difference is that the rolling down would occur in the present age because

$$m_\varphi \sim H_o \sim 10^{-33}\ eV \ , \tag{39}$$

using Eq.(5) and the Appendix. If the potential is also of the type corresponding to the axion

$$V(\varphi) = \rho_c \left(1 - \cos \frac{\varphi}{f_\varphi}\right) \ , \tag{40}$$

where f_φ is the order of the mass of the fermion to which φ is coupled, when $\varphi \simeq$ constant $\sim f_\varphi$ V gives the critical density as a false vacuum energy, and for $\varphi \sim 0$ the mass Eq.(39) appears provided $f_\varphi \sim m_{pl}$.

An explanation of why m_φ has this particular value and therefore the false vacuum is relevant today may be that φ gets its mass through the mixing with pion, as for axion, but with an interaction between the superheavy fermion and ordinary quarks of gravitational order instead of strong one [19] . In this way, instead of Eq.(28) one obtains

$$m_\varphi \sim \frac{m_\pi^3}{m_{pl}^2} \ , \tag{41}$$

consistent with Eq.(39).

Obviously there is still the big interrogation point of why the apparent cosmological constant contribution is of the same order of that of matter which allows our presence to observe it. This can be answered either by an "anthropic principle" related to the conditions for our existence [20] or by models that internally determine that φ locks into a negative pressure state after the onset of matter dominance [21]

In conclusion, the components of the energy of universe seem reasonably determined but the identification of their nature requires still a great astrophysical and theoretical effort beyond the standard model of fundamental interactions.

Acknowledgment

I thank Gerardo Herrera, Miguel Angel Pérez and the Organizing Committee of the IX Mexican School on Particles and Fields for the kind hospitality at Puebla.

A IMPORTANT RELATIONS OF UNITS

We use natural units

$$\hbar = 1, c = 1, k_B = 1. \tag{A1}$$

Relation between length and energy

$$10^{-5} eV \simeq \frac{1}{cm}. \tag{A2}$$

Relation between mass and energy

$$10^{19} GeV \simeq 10^{-5} gr. \tag{A3}$$

Relation between temperature and energy

$$10^{4}\,{}^{o}K \simeq 1 eV. \tag{A4}$$

Astronomical length unit

$$1 pc \simeq 3\ ligth - yr \simeq 3 \times 10^{18} cm. \tag{A5}$$

Macroscopic energy

$$1\ Joule \simeq 10^{7}\ erg \simeq 10^{19} eV. \tag{A6}$$

Solar mass

$$1\ M_{\odot} \simeq 2 \times 10^{30}\ kgr. \tag{A7}$$

REFERENCES

1. E.W. Kolb and M.S. Turner, The Early Universe (Addison Wesley, Redwood City) (1990).
2. B. Madore et al., Nature **395** (1998) 47.
3. D.J. Fixsen et al., Astrophys. J. **473** (1996) 576.
4. S. Burles and D.Tytler, Astrophys. J. **499** (1998) 699.
5. J. Mohr, B. Mathiesen and A.E. Evrard, Astrophys. J. (1998) submitted.
6. Y. Fukuda et al., Phys. Rev. Lett. **81** (1998) 1562.
7. Y. Zeldovich, Zh. Eksp. Teor. Fiz. **48** (1965) 986.
8. R. Peccei and H. Quinn, Phys. Rev. Lett. **48** (1977) 1440.
9. C. Hagmann et al., Phys. Rev. Lett. **80** (1998) 2043.
10. M. Takeda et al., Phys. Rev. Lett. **81** (1998) 1163.
11. V.A. Kuzmin and V.A. Rubakov, Yadernaya Fisika **61** (1998) 1122.
12. V. Berezinsky and A. Vilenkin, Phys. Rev. Lett. **79** (1997) 5202.
13. L. Masperi and M. Orsaria, Nucl. Phys. B (Proc. Suppl.) **75 A** (1999) 362.
14. A. Riess et al., Astron. J. **116** (1998) 1009.
15. P. De Bernardis et al., Nature **404** (2000) 955.
16. L. Masperi and S. Savaglio, Astron. and Astroph. **321** (1997) 1.
17. E. Witten, Int. J. Mod. Phys. **A 10** (1995) 1247.
18. J. Frieman, C. Hill, A. Stebbins and I. Waga, Phys. Rev. Lett. **75** (1995) 2077.
19. J. Estrada and L. Masperi, Mod. Phys. Lett. **A 13** (1998) 423.
20. S. Weinberg, Phys. Rev. Lett. **59**, (1987) 2607.
21. C. Armendariz-Picon, V. Mukhanov and P.J. Steinhardt, astro-ph/0006373.

Disoriented Chiral Condensates in High-Energy Nuclear Collisions

Jørgen Randrup

Lawrence Berkeley National Laboratory, Berkeley, California 94720, USA

Abstract. This brief lecture series discusses how our current understanding of chiral symmetry may be tested more globally in high-energy nuclear collisions by suitable extraction of pionic observables. After briefly recalling the general features of chiral symmetry, we focus on the SU(2) linear σ model and show how a semi-classical mean-field treatment makes it possible to calculate its statistical properties, including the chiral phase diagram. Subsequently, we consider scenarios of relevance to high-energy collisions and discuss the features of the ensuing non-equilibrium dynamics and the associated characteristic signals. Finally, we illustrate how the presence of vacuum fluctuations or the inclusion of strangenesss may affect the results quantitatively.

INTRODUCTION

The advent of ever more powerful heavy-ion accelerators has made it possible to study strongly interacting matter over a wide range of physical conditions. In particular, at the unprecedented energy densities now within reach at RHIC two fundamental phase changes are expected to occur. One is the dissolution of ordinary hadrons into a deconfined quark-gluon plasma phase, which has long been a principal research objective. The other, on which the present lectures are focussed, is the approximate restoration of chiral symmetry. This topic has gained increasing interest in recent years because of the recognition that the rapid non-equilibrium dynamics may produce coherent oscillations of the pion field with observable consequences. Reviews of this phenomenon, commonly referred to as *disoriented chiral condensates* (DCC), are given in Refs. [1–3].

Since the u and d quark masses are fairly small, the basic QCD Lagrangian is approximately invariant with respect to chiral transformations and it is therefore expected that matter at high energy density will display approximate chiral symmetry. However, due to the self-interaction of the fields, this symmetry is spontaneously broken at lower energy. In particular, the ordinary vaccum exhibits a finite value of the order parameter, $\langle q\bar{q}\rangle = f_\pi$, and the associated pionic excitations are approximately massless on the hadronic scale. A pedagogical introduction to the basic concepts of chiral symmetry in nuclear physics is given in Ref. [4].

CP562, *Particles and Fields: Ninth Mexican School,* edited by G. Herrera Corral and L. Nellen

The most popular theoretical tool for studies of DCC phenomena has been the linear σ model [5] which describes the O(4) chiral field $\boldsymbol{\phi} = (\sigma, \boldsymbol{\pi})$ by means of a simple effective quartic interaction (see Refs. [6–14] for some examples),

$$\mathcal{L} = \frac{1}{2}\partial_\mu\boldsymbol{\phi} \circ \partial^\mu\boldsymbol{\phi} - \frac{\lambda}{4}(\phi^2 - v^2)^2 + H\sigma \quad \Rightarrow \quad [\Box + \lambda(\phi^2 - v^2)]\boldsymbol{\phi} = H\hat{\sigma} \ . \qquad (1)$$

The three model parameters, λ, v, and H, can be fixed by specifying the pion decay constant, $f_\pi = 92$ MeV, the free pion mass, $m_\pi = 138$ MeV, and the mass of the nominal σ meson (which is rather uncertain), $m_\sigma = 600$ MeV. In vacuum, the chiral field is aligned with the σ direction, $\boldsymbol{\phi}_{\rm vac} = (f_\pi, \mathbf{0})$, and at low temperatures the agitations of the field represent nearly free σ and π mesons. This simplest version of the model is appropriate only in a baryon-free environment, such as the central rapidity region at RHIC, and it needs to be appropriately extended when baryons are present (for an example of this, see Ref. [8]).

Within this conceptually simple framework, the basic mechanism of DCC formation is readily depicted, as illustrated in Fig. 1: The early violent part of a collision event produces an extended region of space within which the energy density is so high that chiral symmetry is temporarily nearly restored. As the collision evolves further, the energy density drops so rapidly that the chiral degrees of freedom fall out of equilibrium, leading to large-amplitude long-wavelength isospin-directed oscillations of the pion field. This unique phenomenon would have a number of specific observational consequences, including anomalous multiplicity distributions of the soft pions and a significant enhancement of dilepton production.

This lecture series reviews efforts to address the phenomenon quantitatively within the framework of the linear σ model.

STATISTICAL EQUILIBRIUM

In order to understand the key features of DCC formation, which is inherently a non-equilibrium phenomenon, it is useful to know the equilibrium properties of the system. In this section we discuss how statistical equilibrium may be addressed quantitatively within a semi-classical framework that is readily extendable to arbitrary time-dependent scenarios. We follow Ref. [15] where more details are given.

Mean-Field Treatment of the Linear σ Model

The present considerations are restricted to systems confined within a cubic box of volume $\Omega = L^3$; the continuum limit is then approached (rapidly) as $L \to \infty$. A given microscopic state of the system is characterized by the field $\boldsymbol{\phi}(\boldsymbol{r})$ and its time derivative $\boldsymbol{\psi}(\boldsymbol{r})$, both of which are real functions and have periodic boundary conditions. (The separate notation for the time derivative of the field emphasizes that it represents independent information; it is also notationally advantageous.)

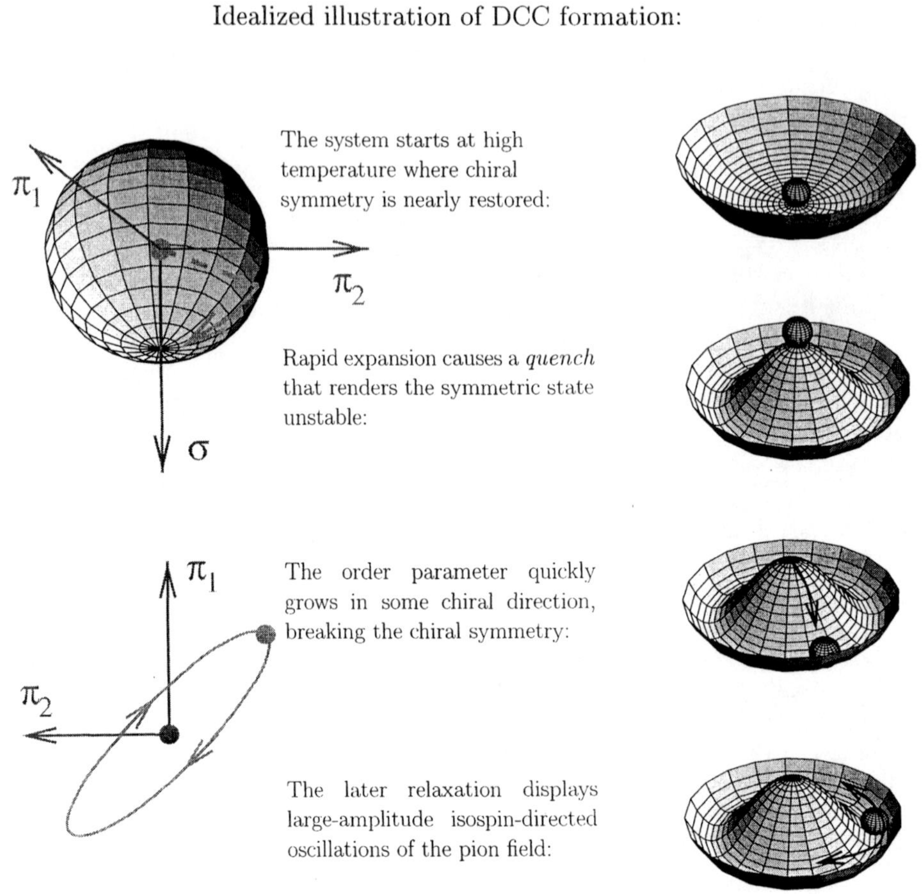

FIGURE 1. The mechanism leading to DCC formation is illustrated for the simplified worlds in which the isospin space has only two (left-hand side) or one (right-hand side) dimension. In the originally conceived DCC quench scenario, the system is initially so highly agitated that the O(4) order parameter $(\sigma, \boldsymbol{\pi})$ is situated at the single minimum of the effective potential near the origin. The system then expands and/or cools so rapidly that the effective potential quickly reverts to its low-temperature form, in which a nearly degenerate minimum appears as a trough on the surface of the 4-hypersphere with radius f_π, $\sigma^2 + \pi^2 \approx f_\pi^2$. Since the early rapid growth of the order parameter is primarily in the radial direction, the generated vector current $\boldsymbol{\pi} \times \dot{\boldsymbol{\pi}}$ is very small and the subsequent oscillatory relaxation towards the ground state is therefore well directed in isospace (lower left illustration). The emitted pion radiation is then correspondingly isospin polarized, resulting in a very broad distribution of the neutral fraction $f = N_0/(N_- + N_0 + N_+)$.

It is useful to separate the field into two parts, $\boldsymbol{\phi}(\boldsymbol{r}) = \underline{\boldsymbol{\phi}} + \boldsymbol{\delta\phi}(\boldsymbol{r})$. The spatial average of the field, $\underline{\boldsymbol{\phi}} = \langle \boldsymbol{\phi}(\boldsymbol{r}) \rangle$ is identified with the O(4) order parameter associated with that particular field configuration. The components of the field fluctuation $\boldsymbol{\delta\phi}(\boldsymbol{r})$ along the O(4) direction of the order parameter is denoted by $\delta\phi_{\parallel}(\boldsymbol{r})$ and $\boldsymbol{\delta\phi}_{\perp}(\boldsymbol{r})$ is its component perpendicular to $\underline{\boldsymbol{\phi}}$. A similar decomposition can be made for the time derivative, $\boldsymbol{\psi}(\boldsymbol{r}) = \underline{\boldsymbol{\psi}} + \boldsymbol{\delta\psi}(\boldsymbol{r})$. This separation makes it possible to interpret the field fluctuations as quasiparticle excitations relative to the environment (or "effective vacuum") characterized by the temperature T and the magnitude of order parameter, ϕ_0.

Moreover, the separation of the field invites the application of the Hartree factorization technique. The first step, taking the spatial average of the full equation of motion (1), leads to an equation of motion for the order parameter $\underline{\boldsymbol{\phi}}$ [16],

$$\left[\partial_t^2 + \mu_0^2\right] \underline{\boldsymbol{\phi}} = H\hat{\boldsymbol{e}}_\sigma \, . \tag{2}$$

The associated effective mass μ_0 for the order parameter is given by

$$\mu_0^2 \;=\; \lambda[\phi_0^2 + \prec \delta\phi^2 \succ + \, 2 \prec \delta\phi_{\parallel}^2 \succ - \, v^2] \, . \tag{3}$$

Here $\phi_0 \equiv |\underline{\boldsymbol{\phi}}|$ is the magnitude of the order parameter. Moreover, $\prec \delta\phi_{\parallel}^2 \succ$ is the thermal average of the field fluctuation along $\underline{\boldsymbol{\phi}}$, while $\prec \delta\phi^2 \succ = \prec \delta\phi_{\parallel}^2 \succ + 3 \prec \delta\phi_{\perp}^2 \succ$ is the total field fluctuation (the three perpendicular fluctuation components have the same thermal average and $\prec \delta\phi_{\perp}^2 \succ$ denotes just *one* of them). Terms vanishing in thermal equilibrium have been ignored, namely correlations between field fluctuations in different $O(4)$ directions, $\prec \delta\phi_{\parallel}\boldsymbol{\delta\phi}_{\perp} \succ = \boldsymbol{0}$, and averages of odd powers of field fluctuations, $\prec \delta\phi^2 \boldsymbol{\delta\phi} \succ = \boldsymbol{0}$.

The above result can be used to determine the phase diagram, since the order parameter experiences no forces in equilibrium, $\partial_t^2 \underline{\boldsymbol{\phi}} = \boldsymbol{0}$. Thus, at a given temperature T, the stationary points of $\underline{\boldsymbol{\phi}}$ are determined by $\mu_0^2 \underline{\boldsymbol{\phi}} = H\hat{\boldsymbol{e}}_\sigma$, which implies that the equilibria are located on the σ axis. If χ_0 is the angle between $\underline{\boldsymbol{\phi}}$ and the σ direction, the aligned component of the order parameter is given by $\sigma_0 = \phi_0 \cos \chi_0$ and the equilibrium condition then amounts to the relation $\mu_0^2 \sigma_0 = H$. When the free pion mass vanishes, so does H. The condition for obtaining an equilibrium for a finite magnitude of the order parameter, $\phi_0 > 0$ (in which chiral symmetry is thus spontaneously broken), then amounts to the requirement that μ_0 be zero. This feature is a manifestation of the Goldstone theorem [17].

In order to calculate the field fluctuations entering in the expression (3) for μ_0^2, it is necessary to know the quasiparticle masses. Subtracting the above equation of motion (2) from the complete equation (1) and applying a suitable Hartree-type factorization, approximate Klein-Gordon equations of motion can be obtained for the field fluctuations $\boldsymbol{\delta\phi} = (\delta\phi_{\parallel}, \boldsymbol{\delta\phi}_{\perp})$ [15],

$$\left[\Box + \mu_{\parallel}^2\right] \delta\phi_{\parallel} = 0 \, , \quad \left[\Box + \mu_{\perp}^2\right] \boldsymbol{\delta\phi}_{\perp} = \boldsymbol{0} \, . \tag{4}$$

The effective quasi-particle masses $\mu_{\parallel}$ and $\mu_{\perp}$ are determined by the gap equations,

$$\mu_{\parallel}^2 = \lambda[3\phi_0^2 + \prec \delta\phi^2 \succ + 2 \prec \delta\phi_{\parallel}^2 \succ - v^2] , \tag{5}$$

$$\mu_{\perp}^2 = \lambda[\ \phi_0^2 + \prec \delta\phi^2 \succ + 2 \prec \delta\phi_{\perp}^2 \succ - v^2] , \tag{6}$$

and the corresponding quasiparticle dispersion relations are $(\omega_k^{\parallel})^2 = k^2 + \mu_{\parallel}^2$ and $(\omega_k^{\perp})^2 = k^2 + \mu_{\perp}^2$. Finally, the thermal fluctuations are given by the usual expressions for bosonic fields,

$$\prec \delta\phi_{\parallel}^2 \succ \;=\; \frac{1}{\Omega}\sum_{\mathbf{k}\neq\mathbf{0}} \frac{1}{\omega_k^{\parallel}} \frac{1}{e^{\omega_k^{\parallel}/T} - 1} \;\asymp\; \frac{\mu_{\parallel} T}{2\pi^2} \sum_{n>0} \frac{1}{n} K_1(\frac{n\mu_{\parallel}}{T}) , \tag{7}$$

$$\prec \delta\phi_{\perp}^2 \succ \;=\; \frac{1}{\Omega}\sum_{\mathbf{k}\neq\mathbf{0}} \frac{1}{\omega_k^{\perp}} \frac{1}{e^{\omega_k^{\perp}/T} - 1} \;\asymp\; \frac{\mu_{\perp} T}{2\pi^2} \sum_{n>0} \frac{1}{n} K_1(\frac{n\mu_{\perp}}{T}) , \tag{8}$$

where the last relations hold in the continuum limit ($L \to \infty$). We note that $\mu_0^2 \leq \mu_{\perp}^2 \leq \mu_{\parallel}^2$, with the equalities holding for $\phi_0 = 0$. Moreover, the gap equations do not contain the parameter H, so the resulting effective masses are independent of the disorientation angle χ_0.

Utilizing the expressions (7-8), the coupled equations (5–6) for the effective masses can be solved for specified values of the temperature T and the magnitude of the order parameter, ϕ_0, provided that these quantities are sufficiently large. The critical boundary on a $\phi_0 - T$ diagram inside which no solution exist, is determined by the vanishing of the transverse mass, $\mu_{\perp} = 0$ (see Eq. 9 below).

Figure 2 shows the resulting effective masses $\mu_{\parallel}$ and $\mu_{\perp}$ as functions of ϕ_0, for temperatures T up to well above critical. At any temperature, there is always a physical solution to the coupled equations (5-6) for the effective masses when $\phi_0 \geq v$. This is easy to see from Eq. (6): At $T = 0$, when the fluctuations vanish, we have $\mu_{\perp}^2 = \lambda(\phi_0^2 - v^2)$ and so $\mu_{\perp}^2$ vanishes at $\phi_0 = v$ and is positive for larger ϕ_0; an increase of T will always increase the fluctuations, and hence the mass. Moreover, we always have $\mu_{\parallel} \geq \mu_{\perp}$.

Since the field fluctuations and the magnitude of the order parameter contribute to the effective masses in qualitatively similar ways, an increase of the temperature (and thus the fluctuations) will permit a further decrease of ϕ_0, so that the point at which $\mu_{\perp}$ vanishes is moved to ever smaller values of ϕ_0. The appearance of $\mu_{\parallel}$, considered as a function of ϕ_0, is nearly independent of temperature, except that each curve terminates at the point where the corresponding $\mu_{\perp}$ vanishes. This limiting curve is indicated by the dotted curve on the interval $(0, v)$ and it is elementary to calculate,

$$\mu_{\perp}{=}0: \ \phi_0^2 = v^2 - \frac{5}{12}T^2 - \prec \delta\phi_{\parallel}^2 \succ \approx v^2 - \frac{T^2}{12}(5 + e^{-\mu_{\parallel}/T}),\ \mu_{\parallel}^2 = \lambda(2v^2 - T^2). \tag{9}$$

This behavior continues until the temperature reaches the value T_0, the lowest temperature for which there is a solution to eqs. (5-6) for all values of ϕ_0. For this particular temperature both effective masses vanish for $\phi_0 = 0$. Consequently, we have $\prec \delta\phi_j^2 \succ = T_0^2/12$ and so $T_0^2 = 2v^2$, *i.e.* $T_0 = 122.63$ MeV with the adopted

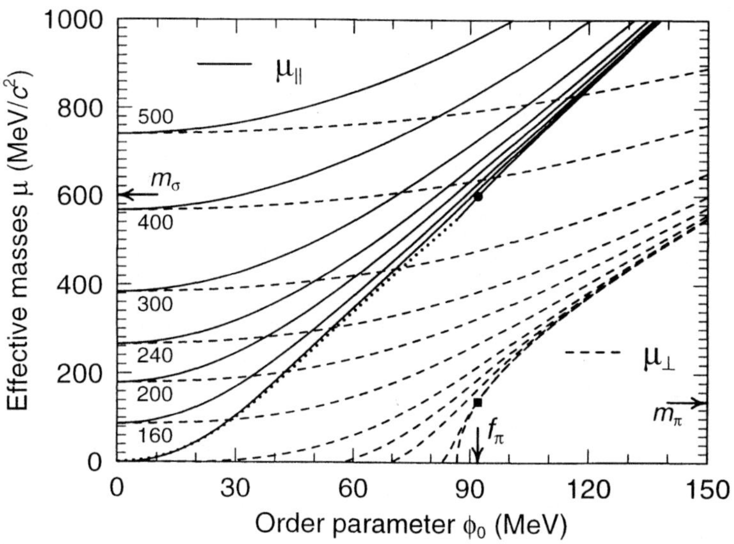

FIGURE 2. Solution to the gap equations. The effective masses $\mu_\parallel$ (solid) and $\mu_\perp$ (dashed), as functions of the magnitude of the order parameter, ϕ_0, for a range of temperatures: $T = 0$, 40, 80, 100, 122.63 ($=T_0$), 160, 200, 240, 300, 400, 500 MeV, calculated in the thermodynamic limit where the box size is large, $L \to \infty$. For a temperature above T_0, the two effective-mass curves start out at $\phi_0 = 0$ with degenerate values, whereas below T_0 they only exist if ϕ_0 is sufficiently large. The corresponding starting points for $\mu_\parallel$ are connected by the dotted curve and, since $\mu_\parallel$ is then nearly independent of T, only the curve for $T = 0$ has been shown. The vertical arrow points to the vacuum value of the order parameter, $\phi_{\rm vac} = f_\pi = 92$ MeV, and the free mass values $\mu_\parallel = m_\sigma = 600$ MeV and $\mu_\perp = m_\pi = 138$ MeV are indicated by the horizontal arrows. The locations of the corresponding points in the diagram are shown by the two solid symbols. Since the field fluctuations are rather insensitive to the box size, except near the critical point, the effective masses exhibit only very little size dependence.

parameter values. [T_0 is here used to denote that unique value of T for which the effective masses are zero when the order parameter vanishes. It is occasionally referred to as the "critical temperature" and denoted T_c but we find this nomenclature unfortunate, since the transition from approximate chiral symmetry to a broken phase generally occurs at significantly higher temperatures, as we shall illustrate later on.] The degeneracy in the masses, $\mu_\parallel = \mu_\perp$, is a general consequence of the $O(4)$ rotational symmetry that emerges for $\phi_0 = 0$ and it therefore remains as T is further increased, with the common mass value μ_0 increasing steadily. Since the effective mass at $\phi_0 = 0$ is given by $\mu_0^2 = \lambda(6\prec\delta\phi_j^2\succ - v^2)$, it becomes proportional to T at high temperatures, $\mu_0 c^2 \approx 1.59\ T$ for $T \gg v$.

Partition Function

The statistical properties of the chiral field are governed by the partition function which, in the semi-classical treatment, is given by

$$\mathcal{Z}_T = \int \mathcal{D}[\boldsymbol{\phi}(\boldsymbol{r}), \boldsymbol{\psi}(\boldsymbol{r})]\, \mathrm{e}^{-\frac{\Omega}{T} E[\boldsymbol{\phi}(\boldsymbol{r}), \boldsymbol{\psi}(\boldsymbol{r})]} . \tag{10}$$

The functional integral is over all possible field configurations and E is the mean energy density of any such state of the system,

$$E[\boldsymbol{\phi}(\boldsymbol{r}), \boldsymbol{\psi}(\boldsymbol{r})] = \langle \tfrac{1}{2}\psi^2 + \tfrac{1}{2}(\nabla\phi)^2 + \tfrac{\lambda}{4}(\phi^2 - v^2)^2 - H\sigma \rangle = E_0 + E_{\rm qp} + \delta V , \tag{11}$$

with $\psi_0 \equiv |\underline{\boldsymbol{\psi}}|$. In the last relation, the energy has been decomposed into terms having instructive physical interpretations [15]. The first term in this decomposition is the energy density that would result if all the field fluctuations were put to zero,

$$E_0 = \tfrac{1}{2}\psi_0^2 + \tfrac{\lambda}{4}(\phi_0^2 - v^2)^2 - H\phi_0 \cos\chi_0 = K_0 + V_0 . \tag{12}$$

It consists of the bare kinetic energy density $K_0 = \frac{1}{2}\psi_0^2$ and the bare interaction energy density V_0.

The second term in E is the energy associated with the gas of independent quasiparticles described by the above Klein-Gordon equations of motion (4),

$$E_{\rm qp} = \tfrac{1}{2}\langle \boldsymbol{\delta\psi} \circ \boldsymbol{\delta\psi} + \nabla\boldsymbol{\delta\phi} \circ \nabla\boldsymbol{\delta\phi} + \boldsymbol{\delta\phi} \circ \boldsymbol{M} \circ \boldsymbol{\delta\phi} \rangle , \tag{13}$$

where $\boldsymbol{M}$ is the O(4) quasiparticle mass tensor. In thermal equilibrium its major principal axis is oriented along the order parameter, the corresponding eigenvalue is $\mu_{\parallel}^2$, and the other three eigenvalues are equal to $\mu_{\perp}^2$.

The last term in E corrects for the fact that the interaction is non-linear,

$$\delta V = \tfrac{\lambda}{4}\langle\delta\phi^4\rangle - \tfrac{\lambda}{2}\langle\delta\phi^4\rangle_G \approx -\tfrac{\lambda}{4}\langle\delta\phi^4\rangle_G = \langle\delta\phi^2\rangle^2 + 2\,\mathrm{Tr}(\langle\boldsymbol{\delta\phi\delta\phi}\rangle \circ \langle\boldsymbol{\delta\phi\delta\phi}\rangle) \tag{14}$$

$$\approx -\tfrac{3}{4}\lambda\, [\prec \delta\phi_{\parallel}^2 \succ^2 + 2 \prec \delta\phi_{\parallel}^2 \succ\prec \delta\phi_{\perp}^2 \succ + 5 \prec \delta\phi_{\perp}^2 \succ^2] \equiv \delta V_T \tag{15}$$

Here $\langle\cdot\rangle_G$ denotes the gaussian approximation to the evaluation of the average. The last line arises if the spatial averages are approximated by the corresponding thermal averages, as is expected to be accurate when L exceeds the correlation length. The resulting quantity, δV_T, then depends on the particular state only through the magnitude of its order parameter, ϕ_0.

With the above preparations, we are now in a position to simplify the expression (10) for the partition function. We first note that the functional integral factorizes into a regular (8-dimensional) integral over the O(4) order parameter $\underline{\boldsymbol{\phi}}$ and its time derivative $\underline{\boldsymbol{\psi}}$ and a functional integral over the field fluctuation $\boldsymbol{\delta\phi}(\boldsymbol{r})$ and its time derivative $\boldsymbol{\delta\psi}(\boldsymbol{r})$,

$$\mathcal{Z}_T = \int d^4\underline{\boldsymbol{\psi}}\, \mathrm{e}^{-\frac{\Omega}{T}K_0} \int d^4\underline{\boldsymbol{\phi}}\, \mathrm{e}^{-\frac{\Omega}{T}(V_0 + \delta V_T)} \int \mathcal{D}[\boldsymbol{\delta\phi}(\boldsymbol{r}), \boldsymbol{\delta\psi}(\boldsymbol{r})]\, \mathrm{e}^{-\frac{\Omega}{T}E_{\rm qp}} . \tag{16}$$

The integration over $\underline{\psi}$ yields the factor $(2\pi T/\Omega)^2$ which depends only on temperature. The integral over the quasiparticle degrees of freedom yields the conditional quasiparticle partition function $\mathcal{Z}_{\rm qp}$. Using the Bose-Einstein values for the mode occupancies $n_k^{(j)}$, we find $\ln \mathcal{Z}_{\rm qp} = \ln \mathcal{Z}_{\rm qp}^{\parallel} + 3 \ln \mathcal{Z}_{\rm qp}^{\perp}$, where

$$\ln \mathcal{Z}_{\rm qp}^{(j)} = \sum_{\mathbf{k}\neq\mathbf{0}} \ln \bar{n}_k^{(j)} = -\sum_{\mathbf{k}\neq\mathbf{0}} \ln[1 - {\rm e}^{-\omega_k^{(j)}/T}] \tag{17}$$

$$= -\frac{1}{T}\sum_{\mathbf{k}\neq\mathbf{0}} \omega_k^{(j)} n_k^{(j)} + \sum_{\mathbf{k}\neq\mathbf{0}} [\bar{n}_k^{(j)} \ln \bar{n}_k^{(j)} - n_k^{(j)} \ln n_k^{(j)}] \tag{18}$$

$$= -\frac{\Omega}{T}\,(V_{\rm qp}^{(j)} - T S_{\rm qp}^{(j)}) \tag{19}$$

for each chiral component j (with $\bar{n}_k^{(j)} \equiv 1 + n_k^{(j)}$). The total energy density of the quasiparticles is then $V_{\rm qp} = V_{\rm qp}^{\parallel} + 3V_{\rm qp}^{\perp}$ and their entropy density is $S_T = S_{\rm qp}^{\parallel} + 3S_{\rm qp}^{\perp}$. Consequently, we may write the overall partition function on a simple form,

$$\mathcal{Z}_T = \left(\frac{2\pi T}{\Omega}\right)^2 \int d^4\underline{\phi}\; {\rm e}^{-\frac{\Omega}{T}F_T} = \left(\frac{2\pi T}{\Omega}\right)^2 4\pi \int_0^\infty \phi_0^3\, d\phi_0 \int_0^\pi \sin^2\chi_0\, d\chi_0\; {\rm e}^{-\frac{\Omega}{T}F_T}\,, \tag{20}$$

where $F_T = V_T(\phi_0, \chi_0) - T S_T(\phi_0)$ is the effective free energy density, with the effective potential energy density being $V_T = V_0(\phi_0, \chi_0) + V_{\rm qp}(\phi_0) + \delta V_T(\phi_0)$.

It should be noted that the χ_0 dependence of the free energy is only through the term $-H\phi_0 \cos\chi_0$ in the bare potential V_0. For $H > 0$ this term has its minimum value for $\chi_0 = 0$ and it increases steadily until reaching its maximum for $\chi_0 = \pi$. All stationary points of F_T are therefore located on the σ axis and occur where $(\partial F_T/\partial\sigma_0)_T$ vanishes.

Phase Structure

Once the expression for the partition function has been derived, it is possible to discuss the statistical properties of the chiral order parameter $\underline{\phi}$. The structure of the free energy density is illustrated in Fig. 3 in the thermodynamic limit ($L \to \infty$) for both the O(4) symmetric case (vanishing m_π) and the realistic case. The equilibrium points can be identified as those where the free energy density F_T is stationary. Using the expression for the free energy derived above, together with the self-consistent dispersion relations, it is elementary to show that

$$\left(\frac{\partial F_T}{\partial \sigma_0}\right)_T = \mu_0^2\, \sigma_0 - H\,. \tag{21}$$

Comparing this relation with Eq. (2), we see that statistical equilibrium (stationary free energy) and dynamical equilibrium (no acceleration of the order parameter) indeed occur simultaneously, as required by thermodynamic consistency. Therefore, the phase diagram can be determined already on the basis of Eq. (2).

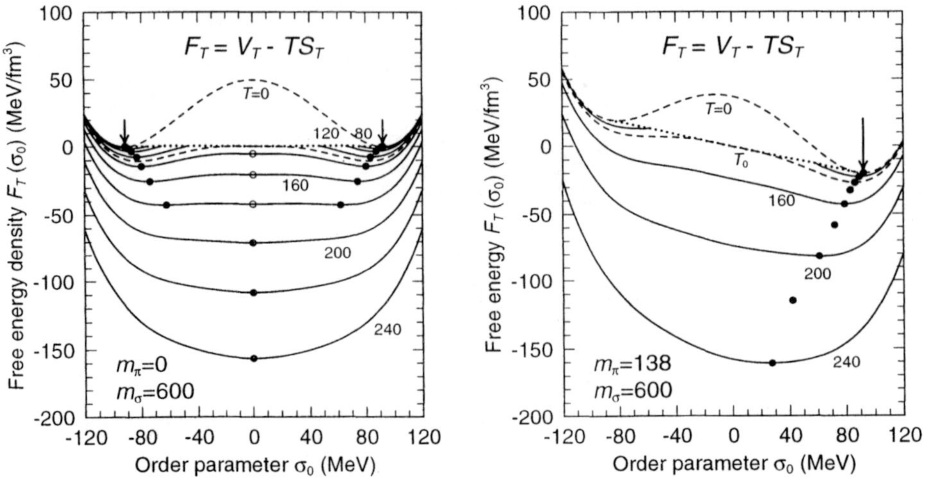

FIGURE 3. *Left panel* (3a): The free energy density F_T as a function of the order parameter σ_0 for various values temperatures T, for the idealized O(4) symmetric case (m_π=0). At each T, the solid (or open) dots indicate the favored (or unfavored) equilibria. The result for $T = T_0 = \sqrt{2}v$ is indicated by a dashed curve. For $T < T_0$, the order parameter must exceed a certain minimum value before all quasiparticle modes are stable and the corresponding end points are connected by the dotted curve. The top dashed curve shows the bare potential V_0 that arises when $T = 0$. *Right panel* (3b): The corresponding result for the realistic case (m_π=138 MeV).

The resulting equilibria are traced out in Fig. 4a for four different values of the free pion mass. In the idealized case of vanishing pion mass, the ground-state minimum in F_T is located at $\sigma_0 = f_\pi$, where F_T vanishes, $F_T^{\rm vac} = 0$. The minimum in F_T moves inwards as T is increased, at first very slowly and then progressively faster, until a certain temperature is reached, $T = T_1$. At this temperature, the stationary points in F_T turn from minima to maxima as the tracing curve continues downwards towards the critical point.

Once the temperature exceeds the critical value T_0, a solution to the gap equations exist for all values of ϕ_0, and the free energy has a minimum at $\phi_0 = 0$. Thus, in the temperature range between T_0 and T_1, there are two minima in F_T, separated by a maximum that is only slightly higher than the symmetric minimum. The outer minimum generally carries a larger statistical weight (the value of F_T is lower). Consequently, it represents the thermodynamcially preferred state and the symmetric configuration is merely metastable (with the maximum representing an unstable equilibrium). We therefore refer to T_1 as the *transition temperature*. Above T_1 there the only stationary point is the minimum at symmety, $\phi_0 = 0$.

This general structure of the free energy, as obtained with the semi-classical treatment, implies that the system will display a first-order phase transition at $T = T_1$, with the preferred magnitude of the order parameter dropping abruptly

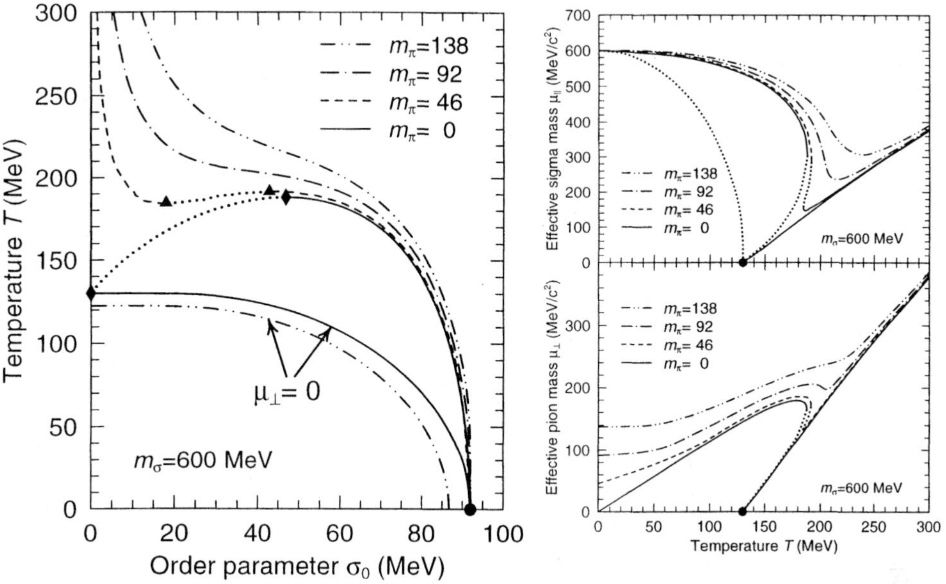

FIGURE 4. *Left panel* (4a): The stable and unstable (dotted) equilibrium values of the chiral order parameter $\sigma_0 = \phi_0 \cos \chi_0$ at a given quasiparticle temperature T, as traced on the chiral phase diagram for the indicated four values of the pion mass m_π. For $m_\pi = 0$ the high-temperature O(4) symmetric equilibrium branch extends upwards along the vertical axis from the critical point marked by the diamond. The critical boundary defined by $\mu_\perp = 0$ (within which the field is supercritical) is delineated for $m_\pi = 0$ and 138 MeV. The solid dot is located at the common ground state, $\sigma_0 = f_\pi = 92$ MeV. *Right panel* (4b): The equilibrium values of the effective masses $\mu_\parallel$ (top) and $\mu_\perp$ (bottom) for σ-like and π-like quasiparticles, respectively, as functions of the temperature T, for the same four different pion masses m_π; the values at the unstable equilibria are traced by the dotted curves. In addition, the top panel shows the value of $\mu_\parallel$ along the critical boundary for $m_\pi = 0$ (dotted curve on far left).

from a fairly large finite value to zero. However, as is evident from Fig. 3a, the metastable minimum at $\phi_0 = 0$ is very shallow and a finite system would therefore display large fluctuations in ϕ_0 (see Fig. 5). Moreover, the use of a finite pion mass will erode this phase structure, as the free-energy curves will be tilted towards the positive σ direction (Fig. 3b). The outer minimum will then occur only on σ axis (at $\sigma_0 = f_\pi$), while there will be a saddle point situated in the opposite direction. As seen in Fig. 4a, when m_π exceeds about half the physical value, the resulting tilting suffices to eliminate the symmetric minimum altogether and there is then only a single minimum at any temperature (located on the positive part of the σ axis). From that value on, the system displays a smooth crossover from a strongly broken to a weakly broken phase, with the most rapid change in order parameter occurring at temperatures somewhat above 200 MeV.

The phase structure discussed above is reflected in the behavior of the equilibrium values of the effective quasiparticle masses, as shown in Fig. 4b. When the specified value of m_π is sufficiently small, the associated phase transition causes a backbend in the mass curves, as the system is traced through the unstable equilibrium branch. For perfect O(4) symmetry ($m_\pi = 0$), the backbend reaches all the way to the critical point where all the effective masses vanish, $\mu_0 = \mu_\perp = \mu_\parallel = 0$, and from there on they are exactly degenerate and grow approximately linearly with T. As noted above, the branch with the large masses is generally the thermodynamically preferred one. For the case of $m_\pi = 0$, Fig. 4b also shows the evolution of $\mu_\parallel$ along the critical path where $\mu_\perp$ vanishes. It drops approximately quadratically from its free value at $T = 0$ to zero at T_0.

For the larger values of m_π, the evolution is smooth, but the basic two-phase structure of the system remains visible as a distinct minimum in the σ mass. (The location of this minimum may be used as an indicator for the effective transition temperature [15].) At higher temperatures, the masses become nearly degenerate as chiral symmetry is approached.

Finite Size

Since the systems available for actual experimental study have a rather limited spatial extension, it is important to assess the importance of finite-size effects. As it turns out [15], the free energy density F_T depends very little on the actual volume Ω, once the side length L is above 5 fm or so, and it makes no practical difference whether it is calculated in terms of the quantized modes, as indicated above, or in the continuum limit. Thus the size dependence of the statistical properties arises primarily through the volume Ω multiplying F_T in the integrand in (20).

When the system is small, the integrand in the partition function is no longer narrowly peaked at the minima in F_T. Rather, it represents a probability density in the O(4) space of the order parameter $\underline{\phi}$. Since F_T depends only on the magnitude of the order parameter, ϕ_0, and its angle of disorientation, χ_0, it is instructive to project onto the plane of those two state variables,

$$\mathcal{Z}_T \sim \int_0^\infty 4\pi\phi_0^3\, d\phi_0 \int_0^\pi \sin^2\chi_0\, d\chi_0\, e^{-\frac{\Omega}{T}F_T(\phi_0,\chi_0)} = \int_0^\infty d\phi_0 \int_0^\pi d\chi_0\, W_T(\phi_0,\chi_0)\ , \quad (22)$$

where W_T is the statistical weight for finding the system with an order parameter having the specified magnitude and disorientation.

Generally speaking, the statistical fluctuations in a finite system tends to wash out the sharp phase structure characteristic of infinite matter. The importance of this effect on the temperature dependence of the order parameter is brought out in Fig. 5. The changes are most significant in the idealized O(4) symmetric scenario. In particular, it can be seen that the first-order transition obtained for m_π=0 becomes less prominent as the volume is decreased to realistic sizes and it has disappeared altogether for the smallest volume considered ($L \approx 6$ fm).

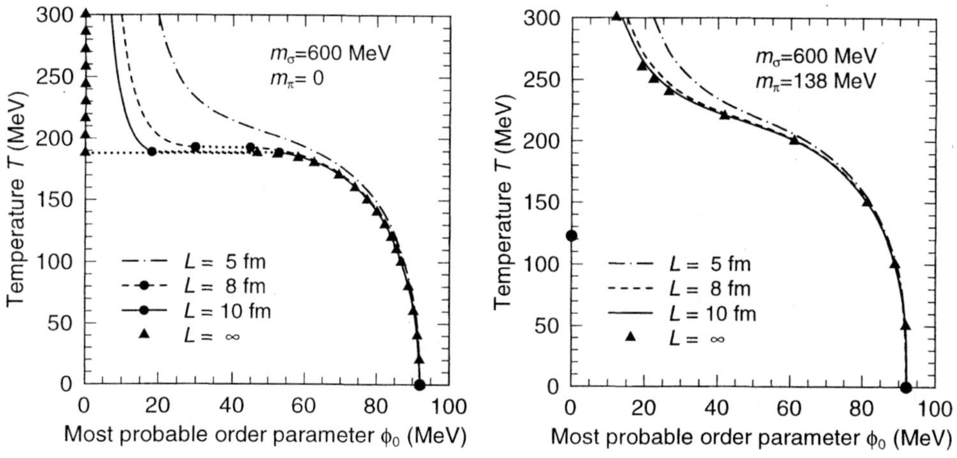

FIGURE 5. The most probable value of the magnitude of order parameter, ϕ_0, as a function of the temperature T, for various values of the side length L of the cubic volume considered, with either m_π=0 (*left panel*: 5a) or m_π=138 MeV (*right panel*: 5b).

Such qualitative change does not occur when a realistic value of m_π is employed, since the behavior is then already smooth for large systems. In fact, the finite-size effect is hardly visible until rather small volumes are reached. Furthermore, for such small volumes the difference between the results obtained for the various specified values of m_π is less noticeable. This finding suggests that a quantitative extraction of the matter equation of state from analysis of the small finite systems involved in actual experiments depends heavily on the availability of reliable models.

Another important finite-size effect is the fluctuation in the O(4) orientation of the order parameter $\underline{\phi}$. However, because of the large entropy carried by the quasi-particles, even a relatively modest change in the disalignment angle χ_0 leads to a strong reduction in the statistical weight. As a result, the equilibrium distribution $P(\underline{\phi})$ remains fairly confined around the positive σ direction, as is illustrated in Fig. 6. Thus, the idealized "sombrero" picture in which the order parameter has a fairly isotropic distribution at high temperatures may be somewhat misleading.

Sampling of Thermal Field Configurations

For actual numerical computations, it is often of interest to sample the field from a suitably characterized ensemble and we outline how it is possible to devise a fast, efficient, and robust method for sampling from the thermal ensemble [15].

The first task is to sample the order parameter in accordance with the statistical weight $W_T(\underline{\phi}, \underline{\psi})$. Since the time derivative $\underline{\psi}$ is governed by a four-dimensional normal distribution which is isotropic and entirely decoupled from the other degrees of freedom (see Eq. (16)), it is elementary to sample this quantity.

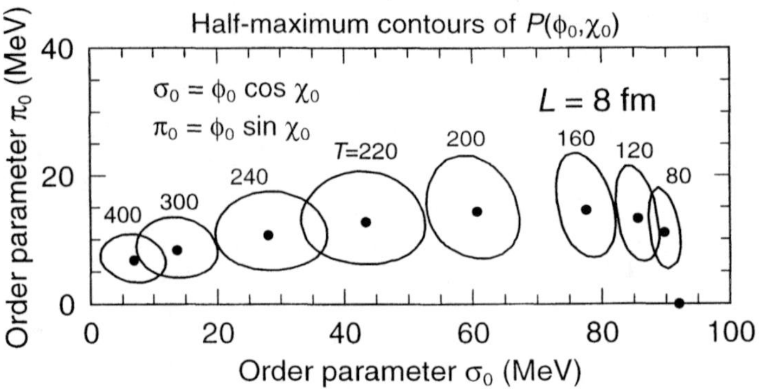

FIGURE 6. The *projected* equilibrium distribution of the chiral order parameter, $P(\phi_0, \chi_0) \sim \phi_0^3 \sin^2 \chi_0 W_T(\underline{\phi})$, displayed as a function of its aligned component, $\sigma_0 = \phi_0 \cos \chi_0$, and the magnitude of its transverse component, $\pi_0 = \phi_0 \sin \chi_0$, for a cubic box of side length $L = 8$ fm. For each temperature T, the solid dot indicates the location of the maximum in $P(\sigma_0, \pi_0)$ and the solid curve traces out the half-maximum contour, as obtained by scaling the continuum results down to the appropriate finite volume $\Omega = L^3$.

It is more complicated to sample the magnitude ϕ_0, due to the intricate structure of its probability distribution, as discussed above. However, the numerical effort required is quite modest. The most efficient method utilizes a precalculation of the effective masses as functions of ϕ_0, for the particular T of interest. The χ_0-independent part of the effective potential, $V_T(\phi_0, 0)$, can then be obtained together with the corresponding entropy $S_T(\phi_0)$. Since the dependence on χ_0 is simple, an exact integration over χ_0 is possible, and so, in effect, the probability distribution for ϕ_0 can be pretabulated. It is then a numerically trivial task to sample ϕ_0.

Once the magnitude ϕ_0 has been selected, it is straightforward to sample the disalignment angle χ_0 (using either its exact form or its Gaussian approximation). In order to orient $\underline{\phi}$ in the $\boldsymbol{\pi}$ subspace, there remains the task of selecting the remaining $O(3)$ spherical angles ϑ_0 and φ_0, upon which the order parameter is given by $\underline{\phi} = (\phi_0 \cos \chi_0,\ \phi_0 \sin \chi_0 \sin \vartheta_0 \cos \varphi_0,\ \phi_0 \sin \chi_0 \sin \vartheta_0 \sin \varphi_0,\ \phi_0 \sin \chi_0 \cos \vartheta_0)$.

Since the different quasi-particle modes can be regarded as effectively decoupled, their sampling is best done by making an expansion into the elementary modes,

$$\delta\phi_{||}(\boldsymbol{r}, t) = \left(\frac{2}{\Omega}\right)^{\frac{1}{2}} \sum_{\mathbf{k} \neq \mathbf{0}} C_{\mathbf{k}}^{||} \cos(\mathbf{k} \cdot \boldsymbol{r} - \omega_k^{||} t - \eta_{\mathbf{k}}^{||}) , \tag{23}$$

and similarly for the three transverse chiral components $\boldsymbol{\delta\phi}_\perp(\boldsymbol{r}, t)$. The phase $\eta_{\mathbf{k}}$ is random in the interval $(0, 2\pi)$ and is thus trivial to sample. Furthermore, the real (and positive) amplitude $C_{\mathbf{k}}$ can be related to the number of quanta $n_{\mathbf{k}}$ by considering the energy carried by the mode, $E_{\mathbf{k}} = n_{\mathbf{k}}\omega_k = \omega_k^2 C_{\mathbf{k}}^2$. By this token, the problem has been reduced to sampling the number of quanta $n_{\mathbf{k}}$ which is elementary. [In fact, $n_{\mathbf{k}}$ can be regarded as counting the number of successive times the sampling of a standard random number yields a value below $\exp(-\epsilon_k/T)$).] The thermal average of $n_{\mathbf{k}}$ is equal to the occupancy f_k employed in the calculation of the entropy, $\prec n_{\mathbf{k}} \succ = f_k$. It may be noted that while the energy relation employed is the classical one (which omits the zero-point contribution), the occupation number is properly quantized and sampled in accordance with quantum statistics. This method eliminates divergencies while retaining the essential quantum fluctuations.

Once the quasiparticle amplitudes and phases have been selected, the expansion (23) readily yields the value of the field fluctuations, $\delta\boldsymbol{\phi}(\boldsymbol{r})$ at a specified time t_0. The corresponding conjugate momentum $\delta\boldsymbol{\psi} \equiv \partial_t \delta\boldsymbol{\phi}$ readily follows,

$$\delta\psi_{\|}(\boldsymbol{r}, t) = \left(\frac{2}{\Omega}\right)^{\frac{1}{2}} \sum_{\mathbf{k}\neq\mathbf{0}} \omega_k^{\|} C_{\mathbf{k}}^{\|} \sin(\mathbf{k}\cdot\boldsymbol{r} - \omega_k^{\|} t - \eta_{\mathbf{k}}^{\|}) \ . \tag{24}$$

When the equations of motion are propagated by a leap-frog method, the field strength $\delta\boldsymbol{\phi}(\boldsymbol{r})$ is calculated at $t_N = N\Delta t$ while the momentum $\delta\boldsymbol{\psi}(\boldsymbol{r})$ is obtained at the mid points. The appropriate initial $\delta\boldsymbol{\psi}(\boldsymbol{r})$ can then easily be obtained by evaluating (24) at $t = t_0 + \frac{1}{2}\Delta t$ after $C_{\mathbf{k}}$ and $\eta_{\mathbf{k}}$ have been selected at t_0.

Finally, the system in which the sampling has been done is aligned with the order parameter $\underline{\boldsymbol{\phi}}$ whose O(4) direction is $(\chi_0, \vartheta_0, \varphi_0)$ (the mass tensor is diagonal in this system). Therefore a corresponding rotation is required to express the sampled field configuration with respect to the chiral reference frame $(\hat{\sigma}, \hat{\pi}_1, \hat{\pi}_2, \hat{\pi}_3)$.

Correlation Function

It is interesting to calculate the correlation function of the chiral field since this quantity determines the spectral distribution of the emitted field quanta. The density matrix for the quasi-particle field is a 4×4 tensor,

$$\boldsymbol{C}(r_{12}, t_{12}) \ \equiv \prec \delta\boldsymbol{\phi}(\boldsymbol{r}_1, t_1)\ \delta\boldsymbol{\phi}(\boldsymbol{r}_2, t_2) \succ \ , \tag{25}$$

where the average is over the thermal ensemble held at the temperature T. Since an ensemble in equilibrium is invariant in time, the correlation function depends only on the time difference $t_{12} = t_1 - t_2$. Moreover, translational symmetry implies that the spatial dependence is via the separation $\boldsymbol{r}_{12} = \boldsymbol{r}_1 - \boldsymbol{r}_2$ and invariance under spatial rotations ensures that only the magnitude $r_{12} = |\boldsymbol{r}_{12}|$ enters.

Utilizing the expansion (23), it is elementary to show that the correlation tensor $\boldsymbol{C}$ is diagonal with the elements $C_{\|}$ and $C_{\perp}$, where

$$C_{\|}(r, t) \ = \ \frac{1}{\Omega} \sum_{\mathbf{k}\neq\mathbf{0}} \frac{1}{\omega_k^{\|}} \frac{\cos(\mathbf{k}\cdot\boldsymbol{r} - \omega_k^{\|} t)}{e^{\omega_k^{\|}/T} - 1} \ \asymp \ \frac{1}{2\pi^2 r} \int_{\mu_{\|}}^{\infty} d\omega \ \frac{\sin kr}{e^{\omega/T} - 1} \cos\omega_k^{\|} t \ , \tag{26}$$

and similarly for $C_{\perp}(\boldsymbol{r}, t)$. Since f_k/ω_k is equal to the thermal average $\prec C_{\mathbf{k}}^2 \succ$, we recognize the familiar result (see Eq. (146.10) in Ref. [18], for example).

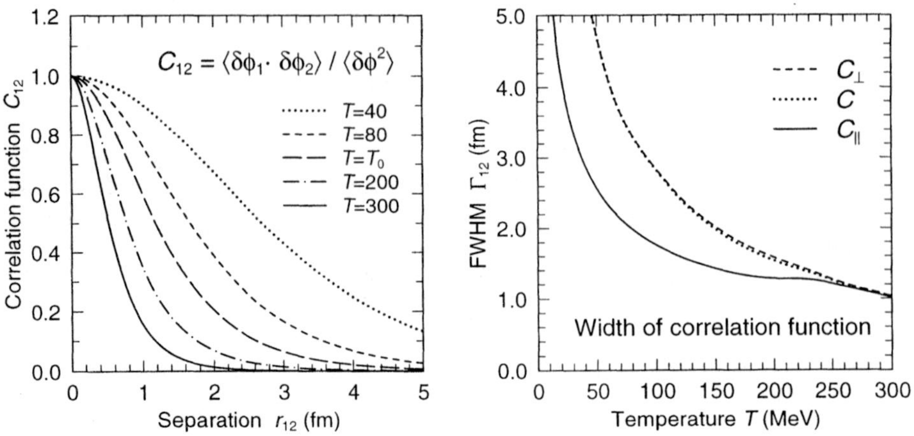

FIGURE 7. Correlation function in equilibrium. *Left* (7a): The reduced correlation function C_{12} for a range of temperatures T, employing for each T the most probable magnitude of the order parameter, ϕ_0. *Right* (7b): The FWHM correlation length Γ_{12} of C_{12} as a function of T, for the total field fluctuations as well as the transverse and parallel components separately.

The usual correlation function is the trace of $\boldsymbol{C}$,

$$C \equiv \prec \boldsymbol{\delta\phi}(\boldsymbol{r}_1, t_1) \cdot \boldsymbol{\delta\phi}(\boldsymbol{r}_2, t_2) \succ = \operatorname{tr} \boldsymbol{C} = C_{\parallel}(r_{12}, t_{12}) + 3C_{\perp}(r_{12}, t_{12}) \ . \quad (27)$$

Its overall magnitude is set by its value at zero, which is simply the corresponding variance in the field strength, $C(0,0) = \prec \delta\phi^2 \succ$. It is instructive to consider the reduced function $C_{12} \equiv C(r_{12}, 0) / \prec \delta\phi^2 \succ$ which is unity for $\boldsymbol{r}_1 = \boldsymbol{r}_2$. In general, $C_{12} \sim (1/r_{12}) \exp(-\mu r_{12})$ in the limit of large separations, $r_{12} \to \infty$, so that $1/\mu c$ provides a simple measure of the correlation length. In the special case when the effective mass vanishes, the reduced equal-time correlation function is given on analytical form, $C_{12} \asymp (3/\zeta)(\coth\zeta - 1/\zeta)$, where $\zeta \equiv \pi T r_{12}$. In this extreme case, the correlation function falls off only as $\sim 1/r_{12}$.

Figure 7a shows how the reduced correlation function evolves with temperature in equilibrium. At high T we have $\mu \sim T$, so then the correlation length tends to zero. For temperatures below critical, the field fluctuations are predominantly associated with the transverse modes, since those have the smallest effective mass, $\mu_\perp \ll \mu_\parallel$, and the correlation length grows ever larger.

It is convenient to characterize C_{12} by its full width at half maximum, Γ_{12}, since this quantity is always possible to extract, even when the mass vanishes. Figure 7b shows this measure of the correlation length as a function of temperature, using again the most probable value of ϕ_0. For temperatures near and below $T \approx 200$ MeV the dominant fluctuations are perpendicular to the order parameter since the corresponding effective mass is relatively small. For higher temperatures the asymptotic regime is approached where the chiral symmetry is approximately restored and the fluctuations are similar in all four chiral directions.

DYNAMICS AND OBSERVABLES

The equilibrium properties of the system discussed above provide a useful reference for understanding the key features of its dynamics, since the system generally seeks to reestablish equilibrium in response to an external disturbance. The environment produced in a high-energy collision is characterized by a rapid cooling, driven primarily by the explosive expansion, and this forces the system away from equilibrium. The overall evolution is then the combined result of those two opposite effects. it is then a delicate task to predict the outcome and one must generally resort to dynamical calculations.

Since the evolving system is in a non-equilibrium state, it is harder to analyze. Fortunately, it is possible to depict a general field configuration on a equal footing with the equilibrium configurations by employing a modified phase diagram in which the degree of agitation is represented by the the average field fluctuation, $\Delta\phi = \langle \boldsymbol{\delta\phi} \circ \boldsymbol{\delta\phi} \rangle^{1/2}$, rather than the quasiparticle temperature T (as in Figs. 4a and 5). Since $\Delta\phi$ is meaningful for any individual field configuration, it is thus possible to project an arbitrary state onto the generalized diagram (which is occasionally referred to as the "chiral road map"). The standard phase diagram (the curves in Fig. 4a obtained for $m_\pi = 138$ MeV) then appears as shown in Fig. 8*a* below.

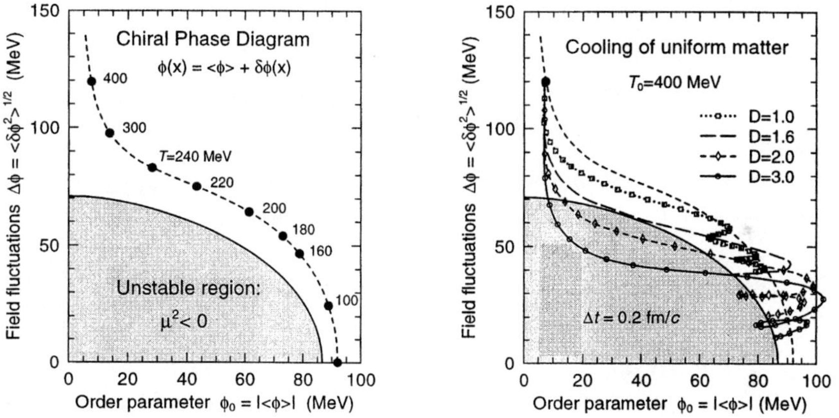

FIGURE 8. *Left panel* (8a): The equilibrium path on the "chiral road map" in which the field configuration is projected onto a single point, $\boldsymbol{\phi}(\boldsymbol{r}) \to (\phi_0, \Delta\phi)$. The adiabatic path of equilibrium points (dashed curve), which represents the major "highway", exhibits a gradual cross over from the broken phase at low temperature to the nearly restored phase at high temperature. The shaded area covers the classically forbidden region within which the quasipions are subject to a supercirtical effective field. *Right panel* (8b): The dynamical paths of systems that have been prepared in equilibrium at $T_i = 400$ MeV and then subjected to a Rayleigh cooling with strengths D adjusted to emulate scaling expansions (D=1 emulates a longitudinal Bjorken expansion). [9]

Emulation of Scaling Expansions

It is expected that the early parton dynamics causes the chiral field to be in a state of rapid expansion. The subsequent evolution may then lead to a supercooled configuration situated inside the unstable region, thus effectively producing a "quench". A number of quenched scenarios have been considered [6,19–21,7,22,23] but they were imposed by *fiat*, thereby reducing the predictive power of the dynamcial calculations (essentially any degree of magnification can be achieved by suitable adjustment of the initial conditions). The degree of arbitrariness can be reduced by elucidating under which conditions a quench-like early scenario may develop dynamically from plausible initial configurations.

Bjorken-like scaling expansions [24] provide especially simple expansion scenarios. Both purely longitudinal [19,7,22,23,13] and fully isotropic scaling expansions [20,25] have been treated. In order to achieve a global impression of the influence of expansion, it suffices to augment the equation of motion (1) by a term emulating a scaling expansion in D dimensions, $-(D/t)\partial_t\boldsymbol{\phi}$. This term has a form akin to the Rayleigh dissipation function in classical mechanics and it acts as a time-dependent damping that reduces the field fluctuations in the course of time. In order to examine its effect, one may ignore the spatial geometry and consider a macroscopically uniform configuration within a large box. The discussion is then simplified and the resulting scenarios can be regarded as idealized representations of chiral matter subjected to an externally prescribed cooling rate, and so the results will have a corresponding general applicability.

Figure 8b shows dynamical trajectories obtained in this manner for $D = 1 - 3$. The effect increases with D, since the dimensionality of the expansion effectively acts as the strength of the damping term. The isotropic expansion (D=3) leads to a significant incursion into the unstable region, while the longitudinal expansion (D=1) keeps the system well within the stable region.

Utilizing Eqs. (5-6) it is possible to extract the time dependence of the effective masses and their evolution will reflect the behavior exhibit by the dynamical paths depicted in Fig. 8b. In particular, the incursions into the supercritical region leads to negative values of $\mu_\perp^2$ and a corresponding exponential growth of those modes for which $\mu_\perp^2 + k^2 < 0$. In fact, the most important effects of the dynamics can be well understood on the basis of the behavior of the quaisparticle mass tensor [9], as will be illustrated later.

It is instructive to see how the various cooling scenarios affect the pion observables. For this purpose, a Fourier decomposition of the pion field is useful,

$$\boldsymbol{\pi}(\boldsymbol{r},t) = \sum_{\mathbf{k}} \boldsymbol{\pi}_{\mathbf{k}}(t)\, e^{i\mathbf{k}\cdot\mathbf{r}} \,. \qquad (28)$$

Due to the presence of the cooling term, the amplitudes tend towards zero as $\boldsymbol{\pi}_{\mathbf{k}}(t) \sim t^{-D/2}$ at late times. Consequently, the field decouples into free modes and it is then meaningful to extract the asymptotic values of the observables.

Of particular interest is the spectral distribution of the pions and Fig. 9a shows their asymptotic power spectrum as a function of their kinetic energy $\omega_k - m_\pi$,

$$P_\pi(E_{\rm kin}) \sim \sum_{\mathbf{k}\neq 0} \omega_k \, |\sqrt{\frac{\omega_k}{2}}\boldsymbol{\pi}_{\mathbf{k}} + \frac{i}{\sqrt{2\omega_k}}\dot{\boldsymbol{\pi}}_{\mathbf{k}}|^2 \; \delta(E_{\rm kin} + m_\pi - \omega_k). \tag{29}$$

It is clear that the incursions into the supercritical region for $D = 2, 3$ lead to dramatic enhancements of the yield in the lowest energy bin, $E_{\rm kin} < 200$ MeV, whereas there is hardly any visible effect for $D = 1$.

The dynamically induced enhancement of the soft pion modes is also reflected in their resulting correlation function $C_\pi(r_{12})$, as shown in Fig. 9b. While the correlation function approximately retains its initial thermal form for D=1, it widens steadily for the larger cooling rates, as the increased strength of the softest pion modes causes it to acquire a pronounced tail.

Such analyses show that the occurrence of instabilities and the associated amplification of pionic modes depend sensitively on the cooling rate, which in turn is intimately related to the character of the expansion. The idealized scenario for D=3 corresponds closely to the isotropic expansion considered in Refs. [20,25] and the results corroborate the conclusion in Ref. [20] that such a scenario leads to amplification. Furthermore, the analysis suggest that a longitudinal expansion alone is insufficient to cause a quench, if the initial fluctuations are of thermal magnitude. This is consistent with what was found in Refs. [19,23] for effectively one-dimensional expansions. This qualitative sensitivity to the collision dynamics underscores the importance of employing physically reasonable initial conditions for the dynamical simulations of DCC formation.

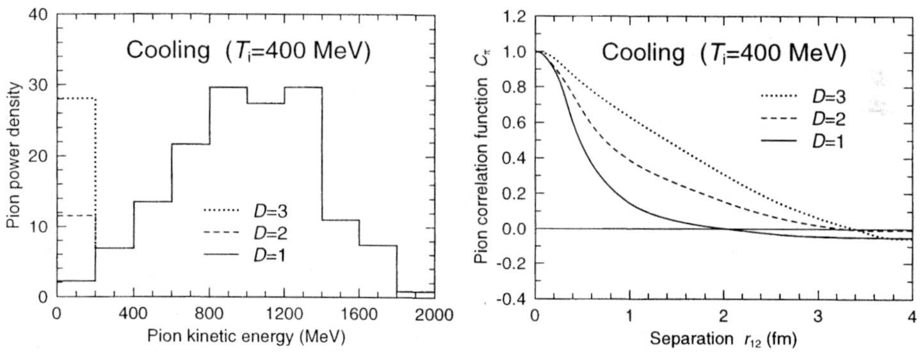

FIGURE 9. *Left panel* (9a): The power spectrum of the pions emerging asymptotically as a result of propagating 100 field configurations (sampled from the thermal distribution at T_i = 400 MeV) with the equation of motion (1) augmented with a Rayleigh dissipation term of strength $D = 1, 2, 3$. *Right panel* (9b): The corresponding pion correlation function normalized to $C_\pi = 1$ at $r_{12} = 0$. (These results are from Ref. [26].)

Neutral Pion Fraction

It was noted early on that isospin-directed oscillations of the pion field will result in an anomalous behavior of the neutral pion fraction $f = n_{\pi_0}/n_\pi$ [27–31]. Indeed, the distribution would be given by $P(f) = 1/(2\sqrt{f})$ in the idealized scenario where all the pions observed arise from a fully aligned source. In practice, the observed pions may originate from unrelated regions and the anomaly is then attenuated. This is illustrated in Fig. 10a which shows the result of combining pions from N independent sources. A Poisson-like distribution peaked near $f = \frac{1}{3}$ emerges when there are many independent sources, or equivalently, when the distance between emission points is large in comparison with the correlation length.

In order to give a quantitative feeling for what this inherent feature amounts to in practice, we show in Fig. 10b the distribution $P(f)$ extracted from an ensemble of 100 events that have been cooled with D=2, as explained above. When all the pions are used for the calculation of f, the resulting distribution looks fairly normal, but when only pions with a kinetic energy below 200 MeV are considered, then $P(f)$ broadens significantly and attains an anomalous form. However, its appearance still differs significantly from the ideal form, which is only reached if all the pions arise from a single mode, such as the one having k=0.

This important feature can also be brought out by viewing a given source at different scales, as illustrated in Fig. 11a where the total source is divided into ever smaller sources, each one leading to a separate value of f. As the source size shrinks, its pion field is increasingly well aligned and the associated $P(f)$ grows correspondingly more anomalous. It has therefore been proposed that wavelet-type analyses may provide a useful tool for the extraction of DCC domain structure from the experimental data [32–34].

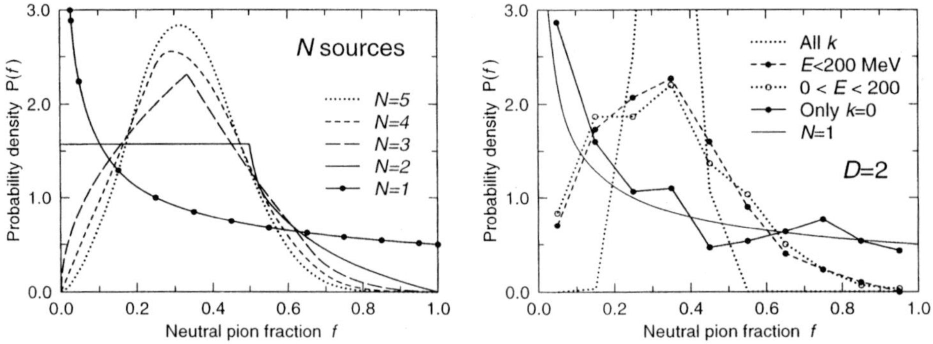

FIGURE 10. *Left panel* (10a): The distribution of the neutral pion fraction, $P(f)$, resulting from combining N similar but independent idealized pion sources with perfect isospin alignments. *Rigt panel* (10b): The distribution $P(f)$ obtained when various energy cuts are applied to events obtained by subjecting field configurations prepared at T_i = 400 MeV to a cooling with D=2. [26]

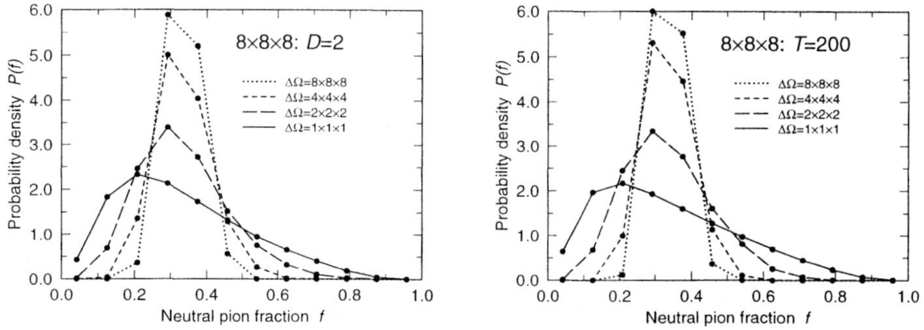

FIGURE 11. The distribution of the neutral pion fraction, $P(f)$, for a cube with side length L=8 fm that has been subdivided into ever smaller cubic sources. The system was either Rayleigh cooled with D=2 starting from $T_i = 400$ MeV (*left panel*: 9a) or it was prepared in equilibrium at T=200 MeV (*right panel*: 9b). (From Ref. [34].)

The key quantity determining the form of the extracted $P(f)$ is the pion correlation length. In the examples above, the growth of the correlation length was caused by the non-equilibrium evolution following a quench. However, a large correlation length can also exist in thermal equilibrium, since it grows steadily as T is reduced, and thus similar results can be produced by a thermal source, as is illustrated in Fig. 11b. Therefore, the appearance of an anomalous neutral pion fraction distribution is not a unique signal of the DCC phenomenon.

Dilepton Production

Electromagnetic observables, namely dileptons [35,36] and photons [37], may provide additional information on the chiral dynamics. It is possible to extend the semi-classical treatment of the linear σ model to the calculation of electromagnetic production processes as well, as illustrated below for the production of dileptons.

The numerical solution of the equations of motion of the chiral fields yields the evolution of the Cartesian components of the pion field, $\boldsymbol{\pi}(\boldsymbol{r}, t)$, in addition to the sigma field $\sigma(\boldsymbol{r}, t)$. The electromagnetic current density coincides with the third component of the isovector current density, $J_\mu(x) = \pi_1(x)\partial_\mu\pi_2(x) - \pi_2(x)\partial_\mu\pi_1(x)$, where the complex fields representing the charged pions are related to the Cartesian components by $\pi_\pm(x) = [\pi_1(x) \pm i\pi_2(x)]/\sqrt{2}$.

The invariant differential dilepton yield may be calculated (to leading order in the fine structure constant $\alpha = e^2/4\pi$), by use of the following expression [36],

$$\frac{d^4N}{d^4q} = \frac{2}{3\pi}\left(\frac{\alpha}{2\pi}\right)^2 \left(\frac{q^\mu q^\nu}{q^4} - \frac{g^{\mu\nu}}{q^2}\right) \int d^4x \int d^4y \; J_\mu(x)\; e^{-iq(x-y)} \; J_\nu(y) \tag{30}$$

$$= \frac{2}{3\pi}(\frac{\alpha}{2\pi})^2 \tilde{J}^*_\mu(q) \left(\frac{q^\mu q^\nu}{q^4} - \frac{g^{\mu\nu}}{q^2}\right) \tilde{J}_\nu(q) \ . \tag{31}$$

This expression ignores the final-state Bose enhancement factors $1+n_{\mathbf{k}}$ which is justified when the occupation number $n_{\mathbf{k}}$ is small, as is typically the case in equilibrium. It may be noted that if the calculated pion field $\boldsymbol{\pi}(\boldsymbol{r},t)$ is assumed to represent a standard coherent state, then the quantal evaluation of the dilepton radiation rate would lead to the above expression (30) when the commutator terms are ignored; if those commutator terms were retained, then the final-state Bose enhancement factors would be recovered.

In general, we consider an entire sample of $\mathcal{N}$ individual evolutions, $\{\boldsymbol{\phi}^{(n)}(\boldsymbol{r},t)\}$, where the label n enumerates the individual "events" in the sample. The resulting ensemble-average dilepton yield is then

$$\prec \frac{d^4N}{d^4q} \succ = \frac{1}{\mathcal{N}} \sum_{n=1}^{\mathcal{N}} \frac{d^4N^{(n)}}{d^4q} \ , \tag{32}$$

where $d^4N^{(n)}/d^4q$ is the contribution from the particular event n, obtained as described above. Since we consider ensembles that have translational symmetry, the current-current correlation function, $\prec J_\mu(x)J_\nu(y)\succ$, will depend only on the spatial separation. Moreover, in the special case of an equilibrium ensemble, its temporal dependence in equilibrium is only via the time difference.

In order to verify that the adopted method indeed leads to physically reasonable results, consider the production of back-to-back dileptons from a thermal gas of free pions. In that special case the four-momentum of the dilepton is of the form $q = (M,\mathbf{0})$ and we are interested in masses M above $2m_\pi$. The current-current contraction in (30) is then especially simple and its ensemble average is given by

$$\prec J_\mu(x) \left(\frac{q^\mu q^\nu}{q^2} - g^{\mu\nu}\right) J_\nu(y) \succ \ = \ \prec \boldsymbol{J}(x)\cdot\boldsymbol{J}(y) \succ \ = \ 2|\nabla C|^2 - 2C\Delta C \ , \tag{33}$$

where we have employed the thermal correlation function of the charged pion fields,

$$\prec \pi_1(x)\pi_1(y) \succ \ = \ \prec \pi_2(x)\pi_2(y) \succ \ = \ C(\boldsymbol{r},t) \ = \ \frac{1}{\Omega}\sum_{\mathbf{k}} \frac{\tilde{n}_k}{\omega_k} \cos(\mathbf{k}\cdot\mathbf{r} - \omega_k t) \ , \tag{34}$$

with $\tilde{n}_k$ being the thermal occupancy, $\tilde{n}_k = 1/(\exp(\omega_k/T)-1)$, and $(\boldsymbol{r},t)$ denoting the difference x-y. [The adopted sampling procedure ensures that the numerically extracted correlation function indeed yields this expression [15].] Since the back-to-back dileptons have vanishing momentum, $\mathbf{q} = \mathbf{0}$, the Fourier transform over the separation $\boldsymbol{r}$ reduces to a spatial average and we readily find

$$\int_\Omega d\boldsymbol{r} \ \prec \boldsymbol{J}(x)\cdot\boldsymbol{J}(y) \succ \ = \ 4\int_\Omega d\boldsymbol{r} \ |\nabla C(\boldsymbol{r},t)|^2 \ = \ \frac{2}{\Omega}\sum_{\mathbf{k}} \frac{\tilde{n}_k^2}{\omega_k^2} k^2 \ [1+\cos 2\omega_k t] \ . \tag{35}$$

The remaining Fourier transformation over the temporal difference then restricts the contributions in the sum to those modes that have frequencies ω_k near half the dilepton mass, $M/2$. Thus, in the continuum limit, when both the box and the time interval are large, we recover exactly the usual expression for production of dileptons by pion annihilation,

$$\frac{d^4N^{th}}{d^4qd^4x} = \frac{4}{3\pi}\left(\frac{\alpha}{2\pi}\right)^2 \int\frac{d\mathbf{k}}{(2\pi)^3}\int dt\,\frac{\tilde{n}_k^2}{\omega_k^2}\frac{k^2}{M^2}\,[1+\cos 2\omega_k t]\,e^{iMt} = \frac{\alpha^2}{3}\frac{n_0^2}{(2\pi)^4}\left(1-\frac{4m_\pi^2}{M^2}\right)^{\frac{3}{2}},$$

where n_0 denotes the occupancy of pion states having the matching frequency $\omega_0 = M/2$. Thus, at the formal level, the semi-classical method is indeed physically reasonable.

Figure 12a shows the dilepton production rate in a source in thermal equilibrium at T=140 MeV, together with the corresponding result obtained when the system is prepared in a quenched configuration where nearly all the thermal energy (about 40 MeV/fm^3) has been converted into potential energy of the displaced order parameter [36]. As is shown in Fig. 12b, the release of the system causes the order parameter to execute large oscillations around its equilibrium value leading to a significant enhancement of the dileptons. In addition to the amplification that occurs whenever the system is inside the unstable region, as discussed above, there is an additional enhancement resulting from the parametric amplification caused by the approximately regular oscillation of the order parameter with a frequency near the σ mass [36,37].

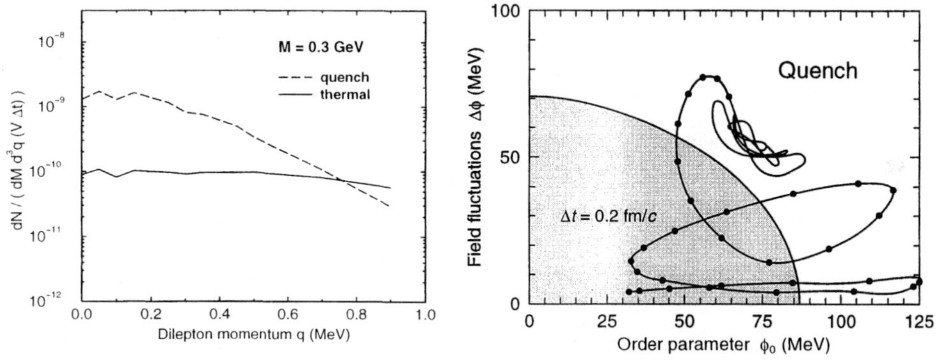

FIGURE 12. *Left* (12a): The dilepton production rate $d^4N/(d^4qV\Delta t)$ as a function of the magnitude of the dilepton momentum, for dilepton masses near M=300 MeV from either a thermal source at T = 140 MeV or as obtained as a result of the quench depicted in the right panel [36]. *Right* (12b): The dynamical path after the same system has been quenched: (the field fluctuations have been nearly eliminated and the order parameter has been placed the supercirtical region with a value that ensures that the energy is the same as in the thermal source [38].

Bjorken Rods

We now turn to the study of a more refined scenario that exhibits some of the most important features expected in real collision events, namely rapid longitudinal expansion and finite transverse extension [13]. Specifically, we shall explore a socalled *Bjorken rod*: a rod-like geometry endowed with a longitudinal scaling expansion (not merely experiencing a Rayleigh cooling). Such systems have a finite (circular) extension in the transverse plane, and the local environment changes smoothly from hot longitudinally expanding matter in the bulk to vacuum outside.

Generally, the numerical propagation of fields exhibiting significant flow patterns, such as rapid expansion, is practically difficult due to the phase oscillations caused by the local boost. However, for idealized scaling expansions this complication can be eliminated by suitable variable transformations. For the longitudinal scaling expansion considered here, it is convenient to replace the usual fixed-frame space-time variables (x, y, z, t) with the comoving variables (x, y, η, τ) [24],

$$t = \tau \cosh \eta \ , \quad z = \tau \sinh \eta \ . \tag{36}$$

Thus, $\tau = (t^2 - z^2)^{1/2}$ is the proper time experienced in a system boosted along the z axis with the local rapidity y equal to the value of $\eta = \frac{1}{2} \ln(\frac{t+z}{t-z})$. The corresponding form of the field equation of motion is then obtained from the transformation of the d'Alembert operator,

$$\Box \equiv \partial_t^2 - \partial_x^2 - \partial_y^2 - \partial_z^2 = \frac{1}{\tau}\partial_\tau \tau \partial_\tau - \partial_x^2 - \partial_y^2 - \frac{1}{\tau^2}\partial_\eta^2 \ . \tag{37}$$

The equation of motion can be readily solved numerically by application of the leapfrog method, once the initial field $\boldsymbol{\phi}(\boldsymbol{r})$ and its time derivative $\boldsymbol{\psi}(\boldsymbol{r})$ are specified (see next page).

When analyzing the asymptotic field for the Bjorken rod, it is natural to perform a Fourier transformation in the transverse plane,

$$\boldsymbol{\phi}_{\mathbf{k}}(\eta) = \int \frac{d^2\boldsymbol{\rho}}{\Omega_\perp}\, \boldsymbol{\phi}(\boldsymbol{\rho}, \eta)\, \mathrm{e}^{-i\mathbf{k}\cdot\boldsymbol{\rho}} \ , \quad \boldsymbol{\psi}_{\mathbf{k}}(\eta) = \int \frac{d^2\boldsymbol{\rho}}{\Omega_\perp}\, \boldsymbol{\psi}(\boldsymbol{\rho}, \eta)\, \mathrm{e}^{-i\mathbf{k}\cdot\boldsymbol{\rho}} \ , \tag{38}$$

where $\Omega_\perp = L_x L_y$ denotes the transverse area of the spatial lattice. Furthermore, the position in the transverse plane is denoted by $\boldsymbol{\rho} = (x, y)$ and the transverse wave number is $\mathbf{k} = (k_x, k_y)$. For large times the longitudinal velocity of a given part of the system is given by $v_z \to z/t = \tanh(\eta)$. Thus, in that limit, the coordinate η equals the rapidity, $\eta \to \mathsf{y} = \tanh^{-1}(v_z)$.

At present, we are only interested in observables based on the pion component of the chiral field, which are expressed in terms of the spherical components, $\boldsymbol{\pi} = (\pi_-, \pi_0, \pi_+)$. The mean number of pions (a given charge state j) emerging with transverse wave vector $\mathbf{k}$ and a rapidity y in the interval $(\mathsf{y}_1, \mathsf{y}_2)$ is then given by

$$\bar{n}_{\mathbf{k}}^{(j)}(\mathsf{y}_1, \mathsf{y}_2) = \Omega_\perp \int_{\mathsf{y}_1}^{\mathsf{y}_2} d\eta\, |\sqrt{\frac{m_k}{2}}\boldsymbol{\phi}_{\mathbf{k}}(\eta) + \frac{i}{\sqrt{2m_k}}\boldsymbol{\psi}_{\mathbf{k}}(\eta)|^2 \ . \tag{39}$$

Preparation of the Rod

In order to prepare the initial field configuration for the rod, we proceed at first in the same manner as for the preparation of the matter scenario addressed above and sample the field configuration $(\boldsymbol{\phi}_{\text{box}}, \boldsymbol{\psi}_{\text{box}})$ from a thermal ensemble describing macroscopically uniform matter within the overall box containing the calculational lattice. This field configuration can be uniquely decomposed into its spatial average, the order parameter, and the remainder, the fluctuating part of the field,

$$\boldsymbol{\phi}_{\text{box}}(\boldsymbol{\rho},\eta) \;=\; \underline{\boldsymbol{\phi}} + \boldsymbol{\delta\phi}(\boldsymbol{\rho},\eta)\;,\; \boldsymbol{\psi}_{\text{box}}(\boldsymbol{\rho},\eta) \;=\; \underline{\boldsymbol{\psi}} + \boldsymbol{\delta\psi}(\boldsymbol{\rho},\eta)\;. \tag{40}$$

In order to obtain the field configuration describing the initial state of the rod, $(\boldsymbol{\phi}_{\text{rod}}, \boldsymbol{\psi}_{\text{rod}})$, we rescale the two parts based on a specified local temperature $T(\rho)$ and then recombine them into the desired initial conditions,

$$\boldsymbol{\phi}_{\text{rod}}(\boldsymbol{\rho},\eta,\tau_0) \;=\; g(\rho)[\underline{\boldsymbol{\phi}} - \boldsymbol{\phi}_{\text{gs}}] + \boldsymbol{\phi}_{\text{gs}} \;+\; \boldsymbol{h}(\rho)\circ[\boldsymbol{\delta\phi}(\boldsymbol{\rho},\eta) - \underline{\boldsymbol{\phi}}] + \underline{\boldsymbol{\phi}}\;, \tag{41}$$

$$\boldsymbol{\psi}_{\text{rod}}(\boldsymbol{\rho},\eta,\tau_0) \;=\; g(\rho)[\underline{\boldsymbol{\psi}} - \boldsymbol{\psi}_{\text{gs}}] + \boldsymbol{\psi}_{\text{gs}} \;+\; \boldsymbol{h}(\rho)\circ[\boldsymbol{\delta\psi}(\boldsymbol{\rho},\eta) - \underline{\boldsymbol{\psi}}] + \underline{\boldsymbol{\psi}}\;, \tag{42}$$

where $\boldsymbol{\phi}_{\text{gs}} = (f_\pi, \mathbf{0})$ and $\boldsymbol{\psi}_{\text{gs}} = (0, \mathbf{0})$ are the vacuum values. The scaling coefficients for the order parameter and the fluctuations, $g(\rho)$ and $\boldsymbol{h}(\rho)$, are obtained by using the corresponding equilibrium values for the specified local temperature $T(\rho)$,

$$g(\rho) = \frac{\phi_0(T(\rho)) - f_\pi}{\phi_0(T_0) - f_\pi}\;,\quad h_\sigma(\rho) \;=\; \left(\frac{\delta\sigma_T^2}{\delta\sigma_{T_0}^2}\right)^{\frac{1}{2}}\;,\quad h_\pi(\rho) \;=\; \left(\frac{\delta\pi_T^2}{\delta\pi_{T_0}^2}\right)^{\frac{1}{2}}\;. \tag{43}$$

By proceeding in this manner we ensure that the local environment, as characterized by the order parameter and the field fluctuations, reflects approximately thermal equilibrium in matter held at the local temperature T, which decreases from its bulk value T_0 to zero according to the prescribed radial profile $T(\rho)$.

The initial system is illustrated in Fig. 13 for $T_0 = 240$ MeV. Since the local temperature drops steadily as a function of the transverse distance ρ, moving out along the abscissa corresponds to reducing the temperature (though not at a steady rate). As one moves out through the surface, the order parameter increases steadily from its reduced value (≈ 27 MeV) in the hot bulk region towards its vacuum value f_π ($= 92$ MeV), and the local thermal fluctuations drop correspondingly towards zero. (Since generally $\mu_\sigma > \mu_\pi$, the fluctuations along a pion direction exceed those in the σ direction.) The resulting profiles of the effective masses also reflect their temperature dependencies: Starting from nearly degenerate values (≈ 300 MeV) in the hot interior, where chiral symmetry is approximately restored, μ_π and μ_σ diverge steadily towards their free values of 138 MeV and 600 MeV, respectively. For higher values of the central temperature T_0, the central value of the order parameter is smaller and the effective masses are larger (and even closer in value) and μ_σ will in fact exhibit a dip in the surface region as the local temperature passes through the critical region.

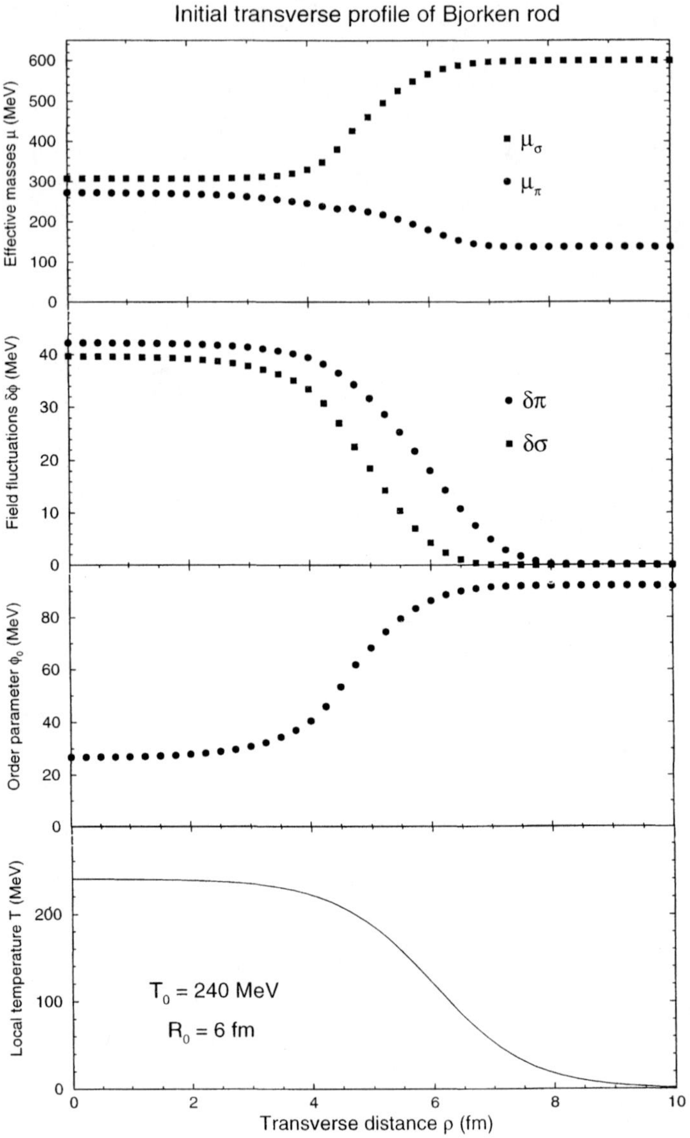

FIGURE 13. Initial transverse profiles of key quantities for a Bjorken rod prepared with a radius of $R_0 = 6$ fm and a central temperature of $T_0 = 240$ MeV: the specified local temperature $T(\rho)$, the order parameter $\phi_0(\rho)$, the mean fluctuation of the field in a given O(4) direction, $\delta\sigma(\rho)$ and $\delta\pi(\rho)$, and the corresponding effective quasiparticle masses, $\mu_\sigma(\rho)$ and $\mu_\pi(\rho)$ (from Ref [13]).

After preparing the field configuration of the rod, the dynamical propagation is readily obtained by solving the field equation (1) in the usual manner. Since the system is no longer macroscopically uniform, it is more complicated to discuss. but it is especially instructive to see how the bulk region develops. For this purpose, we compare with uniform matter prepared at the bulk temperature T_0 and endowed with a corresponding longitudinal expansion. Such a system may be denoted *Bjorken matter* and represents the environment that would be obtained in interior of a rod with a very large radius, $R_0 \to \infty$.

The resulting evolution of the order parameter and the field fluctuations is illustrated in Fig. 14 for a rod with $T_0 = 240$ MeV and $R_0 = 6$ fm. The corresponding quantities for Bjorken matter prepared with the same value of T_0 are also shown. It is seen that the environment in the interior of the rod evolves in a manner quantitatively very similar to that of the corresponding matter scenario throughout the first complete oscillation of the the order parameter, which takes about 4 fm/c. At around that point in time, the decompression wave has arrived from the rod surface and the self-generated transverse expansion now extends over the entire cross section of the rod. As a consequence, the relaxation progresses faster than it would in matter where this effect is absent.

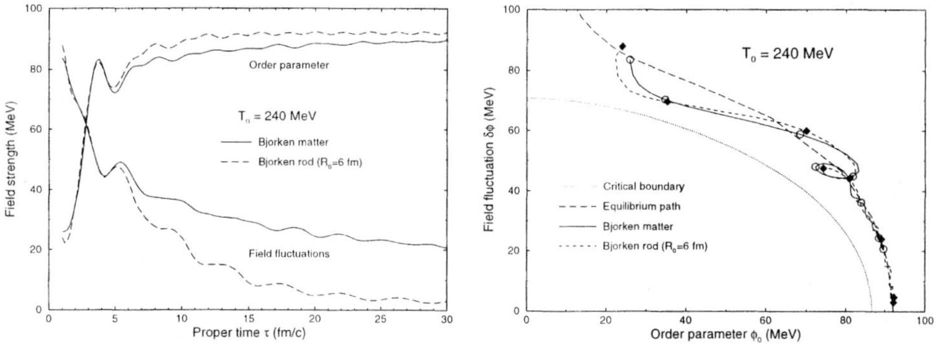

FIGURE 14. *Left* (14a): The time dependence of ϕ_0, the magnitude of the order parameter (increasing curves), and $\delta\phi$, the associated dispersion of the field fluctuations (decreasing curves), in both Bjorken matter prepared with $T_0 = 240$ MeV (solid curves) and the corresponding Bjorken rod with $R_0 = 6$ fm (dashed curves). The information for the rod has been obtained by averaging over a hollow cylindrical volume, $1 < \rho(\text{fm}) < 3$. *Right* (14b): The corresponding phase evolution of the interior of the Bjorken rod (dashed path, filled diamonds), and the associated Bjorken matter (solid path, open circles). The equilibrium path is shown by the short-dashed curve while the dotted curve delineates the region of instability within which the field is supercritical. The symbols give the path locations at successive proper times $\tau(\text{fm}/c) = 1, 2, 3, 4, 5, 10, 20, 30$. [13]

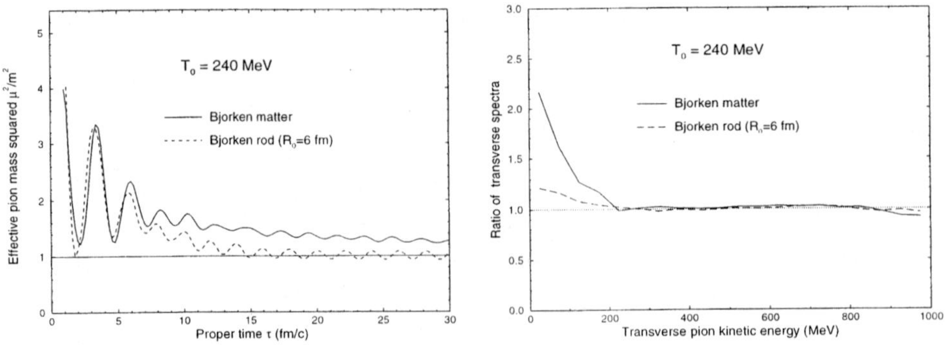

FIGURE 15. *Left panel* (15a): Time evolution of the square of the effective pion mass, $\mu_\pi^2(t)$ (divided by the square of its free mass m_π) for a Bjorken rod (dashed) prepared with $T_0 = 240$ MeV and $R_0 = 6$ fm as well as the corresponding result for Bjorken matter (solid). The information for the rod has been obtained by averaging over a hollow cylindrical volume, $1 < \rho(\mathrm{fm}) < 3$. *Right panel* (15b): The ratio between the final transverse pion spectrum, $d^3N/d^2\mathbf{p}_\perp dy$, and the associated equilibrium spectrum obtained by fitting the dynamical result with a Bose-Einstein form within the energy interval 200-1000 MeV, for the longitudinally expanding rod having an initial radius of $R_0 = 6$ fm and with an initial bulk temperature of $T_0 = 240$ MeV (dashed curve), as well as the corresponding result for Bjorken matter. (These illustrations are from [13].)

The quicker relaxation of the field for the rod is also reflected in the behavior of the effective pion mass, as illustrated in Fig. 15a for the same case. Again we see that through the first several time units the effective mass in the interior of the rod follows closely the evolution of the effective mass extracted for the corresponding matter scenario, but it then drops much faster towards the free value. Furthermore, the oscillations in the effective mass persist for quite a long time, thus making it possible to achieve a significant degree of parametric amplification.

Observables

We now consider a few specific observables that are practically accessible in the analysis of actual experimental data. First we consider the transverse spectra of the emerging pions, $Ed^3N/d^3\mathbf{p} = d^3N/d^2\mathbf{k}dy$. As it turns out, the transverse spectral shape is well approximated by an equilibrium form above kinetic energies of 200 MeV or so and one may thus extract an spectral effective temperature. Moreover, it is instructive to divide by that overall shape. The resulting ratio between the dynamical spectrum and the corresponding Bose-Einstein form is displayed in Fig. 15b together with the corresponding matter result. Clearly, there is a significant enhancement of soft pions, although the effect is considerably smaller in the rod due to the presence of the surface.

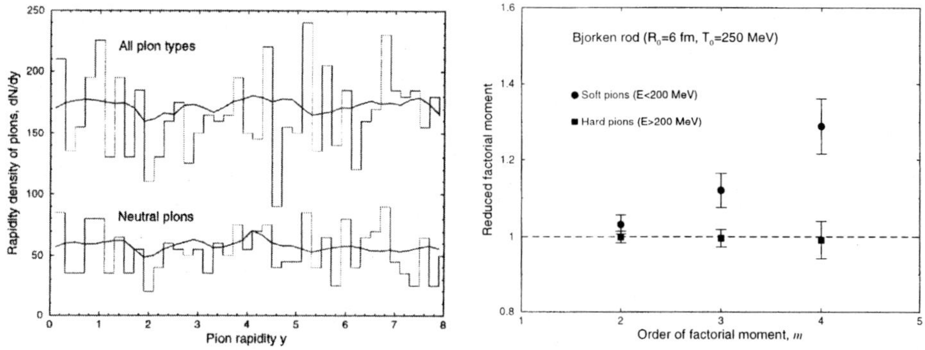

FIGURE 16. *Left panel* (16a): The rapidity density of pions for a single event prepared with $T_0 = 250$ MeV (with rapidity bins $\Delta y = 0.2$). The expected multiplicity based on the asymptotic field is indicated by the smooth solid curve, while the histogram shows the actual number as picked from the corresponding Poisson distributions. The top curves include all three pion types, whereas the bottom curves show the neutras only. *Right panel* (16b): The corresponding reduced factorial moments $\mathcal{M}_m/\bar{N}^m$ as a function of the order m, for either soft or hard pions. [13]

By invoking the relation (39), it is straightforward to calculate the expected number of pions in a given rapidity interval and within a specified transverse energy bin. This result varies from event to event, since the field configurations vary at the microscopic level as a result of the fluctuations in the ensemble of initial states. Moreover, for a given value of the expected number $\bar{n}$, the actual number of particles emerging is a stochastic variable n, since the state described by a given field has no well-defined particle number. For an exploratory study it may suffice to employ a Poisson form, $P_{\bar{n}}(n) = \bar{n}^n \exp(-\bar{n})/n!$. Figure 16a shows both the expected and the actual multiplicity ($\bar{n}$ and n, respectively), as obtained for a single event.

In order to investigate whether the resulting multiplicity distributions contain deviations from Poisson statistics, it is useful to extract the factorial moments.

$$\mathcal{M}_m \equiv \prec N(N-1)\cdots(N-m+1) \succ \,, \tag{44}$$

where N is the number of pions emitted in a given rapidity interval and the average is over all such intervals and events. They are shown in Fig. 16b. While the hard pions appear to be perfectly consistent with pure Poisson statistics, the soft pions exhibit a significant non-poissonian behavior. The character of the deviation of the soft factorial moments suggests that the source occasionally emits anomalously many pions, as one would expect if some modes are especially amplified. This feature is consistent with the enhancement of the soft spectrum and it can also be explored by other means of analyzing the data for anomalous fluctuations. Furthermore, standard event generators (UrQMD and HIJING) do not produce any such difference between soft and hard multiplicity fluctuations [39].

CONCLUDING REMARKS

In this brief lecture series, we have tried to give an impression of current efforts towards exploiting high-energy nuclear collisions to elucidate our understanding of chiral symmetry in strongly interacting systems. We have especially discussed certain key aspects of the treatment (while others had to be left out), hoping thereby to make the rapidly growing literature on the subject more accessible. We have also made contact with experiment by illustrating how certain observables may carry signals of the expected non-equilibrium DCC dynamics.

Our discussion has been carried out within the framework of the linear σ model which was treated in a semi-classical manner. Comparisons with more sophisticated approaches (such as that of Ref. [41]) as well as dynamical self-consistency tests [15] have suggested that this simple treatment is in fact semi-quantitatively reasonable. Nevertheless, refinements would be desirable in a number of respects.

One obvious challenge is to go beyond the semi-classical level and invoke actual quantum field theory. The practical treatment of real-time quantum field theory in an environment that is neither uniform nor thermal presents an interesting but formidable challenge. Fortunately, it appears possible to incorporate the quantitatively most important effects into the numerical DCC simulations, as is briefly illustrated in Appendix A. The principal lesson is that the ever present vacuum fluctuations act on an equal footing with the statistical fluctuations as seeds for dynamical amplification and, in this regard, they appear to be at least as important.

Another direction of desirable improvement concerns the incorporation af additional degrees of freedom, since the limitation to the $(\sigma, \boldsymbol{\pi})$ space is a drastic simplification, especially at the high energy densities occuring at the early stage. Particularly important is the inclusion of strangeness, as is briefly illustrated in Appendix B. This extension leads to significant modifications of the results but it also enriches the dynamics and offers additional possible signals.

Thus, many theoretical challenges remain, with regard to both the formal refinement of the treatment and the implementation of the theory into a practical tool that can provide specific guidance for the design and analysis of the experiments. With the recently completed RHIC facility now beginning to yield data, the need for further advances has intensified.

This work was supported by the Director, Office of Energy Research, Office of High Energy and Nuclear Physics, Nuclear Physics Division of the U.S. Department of Energy under Contract No. DE-AC03-76SF00098.

A: QUANTUM FIELD EFFECTS

A full quantum-field treatment of the chiral dynamics is beyond current reach and the dynamical simulation studies have therefore employed classical fields. Although much valuable insight can been gained in this manner, it is important to recognize that such treatments are not always quantitatively accurate. This is perhaps best illustrated by considering a free pionic mode **k** with a time-dependent frequency, $\omega_k^2(t) = k^2 + \mu^2(t)$, where $\mu^2(t)$ has a given form. The equation of motion for the corresponding field operator is $[\Box + \omega_k^2(t)]\hat{\phi}_{\mathbf{k}}(t) = 0$, and the associated time-evolution operator can be determined (see Ref. [40]).

If the initial occupancy of the mode is $n_{\mathbf{k}}^{\rm init}$, then the final occupancy becomes

$$n_{\mathbf{k}}^{\rm final} = X_k\left[n_{\mathbf{k}}^{\rm init} + \tfrac{1}{2}\right] - \tfrac{1}{2} > X_k n_{\mathbf{k}}^{\rm init}\,, \tag{45}$$

after averaging over the initial state phase. This exceeds the classical result $X_k n_{\mathbf{k}}^{\rm init}$, since the amplification coefficient is generally larger than unity, $X_k \geq 1$. The above expression brings out the fact that the quantum fluctuations and the statistical fluctuations combine in a democratic fashion as seeds for the parametric amplification. A quantitative impression of their relative importance can be gained from Fig. 17a. It is evident that the thermal occupancies are never large compared to one half. Consequently, the vacuum fluctuations are never negligible.

Efforts to develop a suitable quantum-field treatment are underway [42] and an illustration is given in Fig. 17b. It is seen that pions with momenta around 200 MeV/c are produced, even if the initial state contained no quasiparticles at all (the initial presence of thermal excitations will then increase the yield further). This spectral shape reflects the regular temporal modulation of the pion mass.

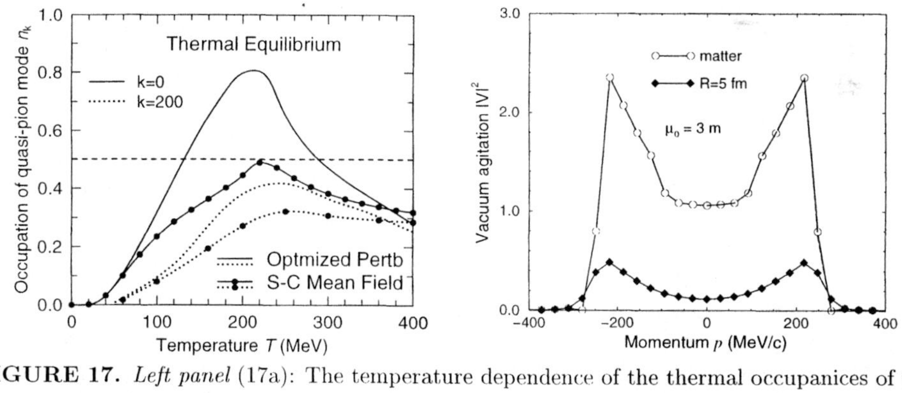

FIGURE 17. *Left panel* (17a): The temperature dependence of the thermal occupanices of the quasipions as obtained with either a semi-classicial treatment [15] or an optimized perturbation method [41]. *Right panel* (17b): The final pion spectrum arising solely from amplification of the vacuum fluctuations in one-dimensional systems with given effective mass functions $\mu_\pi^2(x,t)$ of forms similar to those obtained for Bjorken rods (see fig. 15a) [42].

B: INCLUSION OF STRANGENESS

Since the temperatures of interest easily exceed the mass of the s quark, it is expected that the strange degrees of freedom are agitated as well and thus strangeness should be incorporated into the description. The extension from ud to uds enlarges the meson group from SU(2) to SU(3), which contains a total of 18 fields,

$$\mathrm{SU}(3)[uds]:\left\{\begin{array}{lllllllll}\sigma, & a_0^0, & \zeta; & a_0^-, & a_0^+, & K^{*-}, & K^{*+}, & K^{*0}, & \bar{K}^{*0}; \\ \pi^0, & \eta_8, & \eta_0; & \pi^-, & \pi^+, & K^-, & K^+, & K^0, & \bar{K}^0\end{array}\right. . \tag{46}$$

In equilibrium, both the σ and the ζ fields have finite values, and so the order parameter acquires a strange component, $\sigma \to (\sigma, \zeta)$, where $\zeta = \langle \bar{s}s \rangle$. It is possible to apply the semi-classical treatment to this more complicated case as well [43].

As is evident from Fig. 18 (*left*), the inclusion of strangeness severely impedes the restoration of chiral symmetry as the temperature is raised, thus casting doubt on the standard scenario in which approximate restoration is assumed to occur once T exceeds a few hundred MeV.

On the other hand, as illustrated in Fig. 18 (*right*), the dynamics of the order parameter becomes more intricate. As a result, the kaon modes may experience parametric amplification in analogy to what happens for the pions. For an idealized source, the neutral pion fraction $f_\pi = \pi^0/(\pi^+ + \pi^- + \pi^0)$ retains its SU(2) distribution, $P_\pi(f_\pi) = 1/2\sqrt{f_\pi}$, and the neutral kaon fraction $f_K = (K^0 + \bar{K}^0)/(K^+ + K^- + K^0 + \bar{K}^0)$ has a uniform distribution, $P_K(f_K) = 1$.

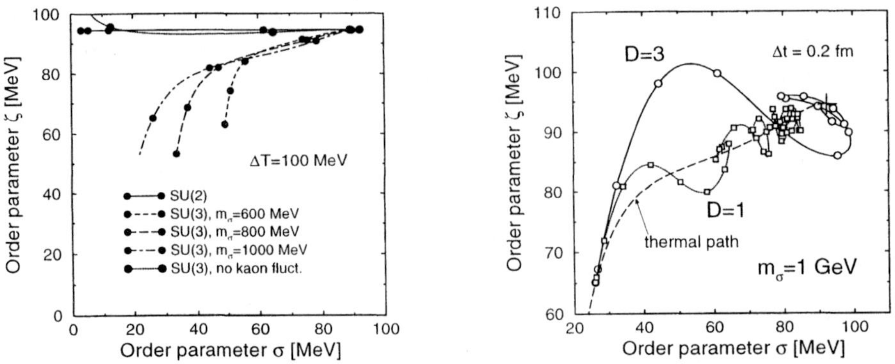

FIGURE 18. *Left panel* (18a): The thermal path of the two order parameter (σ, ζ) for various cases of interest, as indicated. The solid dots are plotted in temperature steps of 100 MeV. All paths start from the vacuum point at the upper-right corner and the points for T=100 MeV are still very close to the vacuum point. *Right panel* (18b): The dynamical path of the order parameter (σ, ζ) as a result of a pseudoexpansion in one and three dimensions, starting from equilibrium at T=400 MeV. Equidistant time steps with $\Delta\tau$=0.2 fm/c are indicated. The thermal path is shown by the long-dashed curve, while the cross indicates the vacuum point, towards which all trajectories converge in time. (These illustrations are from Ref. [43].)

REFERENCES

1. K. Rajagopal, in *Quark-Gluon Plasma 2*, Ed. R. Hwa, World Scientific (1995).
2. J.-P. Blaizot and A. Krzywicki, *Acta Phys. Polon.* **B27**, 1687 (1996).
3. J.D. Bjorken, *Acta Phys. Polon.* **B28**, 2773 (1997).
4. V. Koch, *J. Mod. Phys.* **E6**, 203 (1997).
5. M. Gell-Mann and M. Levy, *Nuov. Cim.* **16**, 705 (1960).
6. K. Rajagopal, F. Wilczek, *Nucl. Phys.* **B404**, 577 (1993).
7. M. Asakawa, Z. Huang, and X.-N. Wang, *Phys. Rev. Lett.* **74**, 3126 (1995).
8. L.P. Csernai and I.N. Mishustin, *Phys. Rev. Lett.* **74**, 5005 (1995).
9. J. Randrup, *Phys. Rev. Lett.* **77**, 1226 (1996).
10. G. Amelino-Camelia, J.D. Bjorken and S.E. Larsson, *Phys. Rev.* **D56**, 6942 (1997).
11. A. Abada and M.C. Birse, *Phys. Rev.* **D57**, 292 (1998).
12. D. Kaiser, *Phys. Rev.* **D59**, 117901 (1999).
13. T.C. Petersen and J. Randrup, *Phys. Rev.* **C61**, 024906 (2000).
14. Z. Xu and C. Greiner, *Phys. Rev.* **D62**, 036012 (2000).
15. J. Randrup, *Phys. Rev.* **D55**, 1188 (1997).
16. G. Baym and G. Grinstein, *Phys. Rev.* **D15**, 2897 (1977).
17. J. Goldstone, *Nuov. Cim.* **19**, 154 (1961).
18. L.D. Landau and E.M. Lifshitz, *Statistical Physics*, Pergamon, New York (1980).
19. Z. Huang and X.-N. Wang, *Phys. Rev.* **D49**, 4335 (1994).
20. S. Gavin and B. Müller, *Phys. Lett.* **B329**, 486 (1994).
21. D. Boyanovsky and H.J. de Vega, *Phys. Rev.* **D51**, 734 (1995).
22. F. Cooper, Y. Kluger, E. Mottola, and J.P. Paz, *Phys. Rev.* **D51**, 2377 (1995).
23. Y. Kluger, F. Cooper, E. Mottola, J. Paz, A. Kovner, *Nucl. Phys.* **A590**, 581c (1995).
24. J.D. Bjorken, *Phys. Rev.* **D27**, 140 (1983).
25. M.A. Lampert, J.F. Dawson, and F. Cooper, *Phys. Rev.* **D54**, 2213 (1996).
26. J. Randrup, *Nucl. Phys.* **A616**, 531 (1997).
27. A.A. Anselm, *Phys. Lett.* **B217**, 169 (1989).
28. A.A. Anselm and M.G. Ryskin, *Phys. Lett.* **B266**, 482 (1991).
29. J.D. Bjorken, K.L. Kowalski, C.C. Taylor, SLAC-PUB-6109; *hep-ph/9309235* (1993).
30. K. Rajagopal and F. Wilczek, *Nucl. Phys.* **B399**, 395 (1993).
31. J.P. Blaizot and A. Krzywicki, *Phys. Rev.* **D46**, 246 (1992).
32. Z. Huang, I. Sarcevic. R. Thews, and X.N. Wang, *Phys. Rev.* **D54**, 750 (1996).
33. K. Rajagopal, Hirschegg Winter Workshop XXV; *hep-ph/9703258* (1997).
34. J. Randrup and R.L. Thews, *Phys. Rev.* **D56**, 4392 (1997).
35. Z. Huang and X.N. Wang, *Phys. Lett.* **B383**, 457 (1996).
36. Y. Kluger, V. Koch, J. Randrup, and X.N. Wang, *Phys. Rev.* **C57**, 280 (1998).
37. D. Boyanovsky, H.J. de Vega, R. Holman, S.P. Kumar, *Phys. Rev.* **D56**, 5233 (1997).
38. J. Randrup, *Nucl. Phys.* **A630**, 468c (1998).
39. M. Bleicher, J. Randrup, R. Snellings, and X.-N. Wang, *Phys. Rev.* **C62** (in press).
40. J. Randrup, *Heavy Ion Phys.* **9** (1999) 289.
41. S. Chiku and T. Hatsuda, *Phys. Rev.* **D58**, 076001 (1998).
42. J. Randrup, *Phys. Rev.* **C62** (in press); LBNL-46265 (2000).
43. J. Schaffner-Bielich and J. Randrup, *Phys. Rev.* **C59**, 3329 (1999).

The Experiment Road to the Heavier Quarks and Other Heavy Objects

Jeffrey A. Appel

Fermilab, PO Box 500, Batavia, IL 60510, USA
E-mail: appel@fnal.gov

Abstract. After a brief history of heavy quarks, I will discuss charm, bottom, and top quarks in turn. For each one, I discuss its first observation, and then what we have learned about production, hadronization, and decays - and what these have taught us about the underlying physics. I will also point out remaining open issues. For this series of lectures, the charm quark will be emphasized. It is the first of the heavy quarks, and its study is where many of the techniques and issues first appeared. Only very brief mention is made of CP violation in the bottom-quark system since that topic is the subject of a separate series of lectures by Gabriel Lopez. As the three quarks are reviewed, a pattern of techniques and lessons emerges. These are identified, and then briefly considered in the context of anticipated physics signals of the future; e.g., for Higgs and SUSY particles.

A BRIEF HISTORY OF HEAVY QUARKS

Today's Elementary Particles

Today's picture of the most elementary particles is composed of six quark types, six lepton types, and four force carriers. The quarks and leptons come in three generations, each generation with a pair of quarks (one with charge + 2/3, and one with charge - 1/3) and a pair of leptons (one charged, and one neutral). See Table 1. The quarks and leptons have spin 1/2, and couple variously to the spin 1 force carriers: gluons, photons, and charged and neutral weak bosons (W^+, W^-, and Z). Unlike the leptons, the quarks appear to come in three varieties, called colors. Perhaps the number of colors is related to the quarks' third-integer charges. Also, quarks never appear in isolation, being permanently confined, for example, in meson and baryon combinations (quark-antiquark and three-quark combinations, respectively).

These quark, lepton, and force-carrying particles are the foundation of the so-called Standard Model of particle physics. It is widely and completely accepted by the community. However, it was not always that way. The acceptance of the quark

CP562, *Particles and Fields: Ninth Mexican School,* edited by G. Herrera Corral and L. Nellen

TABLE 1. Today's elementary particles, by type vs generation.

Type	charge	1st Generation	2nd Generation	3rd Generation
up-type quarks	+ 2/3	u up	c charm	t top
down-type quarks	- 1/3	d down	s strange	b bottom
neutral leptons	0	ν_e e neutrino	ν_μ μ neutrino	ν_τ τ neutrino
charged leptons	-1	e electron	μ muon	τ tau

picture of elementary particles owes a great deal to the discovery and understanding of the heavier quarks. Quarks were unexpected, even not accepted when I was a student. The idea of "partons" obeying SU(3) symmetry [1–3] was a mathematical tool at best, not a physical reality! [4] There were alternate possibilities for underlying structure; e.g., based on shapes, on how things are put together, and on the "bootstrap" model. [5] Perhaps the hadrons we see in the laboratory are each made of combinations of all the others, with no special subset being the most elementary.

The first confirmation of the idea of quarks came after the prediction by Murray Gell-Mann [4] of what we call the omega minus, understood now to be a baryon made of three strange quarks. This otherwise unheralded particle, was discovered in a 1964 Brookhaven bubble-chamber experiment headed by Nick Samios. [6]

The Revolution of November, 1974

The real watershed in thinking began with the announcement in November, 1974, that teams of physicists had observed a rather narrow resonance at a mass of 3.1 GeV/c^2. The resonance was seen in both hadroproduction [7] and e^+e^- annihilation. [8] The resonance still carries the dual name J/ψ from these concurrent observations. The quickly-accepted model explaining this narrow resonance was that it is made up of a new quark-antiquark pair. These new quarks were characterized by a new quantum number, called "charm." The existence of a new quark implied a whole spectrum of new particles containing at least one charm quark. Examples of these were soon discovered in e^+e^- collisions by the Mark I Collaboration at SLAC. [9]

The Growing Variety of Quarks

With the discovery of charm particles, there was a feeling that the quark picture of matter was complete, perhaps like the feeling at the beginning of the century when the atomic nature of matter was first understood. Nevertheless, only three years after the discovery of charm, an even heavier resonance was observed in proton-nucleus collisions at Fermilab. [10] This resonance, at a mass of about 9.5 GeV/c^2 mass, was called upsilon, Υ, by it's discoverers. There was evidence in the initial data of some excited states of the ground-state resonance. These were quickly confirmed at DESY in Hamburg, [11] where the DORIS storage ring energy was boosted to be able to produce the new states. Thus, the bottom quark came to be an accepted member of the hierarchy of quarks.

As an aside, it might be noted that the τ, the charged lepton of the third generation, appeared just before the upsilon particle. The tau was not widely accepted, though it's discoverers were happy to see the upsilon as a confirmation that the earlier picture of two generations was incomplete. The tau was, then, also soon widely accepted. Direct observation of the tau neutrino has, by the way, also only just been announced this month.

A sixth quark was anticipated to be roughly 3 (or π) times the mass of the bottom quark. After all, there is such a pattern apparent among the strange, charm, and bottom quarks. The TRISTAN accelerator in Japan was even built with that goal in mind. However, in spite of major efforts, the discovery of the sixth, the "top" quark, did not occur until twenty years later, in 1997. The discovery required the dedicated running of the highest energy colliding-beams accelerator in the world, Fermilab's Tevatron Collider. This was because the top quark was not three times as heavy as the bottom quark, but about forty times as heavy; weighing in at about 175 GeV/c^2. The reason for this enormous mass remains a mystery.

As the number of quarks grew, another feature of quarks appeared. That is, the eigenstates relevant to their production in strong and electromagnetic interactions (called flavor eigenstates) are not the same as the weak-interaction eigenstates of their decay (called mass eigenstates). The various eigenstates mix, as related by the Cabibbo-Kobayashi-Maskawa (CKM) matrix. [12]

$$WeakEigenstates = \mathbf{V} \times FlavorEigenstates \tag{1}$$

That is,

$$\begin{matrix} d' \\ s' \\ b' \end{matrix} \quad = \quad \begin{matrix} V_{ud} & V_{us} & V_{ub} \\ V_{cd} & V_{cs} & V_{cb} \\ V_{td} & V_{ts} & V_{tb} \end{matrix} \quad \begin{matrix} d \\ s \\ b \end{matrix}$$

Using the Wolfenstein parameterization, [13] to order λ^3, the CKM matrix is

$$\mathbf{V} = \begin{matrix} 1-\lambda^2/2 & \lambda & A\lambda^3(\rho - i\eta) \\ -\lambda & 1-\lambda^2/2 & A\lambda^2 \\ A\lambda^3(1-\rho-i\eta) & -A\lambda^2 & 1 \end{matrix}$$

The strong and electromagnetic interactions conserve the new quantum numbers, called "flavor:" strangeness, charm, bottom, and top. Thus, production of these quarks in strong and electromagnetic interactions always occurs in pairs; i.e., a quark and an antiquark of the same flavor. This is the origin of the "strange" behavior of the strange particles – and of the heavier quarks which followed. On the other hand, the weak interactions do not conserve flavor. The violation of flavor conservation occurs in a well defined way, however, with the rates governed by the CKM matrix elements, V_{ij}, above.

A Few Comments on Names

The original names, up and down, came from an analogy with spinors, with spin direction pointing up and down. The "strange" quark name was chosen as a reminder that it was supposed to explain the strange behavior of particles containing these quarks; e.g., the K mesons which were always produced in pairs or in association with strange baryons. Stranger still, even with strange quarks, some neutral K meson decay behavior was not explained. The branching ratios to certain decay modes (e.g., $\mu^+\mu^-$) were much smaller than expected. One proposed solution, the GIM mechanism [14], suggested a fourth quark to fix things up. This fourth quark was to have just the right properties, be "charmed" in just the right way for the fix to work. The name "charm" was actually suggested earlier by Bjorken and Glashow in a paper generalizing SU(3) quark symmetry to SU(4). [15] I think that the name stuck, in part, because of the charmed properties of the quark. In any event, the mass of the charm quark was predicted on the basis of the needed properties for its cancellation of other contributions to the neutral K decays.

The third generation returned to more prosaic names, "top" and "bottom," like up and down, but not before flirting with the names "truth" and "beauty." I have preferred these latter names, since I used to describe my personal research as the "search for truth and beauty" – but that was before they were both discovered! This takes us to near the end of the current story. Let's review how the heavy quarks were first observed, their properties, and the basic physics associated with each of them.

EXPERIMENT TECHNIQUES LEADING TO HEAVY QUARK CAPABILITIES

Heavy quarks have been studied in a large number of experiments at quite a range of energies, from near threshold to much higher energies. Nevertheless, it will be evident as each heavy quark is discussed that certain experiment techniques have been critical to the success of the physics program. The same techniques appear and reappear. They will also be important for future physics efforts beyond heavy quarks.

The most important techniques for heavy quarks have been (1) efficient event selection, (2) long data-taking runs with lots of beam, (3) large data sets and lots of computing, and (4) the use of solid-state detectors for precision charged-particle tracking. Each of these will be discussed in turn, though it is the combination of all of them together which has really made the difference.

Open, Efficient Event Selection

For charm, the first extensive open charm particle studies were done at e^+e^- colliders running at the ψ(3S), also called the ψ''. Theses excited states of $c\bar{c}$ are produced copiously relative to an underlying continuum of states, and decay dominantly to $D\,\overline{D}$ mesons. Thus, experiments are able to record all the hadronic events in their studies of the D mesons. Final event selections are made later, off-line.

The early attempts to study charm particles at fixed-target experiments were less successful. They attempted to use special geometries for a single decay mode [16] and specialized triggers for particular production mechanisms [17]. Later experiments used looser event selection, and were more effective. Amid the generally-discouraging first efforts at fixed-target experiments, a list of algorithms was examined for event selection. [18] The more successful experiments made their event selections based on very open, efficient, inclusive triggers using total event transverse energy and evidence for decays of particles with lifetimes in the picosecond range. The first of these was effective to the extent that the new particles were heavy as a fraction of the center-of-mass energy, the second to the extent that the new particles had long lifetimes or decayed to particles with long lifetimes.

As was the case for charm, the first extensive studies of bottom particles were most successful at e^+e^- colliders. For bottom particles, the accelerators run at the energy where Υ(4S) particles are produced. Theses excited states of $b\bar{b}$ are produced copiously relative to an underlying continuum of states, and decay dominantly to $B\,\overline{B}$ mesons. Thus, experiments were again able to record all the hadronic events in their studies of the B mesons. The openness of the on-line event selection led to the ability of each experiment to attack a broad range of decays and of physics topics.

Even for the top quark discovery, the on-line trigger is quite efficient for top-quark events. The properties of the top quark are so extreme, that enormous numbers of less interesting events could be discarded at the on-line trigger level.

Large Data Sets and Lots of Computing

For each technique, technical progress has been critical to the extensive progress made in heavy-quark physics. In the case of data acquisition and analysis, the progress has been adopted from the commercial world where the cost per unit of data storage and computation has dropped amazingly over the last two decades. Table 2 lists the growth of charm samples achieved by increasing use of parallelism and computer power per CPU in one series of experiments, those at Fermilab's Tagged

Photon Laboratory. The basic experiment apparatus (a forward, two-magnet spectrometer) did not change much after the addition of silicon microstrip detectors in E691. Nevertheless, the number of reconstructed charm decays used in final physics publications grew exponentially. The physics signals shown in the table improve by a factor of 2,000, not counting the improvement in the signal-to-background ratio. Such numbers may be taken as a rough measure of the physics reach of the experiments.

The improved numbers of observed particles followed very nearly the increases in data set size, which was made affordable by the change from 9-track open-reel 6250 bpi magnetic tapes of E769 and earlier, to the use of 8 mm video tape in E791. A graphic demonstration of this difference is given by comparing the two images in Fig. 1. A fork lift and truck arrived at the E769 experiment each Monday morning to take the weekend's data tapes to the computing center. Compare that with nearly the same amount of data being held in the one arm-load of 8 mm tapes held by one of the E791 physicists. Perhaps even more impressive is the fact that the arm-load of tapes were filled with data in just three hours, not a full weekend. A similar growth of efficiency and cost effectiveness was needed in offline computing. It occurred via the use of "farms" of cheap, parallel, networked CPUs. [19] Table 3 gives the current status of the numbers of equivalent background free signals in some representative charm experiments, calculated from quoted signal sizes and their errors.

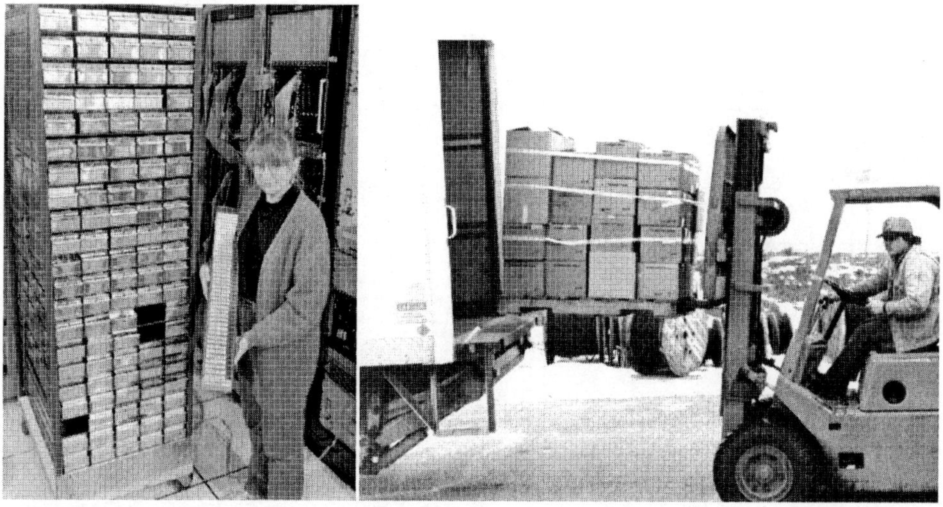

FIGURE 1. A forklift arrives at E769 (right) after a weekend of data taking using 9 track, 6250 bpi, open reel magnetic tapes. E791 physicist Cat James (left) holds an arm-load of 8 mm video tapes next to a storage rack of such tapes. An arm-load of tapes was filled in parallel by 42 tape drives in less than three hours.

TABLE 2. Example of the growth of computing parallelism and power from the series of charm experiments using the Fermilab Tagged Photon Spectrometer.

Time Frame	Exp. #	# Data Streams	# DAQ CPUs	# Output Streams	# Rec'd. Events ($\times 10^6$)	Data Set Size (Gbytes)	# Reconst'ed Charm Decays
1980-2	E516	1	1	1	20	70	100
1984-5	E691	2	1	1	100	400	10,000
1987-8	E769	7	17	3	400	1500	4,000
1990-2	E791	8	54	42	20,000	50,000	200,000

Silicon Microstrip Detectors and Charge-Coupled-Devices

Silicon microstrip detectors [20] (SMDs, shown in Fig. 2) and charge-coupled-devices (CCDs) provide very high precision information about the trajectories of charged particles. From these trajectories, we can obtain the locations where the trajectories overlap, e.g., the primary interaction point (primary vertex). Such a reconstruction from the charm photoproduction experiment E691 is shown in Fig. 3. We see a primary vertex and two secondary vertices where charm particles decay. Given the decay of the two long-lived particles near the primary interaction, the same tracking devices find the decay location as well as the primary vertex. How long lived must be the particles for these observations to be made in the laboratory? The time scale is a picosecond. A particle with such a lifetime will travel 300 microns in the laboratory if its velocity is such that $\beta\gamma$ is equal to 1.0.

TABLE 3. *Numbers of Equivalent Pure Decays Observed by Analysis (scaled to full data sets where needed).*

Physics Topic	Decay Mode	ALEPH	E791	CLEO II.V	FOCUS
Mixing	$D^o_{tag} \to K\pi$	1000	5,400	16,000	
	$D^o_{tag} \to K\pi\pi\pi$		3,300		
Mixing	$D^o_{tag} \to K\mu\nu$		750		7,400
	$D^o_{tag} \to Ke\nu$		760		
$\Delta\Gamma$	$D^o \to KK$		3,150	1,700	
	$D^o \to K\pi$		29,500	30,000	86,000
	$D^o \to K^o_s\phi$			4,100	
CP Vio.	$D^+ \to KK\pi$		1,250		8,100
	$D^+ \to \phi\pi$		800		
	$D^+ \to K^*(890)K^+$		420		
	$D^+ \to \pi\pi\pi$		590		
	$D^+ \to K\pi\pi$		36,000		120,000
CP Vio.	$D^o_{tag} \to KK$		440	2,100	2,200
	$D^o_{tag} \to \pi\pi$		190		
	$D^o_{tag} \to K\pi\pi\pi$		3,000		
	$D^o_{tag} \to K\pi$		10,500	13,500	35,000

This is only a little more than typical longitudinal position resolution values. So, experiments do best when the particles are traveling with higher velocities in the laboratory, and have larger values of $\beta\gamma$ – reaching values of a few tens even. For fixed-target experiments with incident charged-particle beams, one can also include the incident beam track and thin-target locations in fits to find the best estimate of the primary vertex location.

Use of the vertex information in an event provides a double benefit. First, events with evidence of a secondary vertex near the interaction point are highly enriched in heavy quark production. Once a decay vertex is located, the second benefit of precision tracking is evident. When searching for the right combination of observed particles from a single decay, one need only examine the effective mass of those particles coming from that decay vertex, not try all the combinations of all particles in the event. This reduces the backgrounds due to random combinations of tracks by very large factors.

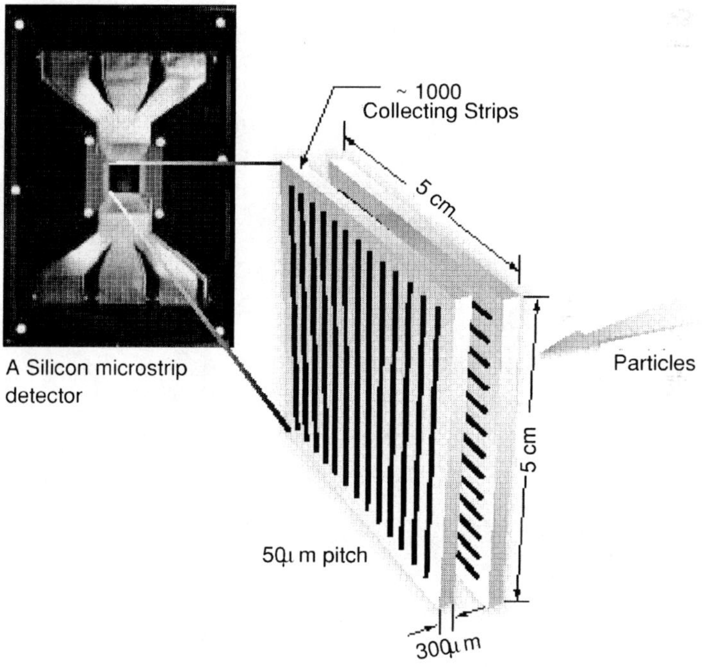

FIGURE 2. A silicon microstrip detector and sketch showing how orthogonal strips are used to map the trajectory of an incident charged particle. The solid-state detectors are fully depleted by application of a reverse bias across their 300 micron thickness, with signals of about 26,000 electrons observed.

OBSERVATIONS AND PHYSICS OF CHARM QUARKS

Observations of Charm

Although one long-lived event in cosmic rays [21] predated the observation of the J/ψ, it was the observation of the J/ψ in both hadronic and e^+e^- interactions that led to the wide acceptance of charm. The hadroproduction resulted in e^+e^- pairs seen in the experiment of Sam Ting and his group at Brookhaven. At SLAC, the group of Burton Richter saw a huge enhancement in the annihilation rate when the center-of-mass energy of the colliding e^+e^- beams was 3.1 GeV. Given that the J/ψ was made of a new quark-antiquark pair, a host of other new particles containing such quarks was expected. Among the new particles were excited states of the quark-antiquark pair. Some of these are listed in Table 4. These so-called charmonium states have no net charm, only "hidden charm." We also expected meson-combinations of charm quarks with lighter quarks and baryon-combinations with three quarks, one or more of which carried charm. The particles with net charm are known as "open charm" particles, and examples are listed in Table 5.

The early developments in charm physics were dominated entirely by experiments at e^+e^- colliders. The fraction of events with charm particles was large at the ψ''. The backgrounds could be well handled for many decay modes, even though it was necessary to examine all combinations of tracks. Once silicon microstrip detectors were introduced into fixed-target experiments, combined with the other features discussed above, the most precise measurements came to be dominated by the Fermilab fixed-target program. Now, we can see the leadership position transferring to the asymmetric e^+e^- collider B factories. This is already evident in

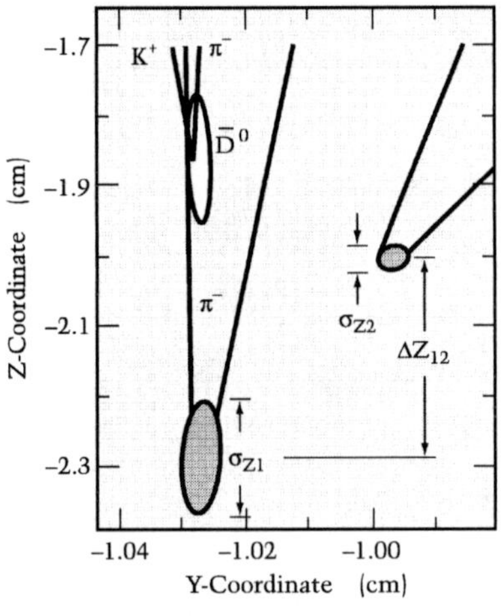

FIGURE 3. The reconstruction of vertices from trajectory information in an example E691 photoproduction event. The ovals represent the uncertainty in the vertex fits. Separation of the primary interaction and the decay points of two charm particles are clearly evident.

the preliminary results just announced at the ICHEP2000 Rochester Meeting held the previous week in Osaka, Japan. Perhaps, eventually, the lead will pass to the forward hadron-collider experiments at Fermilab and CERN – BTeV and LHC-b, respectively. Given the plans evident in the community today, future charm physics will come as a byproduct of the more intensive efforts aimed at understanding the b-quark system.

Techniques in Charm Experiments

The techniques discussed earlier used many charm experiment examples. They will not be repeated here. It is worth noting some additional features, nevertheless. Among these are the so-called D^*-trick, the use of the beam energy in making mass plots of charm candidates from ψ'' data, the use of additional kinematic and topological criteria, the use of high p_t leptons in triggers at fixed-target experiments, and charged particle identification as examples.

One special kinematic feature in charm is the very small kinetic energy available in the decay of the D^*, the first excited state of the D meson. This has been a useful feature, not only to find these D^* mesons in complicated environments like high-energy colliders, but also in selecting events where the nature at birth of the decay D^o is known. Thus, events with a π^+ which comes from a D^{*+} decay means that the accompanying D^o is not a $\overline{D}^o$. One can examine such events for evidence of mixing, that is if the D^o has become a $\overline{D}^o$ before it decays.

At e^+e^- colliders running at the ψ'', backgrounds are reduced and kinematic parameter resolution improves using the beam energy in making mass spectra distributions. The mass calculated this way is called the "beam constrained mass." The requirements used and the calculations are:

$$\Delta E = E_{candidate} - E_{beam} \tag{2}$$

and

$$M_{B\ candidate} = \sqrt{E_{beam}^2 - p_{candidate}^2} \tag{3}$$

In addition to looking for separations of decay vertices from the location of the primary interaction, one can demand for fully charged decay modes that the vector sum of the decay products point back to the primary vertex. This helps reduce backgrounds due to false combinations of reconstructed tracks. The technique even works, though less well, with a missing particle – especially if that particle is light as in the case of a missing neutrino. Of course, one must allow for some mismatch. One can also get to a two-fold-ambiguous momentum of a missing particle if one assumes the mass of that particle as well as the parent particle. The farther from the production point the decay is, the better this reconstruction performs.

TABLE 4. Examples of $c\bar{c}$ hidden-charm particles, their masses, widths, and typical decays.

J^{PC} Assignment	Symbol	Mass (MeV/c^2)	Width (MeV/c^2)	Decay Mode Examples
0^{-+}	$\eta_c(1S)$	2980 ± 2	$13.2^{3.8}_{3.2}$	$\eta\pi\pi$
0^{-+}	$\eta_c(2S)$ η_c'			not observed
1^{--}	$J/\psi(1S)$	3096.87 ± 0.04	0.087 ± 0.005	$2(\pi^+\pi^-)\pi^o$
1^{--}	$\psi(2S)$ ψ'	3685.96 ± 0.09	0.277 ± 0.031	$J/\psi(1S)\pi^+\pi^-$ $J/\psi(1S)\pi^o\pi^o$
1^{--}	$\psi(3S)$ ψ'''	3769.9 ± 2.5	83.9 ± 2.4	$D\overline{D}$
0^{++}	$\chi_{c0}(1P)$	3415.0 ± 0.8	$14.9^{2.6}_{2.3}$	$2(\pi^+\pi^-)$
1^{++}	$\chi_{c1}(1P)$	3510.5 ± 0.1	0.88 ± 0.14	$\gamma J/\psi(1S)$
2^{++}	$\chi_{c2}(1P)$	3556.2 ± 0.1	2.0 ± 0.2	$\gamma J/\psi(1S)$

Spectrum of Charm Mesons and Baryons

The first particles containing charm quarks to be observed were the so-called "onium" states of a charm quark and charm antiquark. These states have no net "charm," and can decay electromagnetically and, for the heavier such states, strongly. Thus, widths of the lower mass states are very narrow, and signals appear clearly above background. On the other hand, such states have unmeasurable decay lengths in the laboratory, making some observations difficult, if not impossible. Some of the observed charmonium states are listed in Table 4. Study of the charmonium states is still an active area. The η_c', for example, has not had a confirmed observation yet, in spite of several attempts. In addition, there are many more excited states of these onium particles.

It is the ground states with open charm that have longer lifetimes, and have been amenable to selection via their observable laboratory flight paths. The ground-state, open, single-charm mesons and baryons are listed in Table 5. No baryons with two charm particles among their three quarks have yet to be observed, though they should certainly appear eventually. Each of the ground-state particles can have a set of excited states, due either to radial or angular-momentum excitations. Many of these have already been observed, and there is a long story to be told here. Let me just note that these states are well described by focusing on the heavy charm quark as defining the coordinates, with an antiquark or diquark system orbiting around it. Spectrum mass-level separations agree with predictions from lattice gauge QCD calculations, and one can now even obtain values for the strong coupling constant from the level splittings in these states.

TABLE 5. Examples of open-charm particles and their decays.

Quark Combination	Symbol	Mass (MeV/c^2)	Lifetime (fs)	Decay Mode Examples
$c\overline{d}$	D^o	1864.5 ± 0.5	413 ± 2.8	$K^-\pi^+, K^-\pi^-\pi^+\pi^+$
$c\overline{u}$	D^+	1869.3 ± 0.5	1051 ± 1.3	$K^-\pi^+\pi^+$
$c\overline{s}$	D_s^+	1968.6 ± 0.6	496^{+10}_{-9}	$K^-K^+\pi^+$
cud	Λ_c	2284.9 ± 0.6	206 ± 12	$pK^-\pi^+$
cuu	Σ_c^{++}	2452.8 ± 0.6		$\Lambda_c^+\pi^+$
cud	Σ_c^+	2453.6 ± 0.9		$\Lambda_c^+\pi^o$
cdd	Σ_c^o	2452.2 ± 0.6		$\Lambda_c^+\pi^-$
csu	Ξ_c^+	2466.3 ± 1.4	330^{+60}_{-40}	$\Lambda K^-\pi^+\pi^+$
csd	Ξ_c^o	2471.8 ± 1.4	98^{+23}_{-15}	$\Xi^-\pi^+, \Omega^-K^+$
css	Ω_c	2704 ± 4	64 ± 20	$\Sigma^+K^-K^-\pi^+$

Physics Issues for Charm Quarks

The physics issues for charm quarks range from searches for clues to new physics to contributions to the understanding and parameters of the Standard Model. The searches for physics beyond the Standard Model include those which would appear as CP violation, oscillations between neutral states, and searches for forbidden and overly copious rare decays. Among the Standard Model parameters to be measured accurately are the CKM matrix elements: $|V_{cs}|$ and $|V_{cd}|$, determination of the strong coupling constant α_s as a test of the flavor independence of QCD, branching ratios and lifetimes to high precision, and a myriad of resonance and other non-perturbative parameters.

Charm, a Unique Window to New Physics

Two features of charm make it unique in the search for new physics: (1) the absence of Standard Model background and (2) the fact that coupling to charm is the only way to see new physics in the up-quark sector. The Standard Model sources of mixing and CP violation for strange and bottom quarks predict significant, even large effects. Yet, for charm, such effects are predicted (so far) to be unmeasurable. Thus, experimental signatures in charm have no SM background, no relevant hadronic uncertainty in background estimates. Any sign of mixing or CP violation in the charm sector is immediate evidence of new physics. [22,23]

As for new physics coupling to the up-quark sector, both the up-quark and the top-quark are prevented from having observable effects in virtually all the standard ways of looking. There is a lack of decay channels for the up quark itself. So, interferences among decay channels are severely constrained. The top quark doesn't live long enough to mix or have the final state interactions needed for CP violation.

Mixing in the Standard Model is a good example of why effects are so small in the charm sector. The Standard Model process is thought to be dominated by contributions of the so-called "box diagram" shown below. Down-type quarks appear in the loop for this mechanism. The amplitude has the form

$$A_{mix}(\text{charm box diagram}) \sim (m_s^2 - m_d^2)/m_W^2 \times (m_s^2 - m_d^2)/m_c^2 \tag{4}$$

where the first term comes from the sum of the two leading terms (GIM suppression), and the second is an off-shell factor. In the limit of SU(3) symmetry, the down and strange quarks have equal masses and the amplitude is zero. Comparing

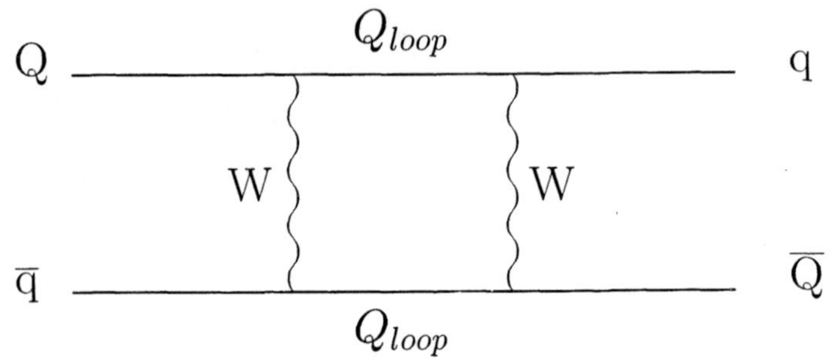

$Q = \text{c} \quad - \quad Q_{loop} = \text{b,s,d}$

$Q = \text{s,b} \quad - \quad Q_{loop} = \text{t,c,u}$

this to the same sort of diagram for mixing of neutral kaons with up-type quarks in the loop shows the orders-of-magnitude difference due to the coupling constants (CKM parameters) and the masses of the quarks involved.

$$A_{mix}(\text{kaon box diagram}) \sim (m_c^2 - m_u^2)/m_W^2 \times (m_s^2 - m_d^2)/m_s^2. \tag{5}$$

Note that for neutral kaon decays to dileptons to be as small as they are, the cancellation of the charm and up quark contributions is required in such loop diagrams. This is what allowed Glashow, Iliopoulos, and Maiani to predict the mass of the charm quark before there was any direct evidence for it. Contributions from these box diagrams, and even those from penguin and long distance effects are of the order of 10^{-7} to 10^{-10}. A rather complete compilation of model-dependent calculations for both Standard-Model and new-physics models is being maintained by Harry Nelson. [24]

We often refer to the gap between Standard Model backgrounds and the current level of experiment limits as an open window. That is, there is an opening for a major discovery. Table 6 lists a number of measurements, the current levels of sensitivity, and the theoretical estimates of the contributions of Standard Model sources to possible signals. There are orders of magnitude available in these open windows. Furthermore, there are many extensions of the Standard Model which might appear in the window. These include fourth-generation quarks, leptoquarks, and various other heavy particles (e.g., Higgs and SUSY particles) which can appear in loops in virtual processes.

Charm Production - Cross Sections

The charm cross section is relatively small for fixed-target photoproduction, where charm is produced in only about a half percent of hadron-producing interactions. The charm cross section is even smaller in fixed-target hadroproduction, where charm is produced only about once per thousand interactions. Once one gets to Tevatron Collider energies, the fractional cross sections rise by an order of magnitude.

The theoretical uncertainties associated with charm hadroproduction predictions are about an order of magnitude at fixed-target energies, even for next-to-leading-order calculations. This is due to the sensitivity to the charm quark mass and the scale dependence of these calculations. In principle, calculations for production at very high transverse momentum might mitigate the scale dependence, but this has yet to be observed. Calculations for photoproduction have somewhat less uncertainty.

The charmonium production is a small fraction of total charm hadroproduction. The comparison with calculations for the J/ψ and ψ' are complicated by the existence of sources like bottom decay and the decay of the higher excited charmonium states. Nevertheless, the levels of these sources have been measured. Once these are subtracted, a mystery remains. At both fixed-target and collider energies, the direct production J/ψ and ψ' mesons are measured to be factors of 7 and 25 larger, respectively, than predicted by the simple color-singlet model. When this is "explained" by including color-octet contributions, the hadron-collider matrix elements are not consistent with the levels observed at the HERA *ep* collider.

Charm Production - Hadronization Effects

The process of produced charm quarks turning into the charm mesons and baryons observed in the laboratory is called hadronization. The process is non-perturbative by its very nature. Nevertheless, some patterns are emerging. For one thing, the longitudinal momentum of observed charm mesons in hadronic interactons looks very much like the predicted distribution calculated for quarks. This appears to come about because of a cancellation of "color drag" and fragmentation. The first of these is an acceleration of the produced quarks in the direction of the incident particles. This is said to be due to the pull of the color strings which attach the charm quark to the forward-going remnants of the incident particle. The fragmentation is the deceleration of the charm quarks as they pick up sea-quarks

TABLE 6. Exmples of the Open Window Sensitivity to Physics Beyond the Standard Model.

Topic	90% CL Limit	SM prediction	Typical Models Tested
CP Violation			
$D^0 \to K^-\pi^+$	$-0.009<a<0.027$ [25]	≈ 0 (CFD)	SUSY,
$D^0 \to K^-\pi^+\pi^-\pi^+$		≈ 0 (CFD)	LR Symm.,
$D^0 \to K^+\pi^-$	$-0.43<a<0.34$ [26]	≈ 0 (DCSD)	Extra Higgs
$D^+ \to K^+\pi^+\pi^-$		≈ 0 (DCSD)	
$D^0 \to K^-K^+$	$-0.026<a<0.028$ [27]		
	$-0.093<a<0.073$ [28]		
	$-0.022<a<0.18$ [25]		
$D^0 \to \pi^+\pi^-$	$-0.002<a<0.094$ [27]		
	$-0.186<a<0.088$ [28]		
$D^+ \to K^-K^+\pi^+$	$-0.006<a<0.018$ [27]		
	$-0.026<a<0.028$ [29]		
$D^+ \to \overline{K}^{*0}K^+$	$-0.092<a<0.072$ [29]	$(2.8\pm0.8) \times 10^{-3}$ [31]	
$D^+ \to \phi\pi^+$	$-0.075<a<0.21$ [30]		
$D^+ \to \pi^+\pi^+\pi^-$	$-0.086<a<0.052$ [29]		
$D^+ \to \eta\pi^+$		$(-1.5\pm0.4) \times 10^{-3}$ [31]	
$D^+ \to K_S\pi^+$		few $\times 10^{-4}$ [32]	
FCNC			
$D^0 \to \mu^+\mu^-$	4×10^{-6} [33,34]	$< 3 \times 10^{-15}$ [35]	4th Gen.,
$D^0 \to \pi^0\mu^+\mu^-$	1.7×10^{-4} [36]		Tree-level
$D^0 \to \overline{K}^0 e^+e^-$	1.1×10^{-4} [37]	$< 2 \times 10^{-15}$ [35]	FCNC
$D^0 \to \overline{K}^0 \mu^+\mu^-$	2.6×10^{-4} [36]	$< 2 \times 10^{-15}$ [35]	
$D^+ \to \pi^+e^+e^-$	6.6×10^{-5} [38]	$< 10^{-8}$ [35]	
$D^+ \to \pi^+\mu^+\mu^-$	1.8×10^{-5} [38]	$< 10^{-8}$ [35]	
$D^+ \to K^+e^+e^-$	2.0×10^{-4} [39]	$< 10^{-15}$ [35]	
$D^+ \to K^+\mu^+\mu^-$	9.7×10^{-5} [39]	$< 10^{-15}$ [35]	
$D \to X_u + \gamma$		$\sim 10^{-5}$ [35]	
$D^0 \to \rho^0\gamma$	2.4×10^{-4} [40]	$(1-5) \times 10^{-6}$ [35]	
$D^0 \to \phi\gamma$	1.9×10^{-4} [40]	$(0.1-3.4) \times 10^{-5}$ [35]	
LF or LN Violation			
$D^0 \to \mu^\pm e^\mp$	8.1×10^{-6} [41]	0	LQ
$D^+ \to \pi^+\mu^\pm e^\mp$	3.4×10^{-5} [41]	0	
$D^+ \to K^+\mu^\pm e^\mp$	6.8×10^{-5} [41]	0	
$D^+ \to \pi^-\mu^+\mu^+$	1.7×10^{-5} [41]	0	
$D^+ \to K^-\mu^+\mu^+$	1.2×10^{-4} [41]	0	
$D^+ \to \rho^-\mu^+\mu^+$	5.6×10^{-4} [36]	0	
Mixing			
$\overset{(-)}{D}{}^0 \to K^\mp\pi^\pm$	$\Delta M_D < 2.8\times10^{-5}$ eV [26]	10^{-7} eV [22,42]	LQ, SUSY,
$\overset{(-)}{D}{}^0 \to K\ell\nu$	$r < 0.005$ [43]		4th Gen., Higgs

to make observable particles. There is some evidence for the attachment of color strings to the charm and valence quarks of projectiles. This is called the "leading particle" effect. In this effect, we see an asymmetry in the forward direction in the numbers of particles with valence quarks in common with the projectile compared to those without. For example, in incident negative-pion-beam experiments, there is a preponderance of D^-'s over D^+'s. For kaon beams, the preponderance is of D_s mesons with the same strange quark as the valence strange quark in the incident kaons.

The details of the production process are also probed by measurements of the correlations of charm and anti-charm particles. One mystery here is why the charm and anti-charm particles do not appear more nearly back-to-back in the plane transverse to the incident particle. Some smearing can come from the intrinsic transverse momentum, k_t, of the partons in the incident particles. However, the amount of k_t required to explain observations in hadroproduction can be as much as 2 to 3 GeV/c. This is rather large, even for a nucleon whose total rest mass is on the order of 1 GeV/c^2.

Charm-Particle Lifetimes

The lifetimes of charm particles would all be the same if they all were the result of a single process, say the color-aligned spectator diagram where the charm quark decays to the Cabibbo-favored W^+ and a strange quark, with the W^+ becoming some final state particle via its virtual decay to $u\bar{d}$. However, there are other possibilities: an annihilation diagram, a W-exchange diagram, and a second (color suppressed) spectator diagram where the quarks from the W decay do not stay together.

Various of the mesons and baryons have differing contributions from each of these diagrams. From the lifetimes listed in Table 5, it is evident that more than a single process must be relevant. Lifetimes of the ground-state charm particles vary by an order of magnitude! A consistent picture of charm decay requires inclusion of all these processes, including coherent interference of diagrams when the final-state particles from two diagrams are the same. [44] There is also evidence of enhancement of the hadronic modes by diagrams with gluons. [45] This gluon participation may account for the fact that the color-suppression in charm decay is not as pronounced as in bottom decays.

Charm Decay - Resonance Dominance and Light Resonance Parameters

Detailed study of the decays of charm mesons and baryons shows a dominance by quasi-two-body modes, that is where the charm particle decays to a low-mass hadron resonance and another particle. The study of such decays is often pursued by examining the distribution of decays in terms of their Dalitz plot, the two-dimensional distribution for three particles in the final state which is uniform for non-resonant s-wave decays, and has various structures evident for other modes. An example is the Dalitz plot for $K\pi\pi$ decay of the D^+ meson shown in Fig. 4. Each dot represents a measured decay; it's location in the plot is a function of the location in the available decay phase-space.

The plot is a beautiful demonstration of basic physics, for example that observations are driven by the square of the more fundamental amplitudes. We see that in the distinct interference pattern between the $K^*(890)$ resonance, observed as a band at the K^*-mass squared, and the more slowly varying other decay contributions. We also see the conservation of angular momentum in this decay of a pseudoscaler particle into a vector particle, the K^*, and a pseudoscalar, the recoiling pion in the decay. This combination of spins, leads to a $(1 + cos(\theta))$ decay angular distribution, which shows up in the Dalitz plot as a concentration of K^* events near the kinematic boundary. Thus, we see that the distributions in the Dalitz plot are characteristic of the resonances involved and their masses, widths, and spins.

As is evident from this one Dalitz plot example, the distribution is sensitive to which resonances contribute to the decay, the mass and width of those resonances, and to the relative phase of the contributions of the decay. Much of the information on low-mass resonances has come from scattering experiments. There, the partial waves which contribute can be quite complex. In the decay of charm particles, the initial state is somewhat better defined, less complex. Thus, charm decay can be used as an independent way of understanding light-resonance physics.

One of the most compelling examples of this is in the three charged pion decays of the D^+ and D_s^+ mesons. In these decays, one observes the contributions from such final states as $\rho^o\pi^+$, $f_o(980)\pi^+$, $f_2(1270)\pi^+$, $f_o(1370)\pi^+$, $\rho^o(1450)\pi^+$, and $\sigma\pi^+$. The decays are dominated by the scalar plus π^+ modes, and the initial quark content of the D's and that of the resonances can be understood to be consistent. This latter point is useful in determining the decay mechanisms at work (e.g., what fraction of the decays come from the annihilation diagram) and the nature of some of the less-well understood resonances such as the $f_o(980)$, the $f_o(1270)$, and the σ. The parameters of these resonances have been determined from the fits to the relevant Dalitz distributions. Although it is not so clear from Fig. 4, a very large part of the slowly varying density of decays comes from the scalar κ resonance. It's mass and width are being determined in a new result from E791. [46]

Study of the Dalitz plots of charm particles can also help in untangling the contributions of the basic quark-level decay diagrams. Various resonances are characteristic of different decay mechanisms for a given charm parent. So far, the contributions of non-spectator diagrams for mesons appears to be somewhat limited. However, baryons have a richer set of possibilities since helicity is no longer a supressing factor. In fact, the variety of baryon lifetimes can be explained by incorporating the variety of possible quark-level diagrams.

Summary of Charm Physics

We have gone very quickly over the enormous range of physics topics where charm quark experiments can contribute. These included (1) new physics searches, (2) Standard Model electroweak parameters, and (3) QCD understanding and parameters in production, spectroscopy, and decay mechanisms.

OBSERVATIONS AND PHYSICS OF BOTTOM QUARKS

As in the case of charm quarks and their physics, I will briefly review the first observations of bottom, the types of experiments which have studied bottom parti-

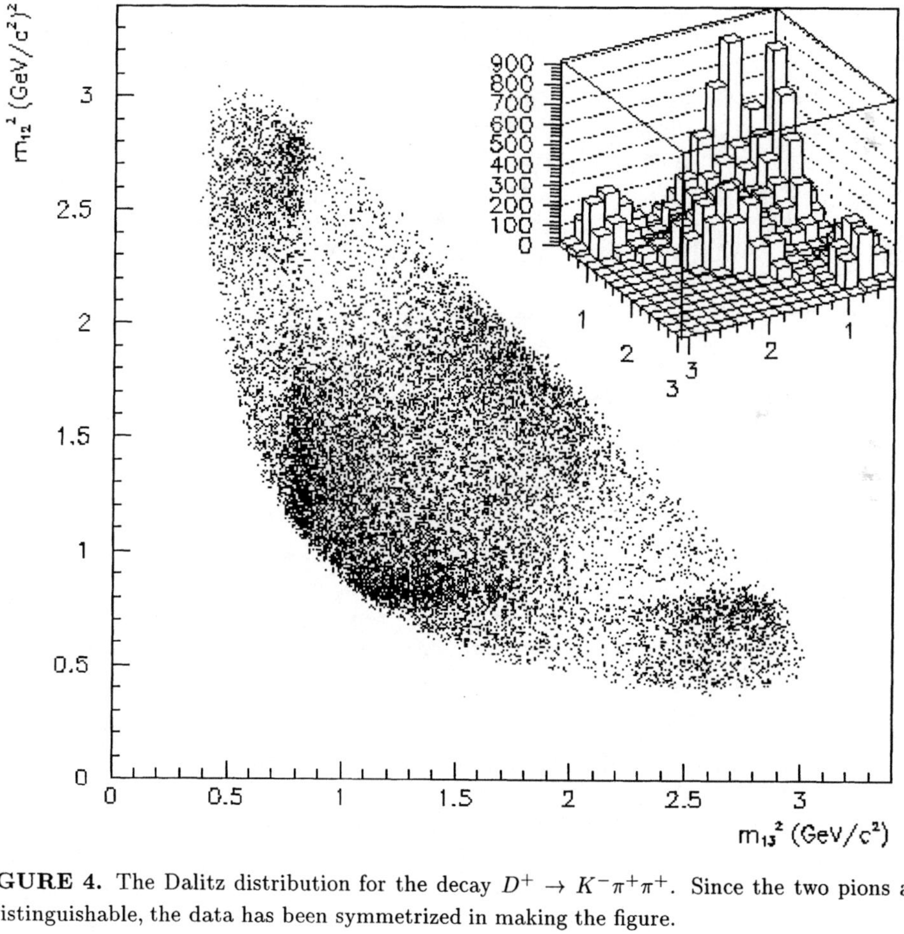

FIGURE 4. The Dalitz distribution for the decay $D^+ \to K^-\pi^+\pi^+$. Since the two pions are indistinguishable, the data has been symmetrized in making the figure.

TABLE 7. Examples of open-bottom particles and their decays.

Quark Combination	Symbol	Mass (MeV/c^2)	Lifetime (fs)	Decay Mode Examples
$\bar{b}d$	B_d or B^o	5279 ± 2	1548 ± 32	$D^-\pi^+\pi^+\pi^-$
$\bar{b}u$	B^+	5279 ± 2	1653 ± 28	$D^o\rho^+$
$\bar{b}s$	B_s	5369 ± 2	1493 ± 62	$D_s^- X$
$\bar{b}c$	B_c	6400 ± 430	460^{+180}_{-160}	$J/\psi\pi^+$
bud	Λ_b	5624 ± 9	1229 ± 80	$J/\psi\Lambda$

cles, the particular techniques used, and finally the range of physics topics of special interest for bottom.

Observations of Bottom Quarks

The first observation of bottom quarks was in a 1977 fixed-target hadroproduction experiment at Fermilab. There, the group headed by Leon Lederman observed $\mu^+\mu^-$ pairs in a complicated peak near a mass of 9.5 GeV/c^2 above a continuum. [10] The experimenters, conscious of the resonances in the J/ψ system, recognized that they might be seeing such a system for a new, third generation quark. The Fermilab results led to an energy upgrade of the DORIS e^+e^- collider at DESY in Hamburg, Germany. There, several experiments were able to offer rather quick confirmation of the discovery. [11] Because of the much better mass resolution based on beam energies at DORIS, the experimenters were able to discern three resonances, just where the Lederman group had said they were.

Spectrum of Bottom Mesons and Baryons

Similarly to the case of charm, but with an even greater potential variety of particles, the bottom system observations are all consistent with the pattern expected for the usual meson and baryon quark combinations. Table 7 gives examples of the better known particles. Many others are anticipated, but have yet to be observed. There are also many states of hidden "beauty," not shown, and of radial and angular-momentum excitations of all the ground states. Even for the observed particles, however, the full demonstration of their expected spin and parity has yet to be accomplished.

Special Techniques Used for Bottom

Particles containing bottom quarks have been observed in a great variety of experiments, from threshold to very high energies – more so than any of the other heavy quarks. At symmetric e^+e^- colliders such as CLEO at Cornell, the machines are operated at the $\Upsilon(4S)$ where the resonance is just above $B\overline{B}$ threshold. Thus, since the center of mass is nearly at rest in the laboratory, the B's are also nearly at rest in the laboratory. In order to produce B's with significant laboratory lifetimes, Pier Oddone and collaborators proposed building an asymmetric e^+e^- collider. [47] At such a machine, since one beam has higher energy than the other, the center of mass is moving in the laboratory. Thus, the Υ(4S) and its decay B's are moving in the laboratory. In this environment, the observation of isolated secondary vertices is again a most useful tool. Two major facilities of this sort have just turned on successfully, and the experiment at each facility has reported its first physics results at the ICHEP 2000 meeting in Osaka, Japan. The experiment is called BaBar at the PEP-II machine at SLAC, and BELLE at the KEK-B machine in Japan.

At fixed-target experiments, where the b was first discovered, the available energy is not as well controlled as at e^+e^- colliders, but the typical energies also produce b's nearly at rest in the center of mass. However, because of the beam momentum, the B's travel with several tens of GeV/c in the laboratory. Several experiments have observed bottom particles, especially the Υ and B-decays to J/ψ's. E288, E605, E653, E771, and E789 at Fermilab are described briefly in a commemorative book available on the web [48]. There are also HERA-B at DESY and BEATRICE (WA92) at the CERN SPS. Only HERA-B among all these is still in the data taking mode.

Much higher energy B's and b-baryons are seen at high-energy hadron collider experiments – e.g., CDF, and DZero already, and at LHCb, BTeV in the future. Here the bottom particles have significant lifetimes in the laboratory, making their observation easier – except that the b quarks often turn into full-scale jets of many particles, not just the pristine B's seen at lower energies. Symmetric e^+e^- colliders running at the Z^o mass and above have similar access to b physics, again with b's dominantly in jets of particles. We have seen interesting results from the SLD experiment at SLAC and four LEP experiments at CERN: ALEPH, DELPHI, L3, and OPAL.

Across this range of experiments, there are a few special techniques which come into play. At the e^+e^- colliders running at the $\Upsilon(4S)$, backgrounds are reduced and kinematic parameter resolution improves using the "beam constrained mass" – as was the case for charm from the ψ" experiments at e^+e^- colliders. In addition, at fixed-target experiments and within jets at higher energies, leptons with high p_t play a role, and visible lifetimes for open bottom mesons and baryons are again crucial.

In looking for mixing of neutral B mesons in hadronic interactions, it is again necessary to know the nature of the B at birth, as well as its nature when it decays. In the case of the e^+e^- colliders, the difference in the nature of the two B's as they decay is needed. This sort of knowledge is obtained by observing one B as it decays, and using information either about the other B in the event or about the produciton of the first B. Using partial information about the other B is called "tagging." Tagging usually involves incomplete knowledge, in order to have as many tagged events as possible. Thus, experimenters examine the net charge, weighted by momentum, or observed decay leptons from the other B or

b-jet. For information about the fully reconstructed B, one can use information on its production angle when there is a strong asymmetry, as when highly polarized electron or positron beams are used at e^+e^- colliders operating at the Z^o with its weak decay asymmetry – so far only at the SLC facility at SLAC. At hadron colliders one can also examine other particles produced in the jet with the B. The most information is available when examining the single other particle nearest in phase space to the B. Detailed information about such production correlations is then needed to fully understand the tagging rate and backgrounds.

Physics Issues for Bottom Quarks

The bottom quark is part of a standard (left-handed) quark isospin doublet (b, t), although there were many early attempts of think of it as a singlet. The dominant b decay is to c quarks, via emission of a virtual W, a 3rd to 2nd generation transition. This coupling is much less strong than the s to u transition, a 2nd to 1st generation transition. The coupling of the b to the u quark, 3rd to 1st generation, is even weaker. The GIM mechanism breaks down since the t quark is so massive. The massiveness of the t quark also leads to very large $B^o\overline{B}^o$ mixing. This was observed very early via a surprisingly large same-sign di-lepton signal, [49] though it was not so widely accepted at first. It was the first indication that the top quark was so extraordinarily heavy.

Today, there are extensive efforts to measure the parameters of the B meson system. While we refer to this as being done to test the Standard Model, most hope that we will find evidence for physics beyond the Standard Model – that is, evidence that there is no single set of parameters within the Standard Model that can explain all observations. Thus, it is important to observe each parameter in more than a single way. This is often called overdetermining the CKM matrix. Experiments are focusing on:

CP violation

Oscillations in neutral states, both B_d and B_s

Rare and nominally-forbidden decays

There are still issues in the details of decay dynamics, since there are discrepancies between expectations and some lifetime measurements. An additional wrinkle in the picture is the relationship between the semileptonic decay rate and the number of charm particles observed in b decay. The semileptonic decay rate is 1 to 2% below the rate (more like 12%) expected from theory using standard CKM matrix elements. Considering $b \to W^-c$ and W^-u, with $W^- \to \overline{u}d$, $\overline{c}s$, $e^-\overline{\nu}_e$, $\mu^-\overline{\nu}_\mu$, and $\tau^-\overline{\nu}_\tau$, the semileptonic branching ratio should be $\sim 1/9$, given 3 colors for the quarks. Phase space for c and t quarks and final state interactions for $W \to$ quarks modify this somewhat.

In addition to the electroweak theory which is the main part of the Standard Model being tested above, there are also issues for QCD. The flavor independence of the strong coupling constant has been established by looking at the rate of gluon emission in e^+e^- interactions involving the light, the charm, and the bottom quark. One might expect that the bottom quark is heavy enough that NLO perturbative QCD would suffice to explain b quark production. However, again there are mysteries to be unraveled here. The importance of the hadronization process for tagging has been mentioned already.

Bottom Production - Cross Sections

Given that the b mass is much larger than Λ_{QCD}, the expectation is that NLO perturbation theory will reliably predict $\sigma_{b\bar{b}}$. At the Tevatron Collider, where the high transverse momentum of the observed B's should also help, the measurements give $\sigma_{b\bar{b}}$ too large by a factor at least two. [50,51]

The CDF and DZero measurements at the Tevatron are dominantly in the central region of phase space. This will also be true for the LHC experiments ATLAS and CMS. However there are two new experiments which will explore the forward regions, BTeV at the Tevatron and LHC-b at the LHC. These latter experiments will take advantage of the higher momentum and longer laboratory lifetimes of bottom particles in their kinematic region. See the left side drawing of Fig. 5. Also shown in Fig. 5 is the strong correlation in produciton of the b and $\bar{b}$ in the forward and backward directions predicted by the PYTHIA simulation model. This appears as the strong peaking in the number of events shown on the figure on the right side of Fig. 5. BTeV and LHC-b should be able to take advantage of this correlation in tagging the B's which they will observe.

Bottom Production - Hadronization Effects

Bottom quarks turn into B mesons which decay to charm. The charm distributions depend on the fragmentation (hadronization) functions for the b quarks, the harder the B meson, the harder the charm. Examples are the

Andersson function:

$$\phi(z) \sim z^{-1}(1-z)^a e^{-bm_t^2/z} \tag{6}$$

and the Peterson function:

$$\phi(z) \sim 1/[z[(1-(1/z)-\epsilon_P/(1-z)]^2] \tag{7}$$

The CLEO Collaboration has looked at D_s decays of B mesons as a particularly clean way to determine the best parameters and relative merits of these two

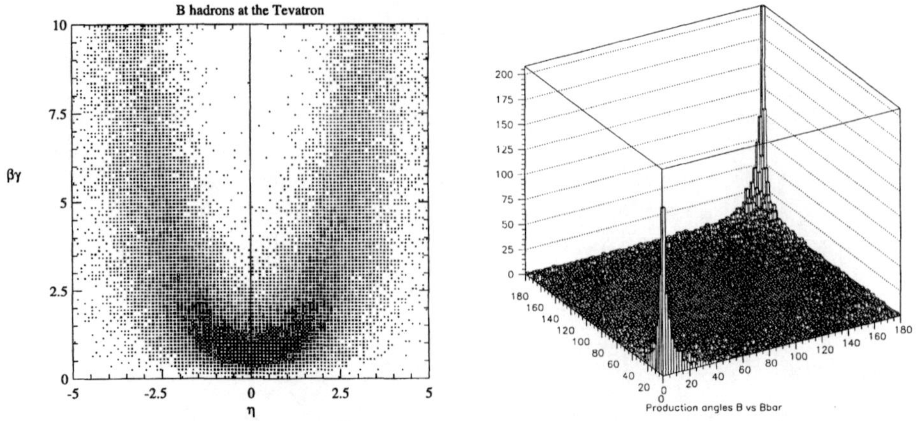

FIGURE 5. Distributions of b and $\bar{b}$ produced in interactions at the Tevatron Collider. On the left is the time dilation factor $\gamma\beta$ vs the pseudorapidity, on the right is the angular correlation of b and $\bar{b}$ quarks.

functions. [52] These distributions are important in understanding experiment apparatus acceptance and in determining backgrounds for top quark and other heavy objects. Yet, we may ask whether the fragmentation functions from e^+e^- are relevant in hadronic environments? [53] If they may be so at high p_t where approximate isolation may be achieved, how low can one go in p_t before the color-field environment is so different that the e^+e^- measurements are not applicable.

Bottom-Particle Lifetimes

The original expectations for bottom particles was that their lifetimes would be short, even with a new quantum number to conserve, because of the large phase space and multitude of decay modes available. However, it turned out that the CKM-matrix parameters were small, thus leading to long lives. As with charm, the expectation was for equality of lifetimes – all ~ 1.5 picosecond. However, as seen in Table 7, there are significant differences. The shortness of the Λ_b lifetime is particularly difficult to explain.

Neutral Bottom-Particle Mixing

Both B_d and B_s particles are expected to have observable mixing. However, the rate of B_s oscillation will be very much more rapid, and correspondingly harder to measure. B_d mixing has been observed by several experiments, in a number of ways – even directly. No B_s mixing has yet been seen. This mixing is expected to lead to CP violation due to the phase in the CKM matrix (i.e., a non-zero value of ρ in Eq. 1).

Additional Bottom Decay Issues

The very hot topic in bottom particle decays is CP violation in B meson decay. The idea is to test the Standard Model prediction for the violation – in the hopes of finding a disagreement which would provide a clue to physics beyond the Standard Model, just as that is the hope in studies of charm decay. We anticipate that very heavy particles may make virtual contributions to the box diagrams in a measureable way. Non-Standard Model "penguin" decay diagrams also are open to virtual particles; e.g., heavy W and charged Higgs in place of the traditional W in loops. See the lectures at this workshop by Gabriel Lopez, "Violación de CP en Mesones B," for more on this subject.

Among the things which complicate experimental work in bottom decays is the large number of decay channels available for each bottom particle. This leads to very small BR's (typically very much less than 1 %) and the difficulty of collecting large samples of a given decay mode. Yet, this is exactly what is needed for high precision measurements and searches; e.g., searches for small CP asymmetries. It is usually the case that the largest such asymmetries are anticipated for the smaller branching-ratio decay modes. Furthermore, only a fraction of the decay channels have actually been observed.

OBSERVATIONS AND PHYSICS OF TOP QUARKS

Again, for the heaviest of quarks, the top quark, I will review its first observation, the techniques used, the measurements made, and make a rapid tour through the physics issues related directly to top quarks.

First Observations of Top Quarks

The discovery of top quarks required seeing interactions with the characteristics expected, and at a rate greater than that which could be expected from other processes. The signal for top quarks so far does not stand out quite as clearly as the signals for charm and bottom quarks did. Given their very high mass, top quarks could only be discovered in proton-antiproton collisions at the 1.8 TeV center-of-mass energy available at Fermilab's Tevatron. Furthermore, the signals appear near the end of the spectra where the rates are low. The cross section turns out to be about $5pb^{-}1$. This corresponds to one such event produced in about 2×10^{10} events.

Observation of the top quarks has depended on the high mass of the top quark, and the high mass and p_t of its decay products. Fortunately, the decay is completely dominated by a single mode:

$$t \to Wb$$

Three topologies result from $t\bar{t}$ production and decay: 6 jets, 4 jets plus a lepton, and 2 jets plus 2 leptons:

$$t\bar{t} \rightarrow [W^-\bar{b}]\ [W^+b] \rightarrow [(q\bar{q}')\ b]\ \ [q''\bar{q}''')\ \bar{b}]$$

$$t\bar{t} \rightarrow [W^-\bar{b}]\ [W^+b] \rightarrow [(\ell\bar{\nu}_\ell)\ b]\ \ [(q\bar{q}')\ b]$$

$$\text{and } [(q\bar{q}')\ \bar{b}]\ \ [(\bar{\ell}\nu_{\bar{\ell}})\ \bar{b}]$$

$$t\bar{t} \rightarrow [W^-\bar{b}]\ [W^+b] \rightarrow [(\ell\bar{\nu}_\ell)\ b]\ \ [(\bar{\ell}'\nu_{\bar{\ell}'})\ \bar{b}]$$

In order to optimize the observation above background, the experiments which made the observations, CDF and DZero, focused on central kinematic region. Most other processes are peaked forward-backward; and the high mass decay daughters benefit from the Jacobian enhancement perpendicular to the incident particles. Only events with very high p_t jets and leptons; and missing, high transverse energy E_t from neutrinos were examined. Furthermore, b-quark jet tagging via leptons and secondary vertices was very helpful. The largest backgrounds came from W plus low-mass q jets. So, the techniques used for bottom quark physics are directly relevant here for top quarks as well. Finally, kinematic constraints on the t-quarks and W^+ and W^- candidates were used. Though the t quark mass was unknown, both t's must have same mass; and W mass was known.

Top Quark Production Cross Section

Table 8 gives the observed cross section results from both CDF and DZero. The events listed in the table come from approximately 100 pb^{-1} of data, meaning that about 500 $t\bar{t}$ events were produced in each experiment. Thus, CDF and DZero saw about 10 % of the top events produced, even after their final selection criteria were applied. That is quite efficient for such rare events. For those signatures with the most observed events, the backgrounds are the greatest. Nevertheless, the various data sets are consistent with each other, and consistently above the expected backgrounds.

Top Quark Mass Measurement

As suggested in the discussion of the cross section, kinematic fits to the events can leave the value of the t-quark mass free, only demanding that both the t and $\bar{t}$ have the same mass. Table 9 shows the results of the fits of essentially the same events as were used for the cross section measurements. Again, agreement of the various sets and between the two experiments gives confidence in the basic result.

TABLE 8. Top Quark Production Cross Section Measurements.

$\sigma_{t\bar{t}}$ (pb)	Source	Ref.	Method	Total Events	Background Events
4.1 ± 2.0	DZero	[54]	lepton plus jets	19	8.7 ± 1.7
8.2 ± 3.5	DZero	[54]	lepton plus jets/μ	11	2.4 ± 0.5
6.3 ± 3.3	DZero	[54]	dileptons and $e\nu$	9	2.6 ± 0.6
5.5 ± 1.8	DZero	[54]	DZero combined	39	13.7 ± 2.2
$6.7^{+2.0}_{-1.7}$	CDF	[55]	lepton plus jets	34	9.2 ± 1.5
				40	22.6 ± 2.8
$8.2^{+4.4}_{-3.4}$	CDF	[56]	dileptons	9	2.4 ± 0.5
$10.1^{+4.5}_{-3.6}$	CDF	[57]	all jets , >0 tags	187	142 ± 12
			all jets , >1 tag	157	120 ± 18
$7.6^{+1.8}_{-1.5}$	CDF	[55]	CDF combined		
$4.7 - 5.8$	Theory	[58]	for m_t 173-175 GeV/c^2		

TABLE 9. Top Quark Mass Measurements.

m_t (GeV/c^2) (pb)	Source	Ref.	Method	Total Events	Background Events
$173.3 \pm 5.6 \pm 5.5$	DZero	[59]	lepton plus jets	76	53.2^{+11}_{-9}
					48.2^{+11}_{-9}
$168.4 \pm 12.3 \pm 3.6$	DZero	[60]	dileptons	6	
$172.1 \pm 5.2 \pm 4.9$	DZero	[59]	DZero combined		
$175.9 \pm 4.8 \pm 4.9$	CDF	[61]	lepton plus jets	76	31 ± 8
$161 \pm 17 \pm 10$	CDF	[62]	dileptons	9	2.4 ± 0.5
$186 \pm 10 \pm 12$	CDF	[63]	all jets , >0 tags	187	142 ± 12
			all jets, > 1 tag	157	120 ± 18
$173.8 \pm 3.5 \pm 3.9$	PDG	[64]	PDG Average		

Relation Between Mass and Cross Section and Other Tests

Given that the top quark mass is above the Wb threshold, the decay width should be about 1 GeV, corresponding to a lifetime of about 10^{-23} seconds. Thus, the t decays before top-hadrons or onia form. This is why the decay is totally dominated by the W b decay mode. The production cross section is calculated in NLO perturbative QCD. The perturbative calculation is expected to be very reliable. It does depend on the mass of the top quark, of course. Any inconsistency of the production rate and measured mass would cast doubt on the discovery. No discrepancy is observed, as seen in Table 8.

On the other hand, discrepancies might indicate exotic production channels, as could an unexpected distribution of the measured $t\bar{t}$ mass distribution and the distribution of the transverse momentum of the t. Angular distributions are sensitive to the presumed V-A coupling and the relative coupling of transverse and longitudinal W's to t. The fraction of decays to transversely and longitudinally polarized W's for 175 GeV/c^2 t is 30 %. A discrepancy would challenge the Higgs mechanism of spontaneous symmetry breaking. Finally, one can also imagine seeing clues to new physics via rare or forbidden decay rates. Finally, the top could be the source of observation of non-Standard Model particles. For example, if the Higgs particle were light enough, the top quark could decay emitting a Higgs particle. One could also be sensitive to decays to light enough SUSY particles.

THE NEXT HEAVY PARTICLES, THEIR OBSERVATION AND NATURE

In the search for heavier objects, heavy quarks will play a critical role as tags for the discovery of those heavier objects, and as a key to identifying what they are. We have seen how this was the case in the top discovery, where heavy quarks in jets separated top events from others among the highest p_t interactions.

Predictions for the Higgs Particle

The Higgs mechanism for electroweak symmetry breaking manifests itself in a particle whose coupling to another particle is proportional to the other particle's mass. Thus, Standard Model predictions of the Higgs mass can be made via virtual loop corrections for mass of the top quark. Note that the dependence of the W mass on the top mass is linear, while the dependence is logarithmic on the Higgs mass. Today, there is a lot of focus on this prediction, as the competition between Fermilab and CERN to discover the Higgs is very intense. The Higgs branching ratio to other particles depends, also, on the Higgs mass – due to threshold and phase space effects. So, identifying a signal at a given mass as a Higgs particle depends on measuring the relative branching ratios to as many of the heavy quarks, WW, WW^*, ZZ, and ZZ^* states as are kinematically allowed. The asterisks in decays to WW^* and ZZ^* refer to virtual, intermediate decay particles, W^* and Z^* whose decay can be to kinematically allowed final states. Many physicists expect that direct observation of a Higgs particle will be the next breakthrough discovery. On the other hand, surprises can happen.

Only Three Generations of Quarks?

Are the three generations of quark pairs truly the complete story? The limits on additional light neutrinos from Z^o decay do not necessarily prove this, since there could be a very heavy neutrino for a fourth generation. There are some limits on this from virtual effects in rare decays of heavy quarks.

Characterizing the Squarks and Higgs of SUSY

There is not enough time in these lectures to give an overview of SUSY. The topic has been well covered in the lectures of Cupatitzio Ramírez and Carlos Wagner. Let me only note that the host of predicted SUSY particles will place a premium on the detailed characterization of any candidate observations, as well as filling out the mass-spectrum of particles discovered. Thus, the heavy squark decays to the corresponding quark, as well as SUSY Higgs couplings are critical to the understanding of what will have been seen.

CONCLUDING COMMENTS

I would like to recommend some reading for more detail than I could give here, in particular: a review article by Harry Lipkin [65] for the state of thinking before the November Revolution of 1974, Ed Thorndike's summary of bottom quark physics [66], and the PDG summary by Michelangelo Mangano and Thomas Trippe [67] on top quarks.

We have seen that heavy quarks have been both interesting in their own right, and useful tools for understanding other things – from light-meson resonances to the high-mass extensions of the current Standard Model. I want to note again the importance played by applying new technology matched to physics opportunities. The precision tracking made possible by microelectronic advances would have been less interesting except for the few mm lifetimes of the charm and bottom quarks in the laboratory. The fantastic increase in computing and storage efficiency made possible the treatment of large amounts of data as required by the low cross sections for heavy quark production. Finally, we should also be a little humble amid all the excitement of the anticipated discoveries of the next heavy objects. In my brief review, we have seen many of the results fly in the face of "common knowledge" and general expectations.

Let me close by reiterating the wealth of rather recent progress in understanding the quark nature of matter, and the enjoyment that has come from being part of this, both from the physics and from technical efforts. I also want to thank our hosts and organizers for a most stimulating and enjoyable workshop.

REFERENCES

1. M. Gell-Mann, "Symmetries of Baryons and Mesons," Phys. Rev. **125**, 1067 (1962).
2. J.D. Bjorken, "Partons," *Proc. of the Int. Conf. on Duality and Symmetry in Hadron Physics*, Tel Aviv, Israel, 1971, pp. 98-115.
3. M.Y. Han and Y. Nambu, "Three Triplet Model with Double SU(3) Symmetry," Phys. Rev. B **139**, 1006 (1965); Y. Nambu and M.Y. Han, "Three Triplets, Paraquarks, and Colored Quarks," Phys. Rev. D **10**, 674 (1974).

4. M. Gell-Mann, "A Schematic Model of Baryons and Mesons," Phys. Lett. **8**, 214 (1964).
5. G.F. Chew, "Bootstrap Theory of Quarks," Nucl. Phys. B **151**, 237 (1979).
6. V.E. Barnes *et al.*, "Observation of a Hyperon with Strangeness -3," Phys. Rev. Lett. **12**, 204 (1964).
7. J.J. Aubert *et al.*, Phys. Rev. Lett. **33**, 1404 (1974).
8. J.-E. Augustin *et al.* Phys. Rev. Lett. **33**, 1406 (1974).
9. Mark I Collaboration, G. Goldhaber *et al.*, Phys. Rev. Lett. **37**, 255 (1976); I. Peruzzi *et al.*, Phys. Rev. Lett. **37**, 569 (1976).
10. E288 Collaboration, S.W. Herb *et al.* Phys. Rev. Lett. **39**, 252 (1977); W.R. Innes *et al.*, Phys. Rev. Lett. **39**, 1240 (1977).
11. PLUTO Collaboration, C. Berger *et al.*, Phys. Lett. B **76**, 243 (1978); DASP-2 Collaboration, C.W. Darden *et al.*, Phys. Lett. B **76**, 246 (1978) and Phys. Lett. B **78** 364 (1978); DESY/Hamburg/Heidelburg/MPI-Munich Collaboration, J.K. Bienlein *et al.*, Phys. Lett. B **78**, 360 (1978).
12. M. Kobayashi and T. Maskawa, "Renormalizable Theory of Weak Interaction," Prog. Theor. Phys. **49**, 652 (1973).
13. L. Wolfenstein, "Parametrization of the Kobayashi-Maskawa Matrix," Phys. Rev. Lett. **51**, 1945 (1983).
14. S.L. Glashow, J. Iliopoulos, and L. Maiani, "Weak Interactions with Lepton-Hadron Symmetry." Phys. Rev. D **2**, 1285 (1970).
15. J.D. Bjorken and S.L. Glashow, "Elementary Particles and SU(4)," Phys. Lett. **11**, 255 (1964).
16. V.L. Fitch *et al.*, "Search for D^* Production in Pion Nucleon Interactions," Phys. Rev. D **33**, 1486 (1986).
17. J. Martin *et al.*, "Use of the ECL-CAMAC Trigger Processor System for Recoil Missing Mass Triggers at the Tagged Photon Spectrometer at Fermilab," in *Topical Conference on the Application of Microprocessors to High-Energy Physics Experiments*, Geneva, Switzerland, May 4-6, 1981, CERN, Microproc. 1981, pp 164ff.
18. J.A. Appel, "Triggering for Charm, Beauty, and Truth," in *The Search for Charm, Beauty, and Truth at High Energies*, November 15-22, 1981, Erice, Italy, G. Bellini and S.C.C. Ting, eds., Plenum Press, New York, USA (1984), pp 555-560.
19. T. Nash *et al.*, "High Performance Parallel Computers for Science: New Developments at the Fermilab Advanced Computer Program," *Workshop on Computational Atomic and Nuclear Physics at One Gigaflop*, Oak Ridge, Tenn., Apr 14-16, 1988, Brazil Exper. Symp. 1987, pp 151; J. Biel *et al.*, Proc. Int. Conf. Computing in High Energy Physics, Asilomar, Feb. 2-6, 1987, Computer Physics Communications **45**, 331 (1987); C. Stoughton and D.J. Summers, Computers in Physics **6**, 371 (1992); F. Rinaldo and S. Wolbers, Computers in Physics **7**, 184 (1993); S. Bracker *et al.*, IEEE Trans. on Nucl. Sci. **43**, 2457 (1996).
20. The earliest successful use of SMDs in high energy physics experiments was by the NA11 collaboration at CERN, who did so much to develop these devices. B. Hyams *et al.*, Nucl. Inst. and Meth. **205**, 99 (1983). J. Kemmer, Nucl. Inst. and Meth. **169**, 499 (1980).
21. K. Niu, E. Mikumo, and Y. Maeda, "A Possible Decay in Flight of a New Type

Particle," Prog. Theor. Phys. **46**, 1644 (1971).
22. G. Burdman, "Potential for Discoveries in Charm Meson Physics," in *Workshop on Tau Charm Factory, Argonne, IL, Jun 21-23, 1995*, J. Repond *ed.*, Argonne National Laboratory, June 21–23, 1995, AIP Conference Proceedings No. 349, pp 409-424, FERMILAB-Conf-95-281, hep-ph/9508349.
23. Ted Liu, hep-ph/9508415.
24. H. Nelson, hep-ex/9908021 for compilation, and for new physics.
25. CLEO Collaboration, J. Bartelt *et al.*, Phys. Rev. D **52**, 4860, (1995).
26. CLEO Collaboration, R. Godang *et al.*, Phys. Rev. Lett. **84**, 5038 (2000).
27. FOCUS Collaboration, J.M. Link *et al.*, Submitted to Phys.Lett.B, hep-ex/0005037, FERMILAB-PUB-00-112-E.
28. E791 Collaboration, E.M. Aitala *et al.*, Phys. Lett. B **421**, 405 (1998).
29. E791 Collaboration, E.M. Aitala *et al.*, Phys. Lett. B **403**, 377 (1997).
30. E687 Collaboration, P.L. Frabetti *et al.*, Phys. Rev. D **50**, R2953 (1994).
31. F. Buccella *et al.*, Phys. Lett. B **302**, 319 (1993);
A. Pugliese and P. Santorelli, "Two Body Decays of D Mesons and CP Violating Asymmetries in Charged D Meson Decays," *Proc. Third Workshop on the Tau/Charm Factory*, Marbella, Spain, 1–6, June 1993, Edition Frontieres (1994), p. 387.
32. Z. Xing, Phys. Lett. B **353**, 313 (1995).
33. E771 Collaboration, T. Alexopoulos *et al*, Phys. Rev. Lett. **77**, 2380 (1996).
34. WA92 Collaboration, M. Adamovich *et al.*, Phys. Lett. B **408**, 469 (1997).
35. J.L. Hewett, "Searching for New Physics with Charm," SLAC-PUB-95-6821, hep-ph/9505246, to appear in *Proc. LISHEP95 Workshop*, Rio de Janeiro, Brazil, Feb. 20–22, 1995.
36. E653 Collaboration, K. Kodama *et al.*, Phys. Lett. B **345**, 85 (1995).
37. CLEO Collaboration, A. Freyberger *et al.*, Phys. Rev. Lett. **76**, 3065 (1996), Erratum-ibid. **77**, 2147 (1996).
38. E791 Collaboration, E.M. Aitala *et al.*, Phys. Rev. Lett. **76**, 364 (1996).
39. E687 Collaboration, P.L. Frabetti *et al.*, Phys. Lett. B **398**, 239 (1997).
40. CLEO Collaboration, D.M. Asner *et al.*, Phys. Rev. D **58**, 2001 (1998).
41. E791 Collaboration, E.M. Aitala *et al.*, Phys. Lett. B **462**, 401 (1999).
42. G. Burdman, "Charm Mixing and CP Violation in the Standard Model," in **The Future of High-Sensitivity Charm Experiments**, *Proc. CHARM2000 Workshop*, Fermilab, June 7–9, 1994, D. Kaplan and S. Kwan *eds.*, FERMILAB-Conf-94/190, p. 75, hep-ph/9407378.
43. E791 Collaboration, E.M. Aitala *et al.*, Phys. Rev. Lett. **77**, 2384 (1996).
44. Harry W.K. Cheung, "Review of Charm Lifetimes," in 8th *International Symposium on Heavy Flavor Physics (Heavy Flavors 8)*, Southampton, England, 25-29 Jul 1999, hep-ex/9912021.
45. I. Bigi, M. Shifman, N.G. Uraltsev, and A. Vainshtein, "Nonperturbative Corrections to Inclusive Beauty and Charm Decays: QCD versus Phenomenological Models," Phys. Lett. B **293**, 430 (1992); Erratum-ibid. **297**, 477 (1993); I. Bigi and N.G. Uraltsev, Z. fur Phys. C **62**, 623 (1994).
46. E791 Collaboration, E.M. Aitala *et al.*, Accepted for publication in Phys. Rev. Lett.

hep-ex/0007027 and hep-ex/0007028.
47. A. Garren *et al.*, "An Asymmetric B Meson Factory at PEP," *Proc. Part. Accel. Conf.*, Chicago, Ill., Mar 20-23, 1989. IEEE Part. Accel. Conf. 1989, pp 1847-1849.
48. J.A. Appel *et al.*, "In Celebration of the Fixed Target Program with the Tevatron," http://conferences.fnal.gov/tevft/book/; hep-ex/0008076 for short version.
49. UA1 Collaboration, C. Albajar *et al.*, Phys. Lett. B **186**, 247 (1987).
50. CDF Collaboration, F. Abe *et al.*, Phys. Rev. Lett. **75**, 1451 (1995); **71**, 2537 (1993); **71**, 2396 (1993); **71**, 500 (1993).
51. D0 Collaboration, B. Abbott *et al.*, hep-ex/0008021; Phys. Rev. Lett. **84**, 5478 (2000); Phys. Lett. B **487**, 264 (2000).
52. CLEO Collaboration, R.A. Briere *et al.*, hep-ex/000428.
53. CDF Collaboration, F. Abe *et al.*, Phys. Rev. D **60**, 092005 (1999).
54. DZero Collaboration, S. Abachi *et al.*, Phys. Rev. Lett. **79**, 1203 (1997).
55. CDF Collaboration, F. Abe *et al.*, Phys. Rev. Lett. **80**, 2773 (1998).
56. CDF Collaboration, F. Abe *et al.*, Phys. Rev. Lett. **80**, 2779 (1998).
57. CDF Collaboration, F. Abe *et al.*, Phys. Rev. Lett. **79**, 1992 (1997).
58. P. Nason, S. Dawson, and R.K. Ellis, Nucl. Phys. B **303**, 607 (1988); W. Beenakker, H. Juijf, W.L. van Neerven, and J. Smith, Phys. Rev. D **40**, 54 (1989); E. Berger and H. Contopanagos, Phys. Lett. B **361**, 115 (1995); E. Laenen, J. Smith, and W. van Neerven, Phys. Lett. B **321**, 254 (1994); S. Catani, M. Mangano, P. Nason, and L. Trentadue, Phys. Lett. B **378**, 329 (1996).
59. DZero Collaboration, S. Abachi *et al.*, Phys. Rev. Lett. **79**, 1197 (1997).
60. DZero Collaboration, B. Abbott *et al.*, Phys. Rev. Lett. **80**, 2063 (1998).
61. CDF Collaboration, F. Abe *et al*, Phys. Rev. Lett. **80**, 2779 (1998).
62. CDF Collaboration, F. Abe *et al*, Phys. Rev. Lett. **79**, 1992 (1997).
63. CDF Collaboration, F. Abe *et al*, Phys. Rev. Lett. **80**, 2767 (1998).
64. Particle Data Group's "Review of Particle Properties," Eur. Phys. J. C **3**, 345 (1998).
65. H.J. Lipkin, "Quarks for Pedestrians," Phys. Rept. **8**, 173 (1973).
66. E.H. Thorndike, "Bottom Quark Physics: Past, Present, Future" in "Symposium on Probing Luminous and Dark Matter, honoring Adrian Melissinos," Rochester, New York, USA, October, 1999, hep-ex/0003027.
67. M. Mangano and T. Trippe, "The Top Quark" in Particle Data Group's "Review of Particle Properties," Eur. Phys. J. C **3**, 343 (1998).

An Experiment in Diffractive Physics

Alberto Santoro [1]

Laboratório de Física Experimental de Altas Energias
Centro Brasileiro de Pesquisas Físicas
Rua Dr. Xavier Sigaud, 150 - 22290-180 Rio de Janeiro - RJ - Brazil

Abstract. The purpose of this talk is to show one of the next future experiment in diffractive Physics which will be installed at the $D\emptyset$ experiment at Tevatron/Fermilab for run II, and the importance for Quantum Chromodinamics (QCD) as the theory of the strong interactions. The apparatus that we have developped is the Forward Proton Detector (FPD) to be introduced on the beam line of the Tevatron at both sides of the $D\emptyset$ detector. The FPD is composed by a set of Roman Pots as we will see in the text below.

I INTRODUCTION

The introduction of a new spectrometer as a set of subdetectors around $D\emptyset$ Detector in additional to the current upgrades for both Tevatron (energy of $\sqrt{s} = 2$ TeV and instantaneous luminosity of $5 \times 10^{31} cm^{-2} sec^{-1}$) and $D\emptyset$ [7] itself (new central tracking with Silicon and Scintillator Fibers detectors, a new solenoid magnet field, new electronics and Data Acquisition), creates new possibilities for Diffractive Physics in the next run II.

Hard diffraction was discovered by UA8-Collaboration [1] and many new results appeared from HERA(H1 and ZEUS) [2], and Tevatron (CDF and $D\emptyset$) (see reference [3] and papers therein for a more complete number of papers about diffraction).

In order to understand strong interactions we have to understand the diffraction component which is caused by the Pomeron. One of the most interesting interpretations of the Pomeron, the Regge trajectory with the quantum numbers of vaccum, is their link with Glueballs. These objects (Glueballs) are constituted by gluons only as a consequence of the non abelian gauge theory or the fact that gluons can interact between them and form a new resonance or a bound state.

Here we will show the progress made with the FPD project and its installation in the Tevatron since the presentation of the proposal to the $D\emptyset$ collaboration. [4]

1) Email: santoro@cbpf.br

CP562, *Particles and Fields: Ninth Mexican School,* edited by G. Herrera Corral and L. Nellen

In conclusion, for the next run II, the set of upgrades submitted by Tevatron, *DØ* detector, plus the additional FPD, we will have a superb possibility to produce new and good results for diffractive physics. To show the interest of diffraction in particle physics, we present the figure 1 from reference [5]. As a complement of this talk I have used the reference [6].

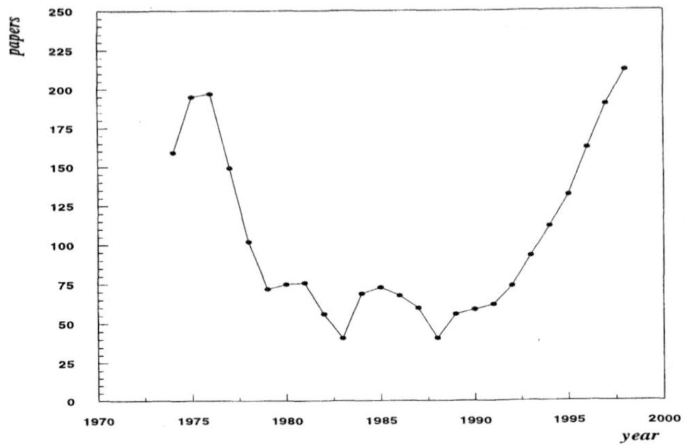

FIGURE 1. Number of papers in Diffractive Physics in the last 25 years.

II MOTIVATION

At present, to have a real progress in diffraction in Particle Physics, we need to have new measurements of several aspects of the diffractive events. We intend to contribute to increase significantly the statistics and present new results for diffraction in particle physics.

Tevatron Upgrade

From one side as we have mentioned above, we have the upgrade of the Tevatron which can be summarized by the table 1,

DØ Upgrade [7]

On the other side the set of upgrades that the DØ detector has planned allow us to say that in fact we have a new detector. The figure 2 summarize these upgrades.

The Addicional Forward Proton Detector

Beyond the upgrade of the Tevatron and DØ as was shown above, we add a set of subdetectors shown in figure 3 to be described later in this paper. Now that

TABLE 1. We show in this table a comparison between the old and new Tevatron and the predictions for TeV33 project.

Tevatron Collider Parameters			
Parameter	Tevatron Run Ib	Tevatron RunII	TeV33
Bunch Spacing (nsec)	3500	396/132	132
Luminosity ($10^3 1cm^-2sec^-1$	1.6	5/20	50
Interactions/Crossing	1 - 2	1-2/1-2	30
Luminous Region (cm)	30	30/15	30
Integrated Luminosity (fb^-1)	0.1	2-4	10-30

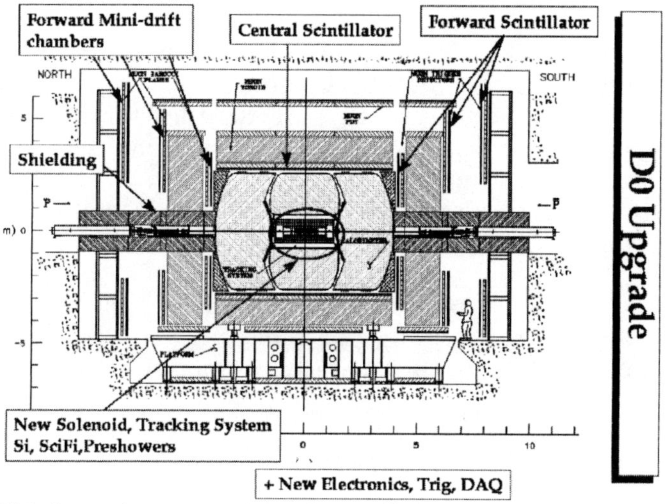

FIGURE 2. This figure shows the sectors of the $D\emptyset$ detector submitted to the upgrade. It is in fact a new detector.

we know what will be our apparatus to investigate the diffraction at $D\emptyset$, let us present a number of topics to be studied before we describe the Forward Proton Detector.

Topics to be studied with FPD

These studies will be generally made on two classes of events: those produced as a result of Soft Diffraction (SoD) and those produced by Hard Diffraction (HD). In SoD and HD we have topologies representing the Single and Double Diffraction and the Double Pomeron exchange. We think that Soft Diffraction is much more well understood in the basis of the Regge phenomenology while Hard Diffraction is being studied at present from the theoretical as well as from the experimental point of view.

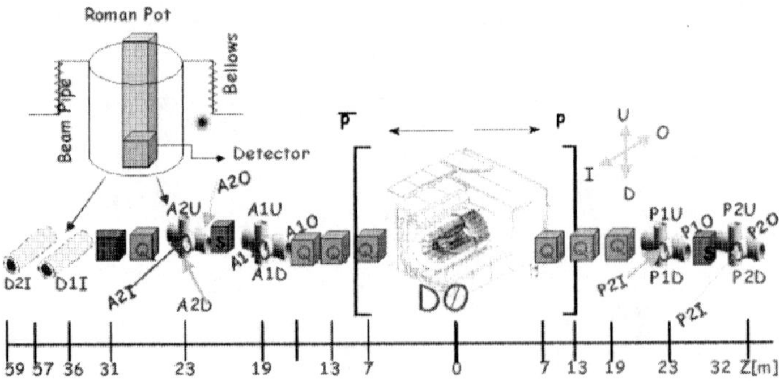

FIGURE 3. This figure shows the sectors of the *DØ* detector submitted to the upgrade. It is in fact a new detector.

In spite of the importance of SoD which will be also studied in our prospects for the FPD, we will mainly focus our interest in Hard Diffraction. We believe that our future results can contribute to improve the understanding of many important subjects, like the hard Pomeron.

Before going to the physics topics we would like to stress that the rapidity gap is an important tool to identify events associated to an especific topology. Rapidity gap is the interval of rapidity without particle production or without color activity. In order to get a diffractive event without Roman Pots we have almost always to do an offline analysis using rapidity gaps techniques. We first plot the multiplicity of the events and count the number of events with multiplicity n $\approx$ 0. After that, we build a lego plot with $\eta \times \phi$ variables, where

$$\eta = -\ln(\tan\frac{\theta}{2})$$

is the pseudo-rapidity, referred here sometimes simply as rapidity; θ and ϕ are the polar and azimuthal angles of the object produced diffractively. In the case of the hard diffraction, they are jets. We associate each topology to a lego plot to well define the class of events that we are looking for.

1. **Diffractive Jet production**

   Jets can not more be seen as a subject of the high p_t physics only. The discovery of diffractively produced jets [1] by UA8 collaboration was very important for diffractive physics since it stablished the definition of the hard diffraction . Due to the copiously sample expected of jet events produced diffractively in several E_T, transverse energies, single and double jet diffractive production will be studied in the next future with the FPD, to have a good separation of those produced by the flux of color interactions.

2. **Low and High $|t|$ elastic scattering**

   With the FPD, DØ will be able to produce both high and low t elastic scattering events . The measurements of elastic cross sections give a direct connection to the total cross section via the Optical Theorem. It is important to know the elastic slope of the differential cross section. The value of the slope characterizes a specific process which can be associated to a particular production (e.g. resonances production). Figure 4 shows the Feynmman graph representation of the elastic scattering and its correspondent lego plot with rapidity gap.

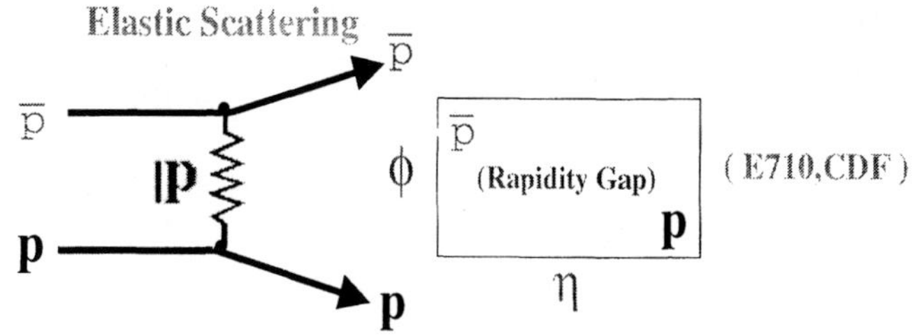

FIGURE 4. This figure shows the topology and the associated lego plot to the elastic diffraction process.

3. **Total cross section**

   We have new results from the Tevatron [8] experiments. From these results we have two possible asymptotic behaviors with the energy. New measurement are important in order to clarify possible violation of the Froissart Bound or not. After the Tevatron only the LHC will offer a new opportunity to make these measurements at higher energies. The results obtained at the Tevatron will be important for the future measurements at higher energies.

4. **Diffractive W/Z boson production**

   The present results from diffractively produced W and Z bosons are not satisfactory, we need more statistics for these events. It is important to understand these processes better, in order to have a comparison with the current boson production. Both CDF [9] and DØ have made progress and the current results are motivating both collaborations to proceed with new measurements.

5. **Diffractive Heavy Flavor Production**

   Heavy Flavor physics, including the Diffractive production, has been extensively studied. We can separate the heavy flavor sector in three types of particles and their corresponding physics. (i) The Charm, (ii) the Bottom and the (iii) Top Physics. Each one has some particularities to be taken into account in the diffractive production.

   Heavy flavor physics has been considered almost as a high p_t physics only. Diffractive heavy flavor production has not been sufficiently studied. This is due mainly to the absence of experimental results in this area for lack of adequate apparatus to observe diffractive production of heavy flavor. One interesting ratio to be measured is studied in reference [10].

$$\frac{Diff.Heavy\,Quark}{All\,Diff.\,events} > \frac{Heavy\,Quark\,Events}{All\,Events}$$

6. **Inclusive single diffraction**

   Many subjects are associated with the inclusive single diffraction particularly for the Tevatron, the Diffractive mass available, for single diffraction events, is M_x = 450 GeV. Inclusive single diffraction has been a good laboratory for several problems in diffraction physics. We intend to use it to study jets, heavy flavors, and to calculate ratios between cross sections of different processes. In figure 5 we show the two topologies, the SoD and HD and its corresponding lego plots.

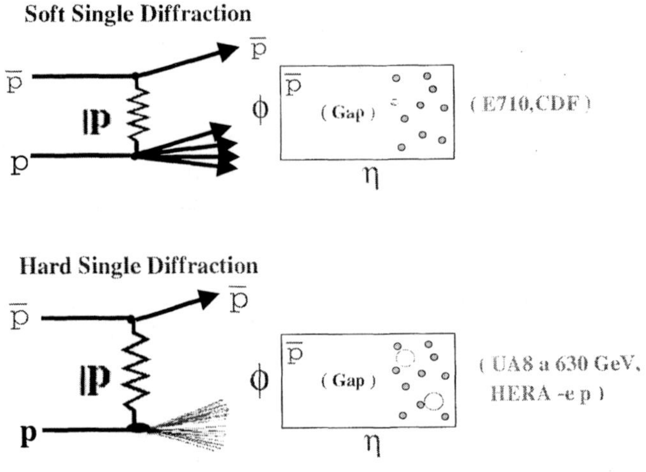

FIGURE 5. This figure shows the topologies and the associated lego plot to the single diffraction.

7. **Hard Double Pomeron Exchange**

   Due to the interesting topology, double Pomeron exchange has been largely discussed as the process for many different types of production. [11]. An advantage of the large Diffractive mass produced at the Tevatron, in this case $M_x \simeq 100\,GeV$, is the possibility to study by direct observation the Pomeron × Pomeron interactions and the associated physics. The instrumentation proposed by FPD/ DØ is appropriated to face the challenges of the double Pomeron mechanism to produce several objects not yet observed. The figure 6 shows the double Pomeron exchange graph and the gaps on its corresponding lego plot.

8. **Glueballs**

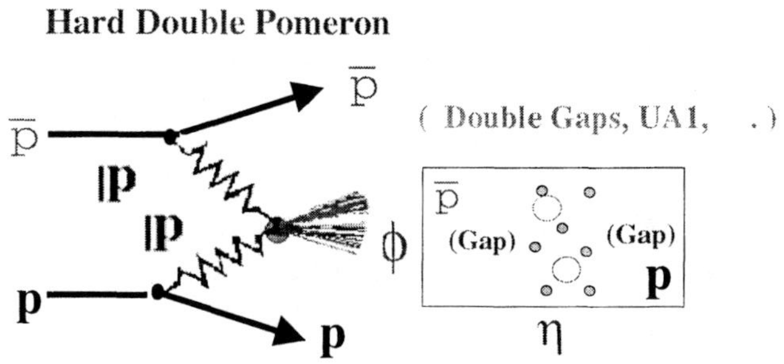

FIGURE 6. This figure shows the topology and the associated lego plot to the double Pomeron exchange event. We have two gaps in both sides and a central jet production.

TABLE 2. This table shows possible glueball state configurations with the mass and the quantum numbers for each one.

Glueballs and Oddballs							
J^{PC}	$(q\bar{q})$	2g	3g	ODD	MASS (GeV)		
					[14]	[15]	[16]
0^{++}	YES	YES	YES	NO	1.58	1.73 ± 0.13	1.74 ± 0.05
0^{+-}	NO	NO	YES	YES			
0^{-+}	YES	YES	YES	NO			
0^{--}	NO	NO	YES	YES	2.56	2.59 ± 0.17	2.37 ± 0.27
1^{++}	YES	YES	YES	NO			
1^{+-}	YES	NO	YES	NO			
1^{-+}	NO	YES	YES	YES			
1^{--}	YES	NO	YES	NO	3.49	3.85 ± 0.24	
2^{++}	YES	YES	YES	NO	2.59	2.40 ± 0.15	2.47 ± 0.08
2^{+-}	NO	NO	YES	YES			
2^{-+}	YES	YES	YES	NO	3.03	3.1 ± 0.18	3.37 ± 0.31
2^{--}	YES	NO	YES	NO	3.71	3.93 ± 0.23	
3^{++}	YES	YES	YES	NO	3.58	3.69 ± 0.22	4.3 ± 0.34
3^{+-}	YES	NO	YES	NO			
3^{-+}	NO	YES	YES	YES			
3^{--}	YES	NO	YES	NO	4.03	4.13 ± 0.29	

Since the origin of QCD, Glueballs has been studied by theoreticians and experimentalists. However, we do not have a significative progress in this subject. We need more experiments dedicated to the discover of glueballs without ambiguity with quark anti-quark competitive states. The family of glueballs is big. Table 2 shows the glueballs (oddballs are also shown). Oddballs should have the priority to be examined due to the fact that they do not have competition with natural q $\bar{q}$ states, mesons, and the qqq states, (baryons) with the same quantum numbers. It is difficult to separate the common hadrons from the glueballs when they appear in the same physical region.

Glueballs are important for QCD, which predicts their existence. Since the Pomeron can be interpreted as a glueball, the study of hard diffraction in the QCD framework is an interesting subject to be developped.

9. **Centauros**

   Centauros were never observed in accelerator particle physics. These objects were discovered in Cosmic Ray Physics as events with several unusual characteristics, like the production of large multiplicity of charged particles accompanied by very few photons. For example, as many as 100 charged particles and no more than 3 π^0. [12] We have enough energy at the Tevatron to produce centauros. Since our diffractive mass is significatively high, we can produce them diffractively. The good calorimetry of the DØ detector can be very useful to observe the absence of electromagnetic activity.

10. **Diffractive Structure Functions**

    The study of Diffractive structure functions at the Tevatron would allow a comparison with the existing Hera results. To understand the structure of the Pomeron one must know its structure function. This type of study has to be pursued exhaustively to get better accuracy and to have a clear interpretation of the Pomeron. One should know how important is the gluon and the quark component of the Pomeron. With these results we can have better calculations of its cross sections. [13] Is the Pomeron the same in electron-proton and proton anti-proton interactions? Are there differences between the Pomeron structure in different reactions?

11. **Correlations between η, t, M_x, b, ξ, x, E_T, θ, ϕ ...**

    It is important to study systematically the correlations between the kinematical variables η, t, M_x, b, ξ, x, E_T, θ, ϕ, ... as well as to obtain the single distributions for each one of these variables. These studies constitute a phenomenological source of investigation of the hidden dynamics of the distributions. η the pseudo-rapidity, (as defined above) is very useful to build the *Lego Plots of* $\eta \times \phi$ where ϕ is the azimuthal angle of the object being studied (e.g. jets); $t = (P_{Beam} - P_{Scattered})^2$ is the transferred momentum between the proton beam and the scattered proton ; $M_x = \sqrt{\xi}\sqrt{s}$ is the diffractive mass ($450.\,GeV$ for single diffraction and $100.\,GeV$ for double Pomeron exchange at the energies of the Tevatron ($\sqrt{2}\,TeV$)); b is the measured slope of the differential cross section, which can be selected globally or for a particular region of the invariant mass produced diffractively ($\frac{d\sigma}{dt} \propto e^{-b(M_x)t}$); $\xi = 1 - x_p = \frac{\Delta P}{P}$ is the fraction of the momentum of the proton carried by the Pomeron; $x_p \geq 0.95$ is the fraction of the momentum of the proton carried by the scattered proton; and E_T = is the transverse energy of the jet produced by hard diffraction.

III THE FORWARD PROTON DETECTOR

We will summarize the FPD since we have already described the studies and the project on reference [4].

The Forward Proton Detector consists of 18 Roman Pots arranged on both sides of the DØ detector as is shown in figure 3 where the Roman Pots are in the beam lineof the Tevatron. We have two castles on the proton side indicated by $P1$ and $P2$ as is shown in figure 3. The orientation is indicated by the additional letter U for up position, D for down position, I for inside position and O for outside position of the pots ($P1U$, $P1D$, $P1I$, $P1O$, same notation for $P2$). On the anti-proton side we have two similar castles labeled by $A1$ and $A2$ followed by the indication of the position similar for the proton side. Two others half castles on the side of the dipole magnet labeled $D1$ and $D2$. The approximated distances of the pots with respect to the interaction point (indicated by 0 on the scale) are shown.

A Tevatron Reconfiguration

We had to modify the beam line to open space for the FPD stations. We show in figure 7 the real position of the Castles. We see that the cryogenic bypass is bigger in the Run II. This was the only place where we could leave the necessary space for the FPD. Figure 7 also shows that the quadrupole magnet Q_1 is no longer present. Other small modifications were necessary, like drilling a hole on the floor to allow the insertion and removal of the bottom detectors. Summarizing the Tevatron modification are: (i) girder modification; (ii) cryogenic bypass, (iii) removal of Q_1 quadrupole.

B Roman Pots

The castle shown in figure 8 is a quadrupole type of one of the mentioned above, P1, P2, or A1, A2. These castles were built by LNLS -Laboratorio Nacional de Luz Synchroton, member of our national collaboration constituted by colleagues from UFRJ, UERJ, UFBa, Unicamp, UNESP, Lafex/CBPF and LNLS.[2] The figure 8 shows only one of the 4 views instrumented by the external tube which is the support of the MAPMTs (Multianode photo multipliers). The photo is taken at NWA/Fermilab, the laboratory where the tests were made for mechanic and vaccum parts.

[2] UFRJ=Universidade Federal do Rio de Janeiro, UERJ=Universidade do Estado do Rio Janeiro, UFBa=Universidade Federal da Bahia, Unicamp= Universidade de Campinas, UNESP= Universidade Estadual de São Paulo

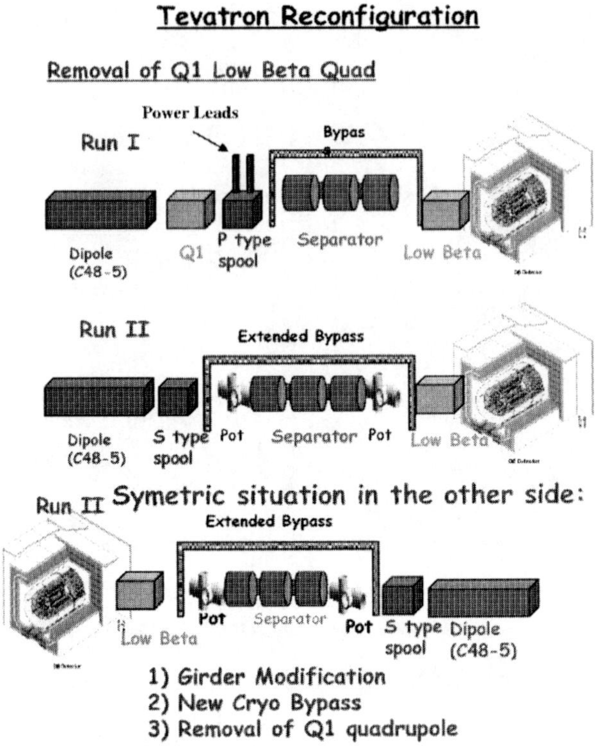

FIGURE 7. The Tevatron Reconfiguration or the Tevatron before and after the introduction of the Castles with the Roman Pots.

FIGURE 8. This photo shows the complete Castle in which the Roman Pots and detectors will be inserted. The photo shows the Castle on the NWA laboratory of the Fermilab to test all mechanic parts and vaccum of the Castle before to go to Tevatron beam line.

C The Detectors

In reference [6] we show the configuration of our detector. The detector is constituted by scintillating fibers placed in the frames for 6 planes X X', U U' and V V' . The scintillating fibers are glued to clear fibers which guide the signal up to the multi-anode photomultipliers. We have 16 channels per plane X X' and 20 channels/plane U U', V V', giving a total of 112 channels per detector and 2016 channels in total. Studies about the signals, efficiency and resolution have been made. Scintillating fibers are the best option for our detectors among many other possible technologies. The frame is made of ordinary plastic. The theoretical resolution is 80 microns. The acceptance of our detectors has been studied in several views. We present the results of these studies in reference [6]

IV CONCLUSION

The primary physics goal of the FPD is to measure hard diffraction, producing new data, turning the study of new diffractive physics possible, as the current measurements do not have enough statistics to get more accurated results. It will also be possible to use the FPD to reduce uncertainties on the luminosity for all DØ physics processes.

We can also obtain results and improve old measurements at lower energies and, in some cases, decide between conflicting results as is the case of the total cross sections. We have given a list of possible topics to be investigated in both hard and soft diffraction.

Our schedule to start the data acquisition is the same of the Dzero Detector, i.e., at the beginning of the year 2001.

This project will give the possibility to upgrade the world diffractive data since many new features will be possible, like the direct observation of the Pomeron Pomeron scattering.

Finally, we would like to end this talk about the FPD project with the photo of the two castle installed in the beam line of the Tevatron as is shown 9.

FIGURE 9. This is the final prototype of our Dipole type Castle built at Laboratório Nacional de Luz Sinchroton, Campinas/Brazil.

I would like to thank the organizer committee of the IX Mexican School on Particles and Fields for giving me the opportunity to give this lecture on Diffractive Physics. All the work behind this paper was done with the effort and contribution of the DØ collaboration, in particular from my colleagues of the FPD group. I would also like to thank our national (Brazilian) collaboration made of colleagues from Universidade Federal da Bahia (N.Oliveira); Universidade Federal do Rio de Janeiro (J.Barreto); Universidade Estadual do Rio de Janeiro (W.Carvalho, C. Martins, V. Oguri and A. Sznajder); Universidade Estadual Paulista (E. Gregores and Sergio Novaes); and Laboratorio Nacional de Luz Synchroton (R.Neuezwander and M. Juni). Without the important contribution and cooperation of Ricardo Rodrigues and Cylon Goncalves from LNLS we could not have built the Roman Pots. Finally I would like to thank my colleagues of our laboratory, (LAFEX/CBPF) G. Alves, H. da Motta, M. Souza and M. Vaz for all the work that they are doing for the development of Diffractive Physics. We thank CBPF, CNPq and FAPERJ for partial financial support.

REFERENCES

1. A.Brandt *et al.*, UA8 Collaboration, Nucl. Inst. Meth.**A327**, 412 (1993); *id.* ,Phys. Lett. **B297**, 93 (1993); *ibid.* , HEP-Ex-9709015, 1997, submitted to PLB.
2. ZEUS collaboration, M. Derrick *et.al.*, Z. Phys. **C65**, 379,(1995); H1 collaboration, T.Ahmed *et.al.*, Nucl. Phys. **B439**, 471 (1995); ZEUS collaboration, M. Derrick *et.al.* Z. Phys. **C68**, 569, (1995); H1 collaboration, T.Ahmed *et.al.* Phys. Lett. **B348**, 681 (1995).
3. A. Brandt, H. Motta and A. Santoro -Editors- LISHEP98- Lafex International School on High Energy Physics - Proceedings of Session C -February 16-20,(1998)
4. A. Brandt et al. *A Forward Proton Detector* Fermilab- Pub-97/377, 1997
5. E.M. Gregores, T.L. Lungov and S. F. Novaes, A.Santoro, - Proceedings of Hadron 2000 - S. Paulo - Brazil.
6. A. Santoro - Proceedings of the XX-ENFPC-Encontro Nacional de Fí sica de Partí culas e Campos -Caxambu-M.Gerais-Brazil (1999)
7. *DØ* Collaboration, Fermilab Pub-96/357-E, (1996).
8. C.Avila, Proceedings of LISHEP98; Thesis:" Measurement the proton - antiproton total cross section at center of mass Energy of 1800 GeV" Cornell University 1997 and references therein; C.Avilez et al., Phys. Lett. **B234**, 158, (1990);*ibid.* , Phys. Rev. Lett. **68**, 2433, (1992); *ibid.* , Phys. Rev. **50**, 5550, (1994); *ibid.* , Phys. Lett. **B234**, 158, (1990); *ibid.* , Nucl. Inst. Meth.**A360**, 80, (1995)
9. F.Abe et al. CDF Collaboration: Phys. Rev. Lett. **78**, 2698, (1997)
10. E.L.Berger, J.C.Collins, D.E. Soper, G.Sterman, Nucl. Phys. **B286**, 704, (1987); A. Kerman and G.Van Dalen, Phys. Rep. **106**, 297, (1984)
11. See for example for Higgs production the paper of D.Kharzeev and E. Levin - "Soft Double-Diffractive Higgs Production at Hadron Collider". -Fermilab-Pub -00/035-T; BNL-NT-00/14; Hep-ph/0005311 (2000)
12. F. Halzen: Felix Home Page, Brazil-Japan Collaboration - Proceedings of the 21st. Int.Conf.-Adelaide-Australia vol.8,259 (1990); C. M. G. Lattes, Y. Fujimoto and S. Hasegawa, Phys. Rep. **65**, 151, 1980; C. E. Navia et al.: Phys. Rev **D40**, 2898, (1989).
13. R. J. M. Covolan and M.S. Soares, IFGW-DRCC 97/04(1997) "A Study on the Pomeron Structure Function"
14. A. B. Kaidalov and Yu. A. Simonov "Glueball masses and Pomeron Trajectory in non perturbative QCD approach"
15. C. Morningstar, M. Peardor, Nucl. Phys. **B63A-C**, 1022, (1998); Phys. Rev. **D60**,034509, (1999)
16. M. Teper, Hep/th981287

The Ups and Downs of J/ψ Suppression

R. Vogt

Nuclear Science Division, Lawrence Berkeley National Laboratory, Berkeley, CA, 94720[1]
and
Physics Department, University of California at Davis, Davis, CA, 95616

Abstract. An overview of the present status of J/ψ suppression in pA and nucleus-nucleus interactions is presented. In both cases, model predictions are summarized and compared to the data. The "anomalous" J/ψ suppression in Pb+Pb collisions is discussed in some detail. Predictions are also shown for quarkonium suppression at collider energies.

INTRODUCTION

Data from the CERN SPS [1–4] and the Fermilab fixed target program [5–8] show that J/ψ production in pA and AB interactions is reduced relative to a linear extrapolation of the pp cross section. Recent claims regarding the production of a new state of matter in Pb+Pb interactions at the CERN SPS [9] illustrate the importance of a thorough understanding of J/ψ production in pA and lighter AB collisions before the strength of the "anomalous" J/ψ suppression in Pb+Pb collisions relative to the Drell-Yan dilepton background can be determined. In these lectures, we begin with a discussion of quarkonium and Drell-Yan production. We then describe the status of J/ψ production in pA interactions with an emphasis on the most recent 800 GeV data [8]. We will then discuss the nucleus-nucleus data and its interpretation in hadronic and phase transition models. Finally, the future of quarkonium suppression in heavy ion colliders is speculated upon.

QUARKONIUM AND DRELL-YAN PRODUCTION

We describe both quarkonium and Drell-Yan production in this section because, in pA interactions, Drell-Yan production is the dominant background to quarko-

---

[1] This work was supported in part by the Director, Office of Energy Research, Division of Nuclear Physics of the Office of High Energy and Nuclear Physics of the U. S. Department of Energy under Contract No. DE-AC03-76SF00098.

CP562, *Particles and Fields: Ninth Mexican School*, edited by G. Herrera Corral and L. Nellen

nium production and, in AB collisions, Drell-Yan production has been used as a reference process to provide a figure of merit for J/ψ suppression. Both processes are detected through the dilepton channel, typically in a muon spectrometer at fixed-target energies. Quarkonium resonances, with $J^{PC} = 1^{--}$, decay to lepton pairs with invariant mass equal to the resonance mass, forming a peak above the continuum which is dominated by Drell-Yan lepton pair production for masses greater than 4 GeV. These pairs contribute to the continuum rather than a resonance peak since the virtual photon exchanged in the Drell-Yan process has no definite mass.

Quarkonium Production

Two basic models of quarkonium hadroproduction have been successfully employed. The first, the color evaporation model, treats *e.g.* all charmonium production identically to $c\overline{c}$ production below the $D\overline{D}$ threshold. The second, nonrelativistic QCD, involves an expansion of quarkonium production in powers of the relative Q-$\overline{Q}$ velocity within the bound state.

In the color evaporation model, CEM, quarkonium production is treated identically to open heavy quark production. However, the invariant mass of the heavy quark pair is restricted to be less than twice the mass of the lightest meson that can be formed with one heavy constituent quark. The upper limit on the $c\overline{c}$ pair mass is $2m_D$. The hadroproduction of heavy quarks at leading order (LO) in perturbative QCD is the sum of contributions from $q\overline{q}$ annihilation and gg fusion. If x_F is the $c\overline{c}$ longitudinal momentum fraction in the AB center-of-mass frame and $\sqrt{s}$ is the center-of-mass energy of a nucleon-nucleon collision, the quarkonium production cross section is [10,11]

$$\begin{aligned}\frac{d\tilde{\sigma}_i}{dx_F} &= 2F_i \int_{4m_c^2}^{4m_D^2} dm^2 \int_0^1 dx_1 dx_2\, \delta(x_1 x_2 s - m^2)\, \delta(x_F - x_1 + x_2) \\ &\times \Big\{ f_g^A(x_1, m^2) f_g^B(x_2, m^2) \sigma_{gg}(m^2) \\ &+ \sum_{q=u,d,s} [f_q^A(x_1, m^2) f_{\overline{q}}^B(x_2, m^2) + f_{\overline{q}}^A(x_1, m^2) f_q^B(x_2, m^2)] \sigma_{q\overline{q}}(m^2) \Big\} \qquad (1)\end{aligned}$$

where x_1 and x_2 are the fractions of the nucleon momentum carried by the projectile and target partons respectively. The parton densities $f_i(x, m^2)$ are evaluated at momentum fraction x and scale m^2, the squared invariant mass of the $c\overline{c}$ pair. F_i is the fraction of the free $c\overline{c}$ cross section that produces the final-state resonance. We use the MRST LO parton distributions [12] for $f_i(x, m^2)$. Once F_i has been determined for each state, *e.g.* J/ψ, ψ' or χ_{cJ}, the model successfully predicts the energy and momentum dependencies. We note that F_ψ includes both direct J/ψ production and indirect production through radiative decays of the χ_{cJ} states and hadronic ψ' decays. In the CEM, the x_F distributions of all states are assumed to be the same. Thus F_ψ in Eq. (1) implicitly includes the χ_{cJ} and ψ' decay contributions. The relative differential and integrated quarkonium production rates are approximately independent of projectile, target, and energy [13], necessary for the model to have any predictive power.

The nonrelativistic QCD, NRQCD, approach to quarkonium production expands the cross section in powers of α_s and Q-$\overline{Q}$ velocity to include not only the leading color singlet contributions but also color octet production [14]. Hadronization occurs through soft gluon emissions. The x_F distribution of a charmonium state, C, in NRQCD is

$$\frac{d\sigma_C}{dx_F} = \sum_{i,j} \int_0^1 dx_1 dx_2 \delta(x_F - x_1 + x_2) f_i^A(x_1, \mu^2) f_j^B(x_2, \mu^2) \sum_n C^{ij}_{Q\overline{Q}\,[n]} \langle \mathcal{O}^C_n \rangle \ , \qquad (2)$$

where the C production cross section is the product of expansion coefficients, $C^{ij}_{Q\overline{Q}\,[n]}$, calculated perturbatively in powers of $\alpha_s(\mu^2)$, and nonperturbative parameters, $\langle \mathcal{O}^C_n \rangle$, describing the hadronization of the charmonium state. We use the parameters determined by Beneke and Rothstein [15] for fixed-target hadroproduction of charmonium with $m_c = \mu/2 = 1.5$ GeV. The CTEQ 3L parton densities [16] were used for $f_i(x, \mu^2)$.

The total J/ψ x_F distribution includes decay contributions from the χ_{cJ} and ψ',

$$\frac{d\sigma_\psi}{dx_F} = \frac{d\sigma_\psi^{\rm dir}}{dx_F} + \sum_{J=0}^{2} B(\chi_{cJ} \to \psi X) \frac{d\sigma_{\chi_{cJ}}}{dx_F} + B(\psi' \to \psi X) \frac{d\sigma_{\psi'}}{dx_F} \ . \qquad (3)$$

Octet production contributes $\approx 60\%$ of the total J/ψ cross section. In NRQCD, the J/ψ, χ_{cJ}, and the ψ' distributions differ from each other, both because the relative octet contributions to each state differ and, in the case of ψ' and directly produced J/ψ, because the fitted parameters $\langle \mathcal{O}^C_n \rangle$ are different. Thus three parameters are needed to fix the ψ' production cross section while eight are needed for the total J/ψ cross section. Only one parameter for each state is needed in the CEM, a considerable reduction.

Drell-Yan Production

Continuum lepton pairs are produced by the Drell-Yan process, $q\overline{q}$ annihilation into a virtual photon, $q\overline{q} \to \gamma^\star \to l^+l^-$. At leading order [17],

$$\begin{aligned} \frac{d\sigma^{\rm DY}}{dx_F dM} &= \frac{8\pi\alpha^2}{9M} \int_0^1 dx_1 dx_2 \, \delta(x_1 x_2 s - M^2) \, \delta(x_F - x_1 + x_2) \\ &\times \sum_q e_q^2 [\{ z_A f_q^p(x_1, M^2) + n_A f_q^n(x_1, M^2)\} \{ z_B f_{\overline{q}}^p(x_2, M^2) + n_B f_{\overline{q}}^n(x_2, M^2)\} \\ &+ \{ z_A f_{\overline{q}}^p(x_1, M^2) + n_A f_{\overline{q}}^n(x_1, M^2)\} \{ z_B f_q^p(x_2, M^2) + n_B f_q^n(x_2, M^2)\}] \ , \end{aligned} \qquad (4)$$

where $z = Z/A$ and $n = N/A$ are, respectively, the proton and neutron fractions in the projectile and target. In pA collisions, $z_A = 1$ and $n_A = 0$. Isospin becomes important in Drell-Yan production because the parton densities are weighted by the square of the parton charge. Since all partons are weighted equally in quarkonium production and gluon fusion dominates over most of the x_F range, isospin effects in quarkonium production are negligible.

When this leading order cross section is compared to data, it falls short by an approximately constant factor K. Experimentally, $K \approx 1.7 - 2.5$, depending on the energy, mass range, and parton distribution functions [17].

Since lepton pairs only interact electroweakly, the A dependence is expected to be weak because no final-state interactions affect the lepton pair.

PA INTERACTIONS

In this section, we address the nuclear dependence of J/ψ and Drell-Yan production in pA interactions. As discussed previously, quarkonium and Drell-Yan production is described by perturbative QCD. In QCD, the nonperturbative parton distributions are separated from the hard interaction by factorization [18]. If factorization is satisfied, then particle production should be independent of the presence of nuclear matter and σ_{pA} would grow linearly with A for quarkonium and Drell-Yan production. Experimentally, the dependence of hard or high mass/transverse momentum particle production on atomic mass number A is parameterized by a power law as [1–8,19,20]

$$\sigma_{pA} = \sigma_{pN} A^{\alpha} \tag{5}$$

where σ_{pA} and σ_{pN} are the integrated particle production cross sections in proton-nucleus and proton-nucleon interactions respectively. The integrated Drell-Yan cross section satisfies $\alpha = 1$ to rather high precision [19]. A less than linear A dependence has been observed for ψ [1–8] and Υ [20] production. Typical values of α are between 0.9 and 1.

Any dependence on the kinematic variables such as projectile energy or longitudinal momentum fraction, x_F, reveals the importance of going beyond a simple A^{α} scaling for production and a constant absorption cross section for J/ψ production. In quarkonium production, α decreases as a function of x_F [4–6,8]. Drell-Yan production shows that α also decreases with x_F [19] but not as strongly as quarkonium.

The x_F dependence of J/ψ production was studied using a two-component model employing concepts developed in Ref. [21]. The first component includes target effects such as shadowing and energy loss and yields an approximately linear A dependence, as in Drell-Yan dilepton production. The second component arises from the projectile alone assuming $c\bar{c}$ pairs are intrinsic to the projectile wavefunction [22,23]. Since the charm quark mass is large, these intrinsic heavy quark pairs carry a significant fraction of the longitudinal momentum and contribute at large x_F whereas the perturbative QCD x_F distribution decreases strongly. The light spectator quarks in the intrinsic $c\bar{c}$ state interact on the nuclear surface, leading to an approximate $A^{2/3}$ dependence [23].

The model was motivated by the NA3 analysis [4] of J/ψ production from hard (low x_F) and diffractive (high x_F) components,

$$\frac{d\sigma_{pA}}{dx_F} = A^{\alpha'} \frac{d\sigma_h}{dx_F} + A^{\beta} \frac{d\sigma_d}{dx_F} \ . \tag{6}$$

We studied the J/ψ and Drell-Yan A dependence as a function of x_F in the context of this model [21,24]. The most recent J/ψ data are from the E866 collaboration, using an 800 GeV proton beam on Be and W targets [8]. They find $\alpha \approx 0.94$ until $x_F \approx 0.25$, decreasing to $\alpha \approx 0.7$ at large x_F. We calculate the cross section per nucleon for each target according to Eq. (6), obtaining from Eq. (5),

$$\alpha(x_F) = 1 + \frac{\ln[(d\sigma_{p\mathrm{W}}/A_{\mathrm{W}} dx_F)/(d\sigma_{p\mathrm{Be}}/A_{\mathrm{Be}} dx_F)]}{\ln(A_{\mathrm{W}}/A_{\mathrm{Be}})} \ . \tag{7}$$

We briefly discuss all the effects considered and compare the results to the E866 data. See Ref. [24] for full details.

Nuclear Absorption and Comover Scattering

We first discuss final-state interactions, occuring after $c\bar{c}$ production. The $c\bar{c}$ pair may interact with nucleons and be dissociated or absorbed before it can escape the target. Comoving secondaries with velocities similar to the $c\bar{c}$ pair or J/ψ, formed after $\tau_0 \sim 1 - 2$ fm, may also scatter with it. Because the J/ψ formation time is less than τ_0, the final-state charmonium is assumed to interact with the comovers.

The effect of nuclear absorption alone on the quarkonium production cross section in pA collisions may be expressed as [11]

$$\sigma_{pA} = \sigma_{pN} \int d^2b \int_{-\infty}^{\infty} dz \, \rho_A(b,z) S_C^{\mathrm{abs}}(b) \tag{8}$$

$$= \sigma_{pN} \int d^2b \int_{-\infty}^{\infty} dz \, \rho_A(b,z) \exp\left\{ - \int_z^{\infty} dz' \rho_A(b,z') \sigma_{\mathrm{abs}}(z'-z) \right\} \tag{9}$$

where b is the impact parameter, z is the $c\bar{c}$ production point, S_C^{abs} is the nuclear absorption survival probability of state C, ρ_A is the nuclear density distribution [25], and σ_{abs} is the nucleon absorption cross section. The J/ψ may interact with nucleons as a fully formed color singlet meson or as a precursor $(c\bar{c})_8 g$ color octet [26]. Expanding the exponent in Eq. (9), integrating, and reexponentiating assuming A is large gives $\alpha = 1 - 9\sigma_{\mathrm{abs}}/(16\pi r_0^2)$ [11].

Three different models of nuclear absorption have been studied. All quarkonium states could be produced either as color octets or color singlets. However, a combination of production in octet and singlet states is more likely. We check all these cases. Our calculation of J/ψ production follows the assumptions regarding the absorption process. When pure octet or pure singlet absorption is considered, J/ψ production is calculated in the CEM. When a combination of octet and singlet absorption is assumed, NRQCD is used to obtain the correct balance between octet and singlet production. Both the CEM and NRQCD parameters are tuned to fit pp production. We include the $\approx 30\%$ contribution from χ_{cJ} decays [27] and the $\approx 12\%$ contribution from ψ' decays [13] decays. Then the total J/ψ survival probability on nucleons, including indirect production, is

$$S_{\psi}^{\rm abs}(b) = 0.58 S_{\psi,\,\rm dir}^{\rm abs}(b) + 0.3 S_{\chi_{cJ}}^{\rm abs}(b) + 0.12 S_{\psi'}^{\rm abs}(b) \ . \tag{10}$$

The first case is pure octet production. Here, the $c\overline{c}$ is produced in a color octet state, $(c\overline{c})_8$, which travels through the nucleus. As it leaves the field of the nucleon that produced it, it neutralizes its color by combining with a collinear gluon in a nonperturbative interaction. The resulting $(c\overline{c})_8 g$ state is in a color singlet as it traverses the nucleus. The final-state charmonium resonance will be formed when the accompanying gluon is absorbed by the colored $(c\overline{c})_8$. When the $(c\overline{c})_8 g$ state interacts with nucleons, a gluon is exchanged which can couple to either the g or the $(c\overline{c})_8$, leaving the remaining state colored. Since this colored state is not yet bound, any interaction can lead to the breakup of the state [26]. There is thus no energy dependent threshold for J/ψ breakup in this description. Because the initial color octet is the same regardless of the eventual final-state resonance, $\sigma_{\rm abs} = \sigma^{\rm o}_{\psi N} = \sigma^{\rm o}_{\psi' N} = \sigma^{\rm o}_{\chi_{cJ} N}$. Therefore $S_C^{\rm abs}$ is identical for all C states and the feeddown contributions to the J/ψ in Eq. (10) do not affect the absorption. We assume $\sigma^{\rm o}_{\psi N}$ is a constant, independent of energy and x_F, and treat absorption as if only the $(c\overline{c})_8 g$ interacts with nucleons, not the final charmonium states.

On the other hand, if the $c\overline{c}$ pair is produced as a color singlet in a time $\tau \propto m_c^{-1}$, the proper time required for the formation of the bound state, $\tau_{\psi_i} \sim 1 - 2$ fm, is considerably longer [28]. The singlet $c\overline{c}$–N absorption cross section may grow as a function of proper time until τ_{ψ_i} when it saturates at the asymptotic value $\sigma^{\rm s}_{\psi_i N}$ [29,30] so that $\sigma_{\rm abs}(z'-z) = \sigma^{\rm s}_{\psi_i N}(\tau/\tau_{\psi_i})^{\kappa}$ for $\tau < \tau_{\psi_i}$. The exponent κ determines the increase of $\sigma_{\rm abs}$ during hadronization of the $c\overline{c}$ pair. If $\sigma_{\rm abs}$ is proportional to the geometric cross section, then we expect $\kappa \sim 2$. The proper time τ is related to the $c\overline{c}$ path length through nuclear matter, $\tau = (z'-z)/\gamma v$. The γ factor introduces x_F and energy dependencies into the growth of the cross section. The J/ψ may be formed inside or outside the target. At 800 GeV, by $x_F = 0$ the final-state meson is produced outside the target so that $\alpha \approx 1$ for $x_F > 0$. Therefore the A dependence of color singlet production is virtually independent of $\sigma^{\rm s}_{\psi_i N}$ for $x_F > 0$ at 800 GeV.

More realistically, J/ψ production is a combination of octet and singlet states, as in NRQCD. The ratio of octet to singlet production is energy and x_F dependent [15]. The relative absorption of each state then depends on x_F since the octet and singlet absorption cross sections are expected to be different [31]. In this case,

$$\frac{d\sigma_{pA}^{\psi,\,\rm tot}}{dx_F} = \left[\frac{d\sigma_{pp}^{\psi,\,\rm dir,\,oct}}{dx_F} + \sum_{J=0}^{2} B_{\chi_{cJ}} \frac{d\sigma_{pp}^{\chi_{cJ},\,\rm oct}}{dx_F} + B_{\psi'} \frac{d\sigma_{pp}^{\psi',\,\rm oct}}{dx_F}\right] \int d^2b\, T_A(b) S_{\rm oct}^{\rm abs}(b)$$

$$+ \int d^2b T_A(b) \left[\frac{d\sigma_{pp}^{\psi,\,\mathrm{dir,\,sing}}}{dx_F} S^{\mathrm{abs}}_{\psi,\,\mathrm{dir,\,sing}}(b) + \sum_{J=0}^{2} B_{\chi_{cJ}} \frac{d\sigma_{pp}^{\chi_{cJ},\,\mathrm{sing}}}{dx_F} S^{\mathrm{abs}}_{\chi_{cJ},\,\mathrm{sing}}(b) \right.$$

$$\left. + B_{\psi'} \frac{d\sigma_{pp}^{\psi',\,\mathrm{sing}}}{dx_F} S^{\mathrm{abs}}_{\psi',\,\mathrm{sing}}(b) \right] \qquad (11)$$

where B_{χ_c} and $B_{\psi'}$ are the branching ratios for χ_{cJ} and ψ' decays to J/ψ and the nuclear density profile is $T_A(b) = \int_{-\infty}^{\infty} dz \rho_A(b,z)$. Note that since σ_{abs} is identical for all charmonium states, the x_F and A dependencies factorize while the different final state radii dictate $\sigma^{\mathrm{s}}_{\psi' N} > \sigma^{\mathrm{s}}_{\chi_{cJ} N} > \sigma^{\mathrm{s}}_{J/\psi N}$ and the x_F and A dependencies of singlet states are intertwined.

The A dependence of J/ψ dissociation by comovers is determined from

$$\sigma_{hA} = \sigma_{hN} \int d^2b \, [0.58 S^{\mathrm{co}}_{\psi,\,\mathrm{dir}}(b) + 0.3 S^{\mathrm{co}}_{\chi_{cJ}}(b) + 0.12 S^{\mathrm{co}}_{\psi'}(b)] \qquad (12)$$

where the comover survival probability for directly produced J/ψ's is [11]

$$S^{\mathrm{co}}_{\psi,\,\mathrm{dir}}(b) \approx \exp\left\{ -\langle \sigma_{\psi \mathrm{co}} v \rangle \frac{dN}{dy} \ln\left(\frac{\tau_I}{\tau_0}\right) T_A(b) \right\} . \qquad (13)$$

Here τ_I is the effective proper time the comovers interact with the J/ψ, τ_0 is the comover formation time, v is the relative J/ψ-comover velocity, the multiplicity in pA collisions is proportional to $T_A(b) dN/dy$, and $\sigma_{\psi \mathrm{co}}$ is the J/ψ-comover cross section. We take $\sigma_{\psi \mathrm{co}} = 0.67$ mb [32] with $\sigma_{\psi' \mathrm{co}} \approx 3.7 \sigma_{\psi \mathrm{co}}$ and $\sigma_{\chi_{cJ} \mathrm{co}} \approx 2.4 \sigma_{\psi \mathrm{co}}$ [33], assuming that the asymptotic charmonium states interact with the comovers.

If a constant absorption cross section is assumed, both the nuclear absorption and comover survival probabilities depend on $T_A(b)$, giving the two processes identical A dependencies. Thus comover contributions to pA interactions, while small, are difficult to rule out entirely.

Nuclear Shadowing

Measurements of the nuclear charged parton distributions, F_2^A, by deep-inelastic scattering off nuclear targets shows that the ratio $R_{F_2} = F_2^A/F_2^D$ has a characteristic shape as a function of x [34], generally referred to as shadowing. This behavior is not well understood for all x. However, the shadowing effect can be modeled by an A-dependent fit to the nuclear deep-inelastic scattering data.

The nuclear parton distributions are assumed to factorize into the nucleon parton distributions, independent of A, and a shadowing function that parameterizes the modifications of the nucleon parton densities in the nucleus, dependent on A, x, and Q^2:

$$f_i^A(x, Q^2, A) = S^i(A, x, Q^2) f_i^p(x, Q^2) .$$

While the location of the parton in the target could influence S^i [35], the impact parameter is difficult to resolve in pA collisions. We studied three different parameterizations of the shadowing function, $S^i(A, x, Q^2)$.

The first, S_1, is a fit to nuclear deep-inelastic scattering data which does not differentiate between quark, antiquark, and gluon modifications and does not include evolution in Q^2 [36]. Therefore it is not designed to conserve baryon number or momentum. The parameterization is available for all A and is designed so that $S_1 \equiv 1$ when $A = 1$.

The second parameterization, $S_2^i(A, x, Q^2)$, modifies the valence quark, sea quark and gluon distributions separately and also includes Q^2 evolution [37], beginning at $Q = Q_0 = 2$ GeV and continuing up to $Q = 10$ GeV. It is based on a fit using the Duke-Owens parton densities [38]. It is assumed that S_2^V and S_2^S are the same for all valence and sea quarks, consistent with the symmetric sea of the Duke-Owens parton distributions. This parameterization conserves baryon number and the parton momentum sum but is only available for $A = 32$ and 200. It is thus applied only to tungsten and the beryllium parton densities are left unmodified.

The third parameterization, $S_3^i(A, x, Q^2)$ [39,40], is based on the GRV LO parton distributions [41]. The initial scale was chosen to equal the charm quark mass in the GRV LO distributions, $Q = Q_0 = 1.5$ GeV. At this scale all sea quark ratios are assumed to be equal, as are both the valence ratios. The parameters are constrained by nuclear deep-inelastic scattering and Drell-Yan data. The gluon ratio is then fixed by the momentum sum rule as well as J/ψ electroproduction data. Above Q_0, the individual quark and gluon distributions are evolved separately.

Effects of Energy Loss

Partons are expected to lose energy when traversing matter [42,43]. Since the projectile parton is typically expected to feel the effects of energy loss, the scaling of the A dependence at different energies with x_F or x_1 [6] suggested that energy loss could be the cause. We will introduce three models of energy loss that have been applied to J/ψ and Drell-Yan production. Two assume that the initial parton loses energy [44,45] while the third assumes the produced $c\bar{c}$ loses energy through its interactions with nucleons [46].

Initial-State Loss

Initial state energy loss, as studied by Gavin and Milana [44] and subsequently developed by Brodsky and Hoyer [45] assumes that the projectile parton momentum fraction, x_1, is depleted by multiple scattering as the parton moves through the nucleus. The projectile parton momentum fraction in the hard scattering is then $x_1' = x_1 - \Delta x_1$ where Δx_1 is the loss in x_1. Thus the shifted value, x_1', enters the partonic cross sections but the parton distributions must be evaluated at the initial

x_1. An additional delta function is added to Eqs. (1) and (2) with the corresponding integral over x'_1 so that Eq. (1) becomes

$$\frac{d\tilde{\sigma}_i}{dx_F} = \frac{F_i}{s}\int dm^2 dx'_1 dx_1 dx_2\, \delta(x'_1 - x_1 + \Delta x_1)\delta(x_F - x'_1 + x_2)\,\delta(x'_1 x_2 s - m^2) \times \Big\{ f_g^A(x_1,m^2) f_g^B(x_2,m^2)\sigma_{gg}(m^2) + \sum_{q=u,d,s} [f_q^A(x_1,m^2) f_{\bar{q}}^B(x_2,m^2) + f_{\bar{q}}^A(x_1,m^2) f_q^B(x_2,m^2)]\sigma_{q\bar{q}}(m^2)\Big\} \quad (14)$$

while Eq. (2) is then

$$\frac{d\sigma^C}{dx_F} = \sum_{i,j}\int_0^1 dx'_1 dx_1 dx_2\, \delta(x'_1 - x_1 + \Delta x_1) \times \delta(x_F - x'_1 + x_2) f_i^A(x_1,\mu^2) f_j^B(x_2,\mu^2) \sum_n C^{ij}_{Q\overline{Q}\,[n]} \langle \mathcal{O}_n^C \rangle \,. \quad (15)$$

The first model of initial-state energy loss applied to J/ψ production was proposed by Gavin and Milana [44], GM, where

$$\Delta x_1 = \epsilon_i x_1 A^{1/3} \,, \quad (16)$$

with $\epsilon_q = 0.00412$ from the Drell-Yan A dependence [44] and $\epsilon_g = 9\epsilon_q/4$ due to the difference in the color factors. This corresponds to $-dE/dz|_q \sim 1.5$ GeV/fm and $-dE/dz|_g \sim 3.4$ GeV/fm [44]. As x_F increases, x_1 grows larger and if the parton densities behave as $\sim (1-x_1)^{n_P}$ when $x_1 \to 1$, a slight decrease in x_1 is magnified. The effect should be stronger for J/ψ than Drell-Yan because $n_g \sim 5$ and $n_{q_V} \sim 3$ in simple spectator counting models [47]. The valence quark distributions dominate Drell-Yan production at large x_F and x_1.

Brodsky and Hoyer [45], BH, argued that the GM energy loss was too large because there is not enough time after the initial QCD bremsstrahlung for the color field of the parton to be regenerated before the next interaction. Therefore, the subsequent interactions of the parton in the target should not lead to a large energy loss [48]. From the uncertainty principle they deduced that the loss should be independent of parton type, giving a bound on Δx_1,

$$\Delta x_1 < \frac{\langle k_\perp^2 \rangle L_A}{2E} \,, \quad (17)$$

where L_A is the path length through the medium, $E = x_1 s/2m_p$, and $\langle k_\perp^2 \rangle$ is the average transverse momentum of gluons radiated by the incoming parton. Because $\Delta x_1 \propto 1/x_1$, at small x_1 (negative x_F) $\Delta x_1 > x_1$ and the model is no longer applicable. The average radiative loss, "original BH loss", is thus expected to be $-dE/dz \sim 0.25$ GeV/fm for quarks and gluons with another 0.25 GeV/fm loss arising from elastic scattering.

The bound on Δx_1 was subsequently refined through the work of Baier *et al.* [43,49]. Their calculations suggest that Δx_1 is

$$\Delta x_1 = \frac{3\alpha_s}{2} \frac{m_p}{x_1 s} L_A \langle p_{\perp W}^2 \rangle \ . \tag{18}$$

Now the average transverse momentum squared of the parton, $\langle p_{\perp W}^2 \rangle$, is proportional to $A^{1/3}$ [43] so that $\Delta x_1 \propto A^{2/3}$ in Eq. (18) rather than $A^{1/3}$ as in Eq. (17) since the BH calculation assumed $\langle k_\perp^2 \rangle$ to be independent of A.

Two estimates of $\langle p_{\perp W}^2 \rangle$ were given in Ref. [43]. The larger value, an upper limit on Δx_1, comes from a single nuclear rescattering of photoproduced dijets [50],

$$\langle p_{\perp W}^2 \rangle = \pi^2 \alpha_s \lambda_{\rm LQS}^2 A^{1/3} \frac{C_A \sigma_g^{\gamma A} + C_F \sigma_q^{\gamma A}}{\sigma^{\gamma A}} \simeq 0.658\, \alpha_s\, A^{1/3}\, {\rm GeV}^2 \ . \tag{19}$$

We assume that $\langle p_{\perp W}^2 \rangle$ is identical for quarks and gluons in this case. When $\alpha_s \sim 0.3$ and $A = 184$, we find $-dE/dz \simeq 1.28$ GeV/fm, the "maximum BH loss". The second estimate depends on the nucleon gluon distribution and contains explicit color factors so that

$$\langle p_{\perp W}^2 \rangle_q = \frac{2\pi^2 \alpha_s}{3} \rho_A x G(x, Q^2) L_A \simeq 0.07\, \alpha_s\, A^{1/3}\, {\rm GeV}^2 \tag{20}$$

$$\langle p_{\perp W}^2 \rangle_g = \frac{9}{4} \langle p_{\perp W}^2 \rangle_q \simeq 0.15\, \alpha_s\, A^{1/3}\, {\rm GeV}^2 \tag{21}$$

where $xG(x) \sim 1-2$ for the x_1 range of E866. This lower estimate is the "minimum BH loss". Now when $\alpha_s \sim 0.3$ and $A = 184$, $-dE/dz|_q \simeq 0.12$ GeV/fm and $-dE/dz|_g \simeq 0.28$ GeV/fm.

We assume that shadowing and energy loss effects are independent. This assumption depends on the ultimate source of nuclear shadowing because the mechanism of initial-state energy loss is multiple parton scattering before the hard collision. If shadowing in deep-inelastic scattering is due to the recombination of high density partons, no rescattering is involved and the effects are independent. If shadowing is caused by $\gamma^\star \to q\overline{q}$ with the $q\overline{q}$ pair free to rescatter as a vector meson, the origin of the two effects is similar. The vector-meson approach is a low Q^2 effect, lower than the effective Q^2 needed for J/ψ production. Therefore we assume that the two effects are independent and include both in our calculations.

Final-State Loss

The second model of energy loss is applicable only to the quarkonium system and not to Drell-Yan production which does not involve color confinement in the final state [46]. When a $c\overline{c}$ pair is produced in a color octet state, it has to emit a soft gluon in order to produce the final-state J/ψ or ψ'. This $c\overline{c}$ can propagate a distance greater than its path through the nucleus before the soft gluon is finally emitted. This is because the Landau-Pomeranchuk-Migdal effect [48] in QCD which regenerates the color field after scattering delays the emission of the third gluon which neutralizes the color of the $c\overline{c}$ state. However, each successive interaction of the $c\overline{c}$ pair degrades its momentum. This final-state loss model, developed by Kharzeev and Satz [46] is applicable only when the $c\overline{c}$ pair interacts in the color octet state, essentially for $x_F \geq 0$. After n interactions along its path length before leaving the target, the pair's momentum is reduced so that a J/ψ observed at a given x_F has actually been produced with a higher value, x_F/δ.

The x_F distribution, $G_A(x_F)$, then has two parts [46],

$$G_A(x_F) \propto S_{\rm oct}^{\rm abs} G_p(x_F) + (1 - S_{\rm oct}^{\rm abs}) \frac{G_p(x_F/\delta)}{\delta} \theta(1 - x_F/\delta) \ , \tag{22}$$

where $G_p(x_F)$ is the x_F distribution in pp interactions and $S_{\rm oct}^{\rm abs}$ is calculated in Eq. (9). The second term includes the scatterings in the target that cause the shift in x_F. The effect of Eq. (22) does not produce an integrated reduction in α, only $\alpha(x_F)$ changes due to the shift in x_F.

Intrinsic Charm

The wavefunction of a proton in QCD can be represented as a superposition of Fock state fluctuations, *e.g.* $|uudg\rangle$, $|uudq\overline{q}\rangle$, $|uudQ\overline{Q}\rangle$, ... of the $|uud\rangle$ state [22,51]. These intrinsic $|uudQ\overline{Q}\rangle$ Fock states are dominated by configurations with equal rapidity constituents so that the intrinsic heavy quarks carry a large fraction of the parent momentum [22].

The frame-independent probability distribution of a 5–particle $c\overline{c}$ Fock state in the proton is

$$\frac{dP_{\rm ic}^5}{dx_i \cdots dx_5} = N_5 \alpha_s^4(m) \frac{\delta(1 - \sum_{i=1}^5 x_i)}{(m_p^2 - \sum_{i=1}^5 (\widehat{m}_i^2/x_i))^2} \ , \tag{23}$$

where N_5 normalizes the $|uudc\overline{c}\rangle$ probability, $P_{\rm ic}^5$. The delta function conserves longitudinal momentum.

While the total intrinsic charm cross section is relatively easy to define, there are some uncertainties in the relative weights of open charm and J/ψ production from an intrinsic charm state. Since these weight factors are not fixed [52], we use an effective intrinsic charm probability, $P_{\rm ic}^{\rm eff}$, proportional to $P_{\rm ic}^5$. The EMC charm structure function data is consistent with $P_{\rm ic}^5 = 0.31\%$ for low energy virtual photons but $P_{\rm ic}^5$ could be as large as 1% for the highest virtual photon energies [53,54].

Including a delta function to combine the x_c and $x_{\overline{c}}$ in the J/ψ, the J/ψ x_F distribution from intrinsic charm is

$$\frac{d\sigma_d}{dx_F} = \sigma_{pN}^{\rm in} \frac{\mu^2}{4\widehat{m}_c^2} \int \prod_{i=1}^5 dx_i \, \frac{dP_{\rm ic}^{\rm eff}}{dx_1 \ldots dx_5} \, \delta(x_F - x_c - x_{\overline{c}}) \ . \tag{24}$$

The factor of $\mu^2/4\widehat{m}_c^2$ arises from the soft interaction which breaks the coherence of the Fock state. We use $\mu^2 \sim 0.1$ GeV^2. Only the 5 particle Fock state is considered. The intrinsic charm contribution is included as in Eq. (6) with $\beta = 0.71$. We show the effect of reducing and/or eliminating the intrinsic charm component.

pA Results

We take the S_3 shadowing parameterization with GM and minimum BH loss and vary $P_{\rm ic}^{\rm eff}$ between 0 and 1% and compare to the E866 data in Fig. 1. The GM loss mechanism alone causes strong reduction in α at large x_F so that including intrinsic charm does not have a large effect. It would appear from Fig. 1(a) that $P_{\rm ic}^{\rm eff} = 0.31\%$ agrees best with the data although the agreement is reasonable in all three cases. The same is true for the combination model but pure singlet absorption would require a larger intrinsic charm probability to agree with the data. On the other hand, the relatively good agreement of the minimum BH loss calculations with the data at large x_F is due to the intrinsic charm contribution. If $P_{\rm ic}^{\rm eff} < 1\%$, the model calculations are far away from the data. Thus increasing the relative intrinsic charm contribution is the only way to produce agreement with the data at large x_F. This is clearly shown in Fig. 1(b). With no intrinsic charm, $\alpha(x_F)$ is relatively flat at large x_F. Similar results are obtained with the pure singlet and combination absorption models. Note that for both loss mechanisms, intrinsic charm only affects the shape of $\alpha(x_F)$ at $x_F > 0.25$.

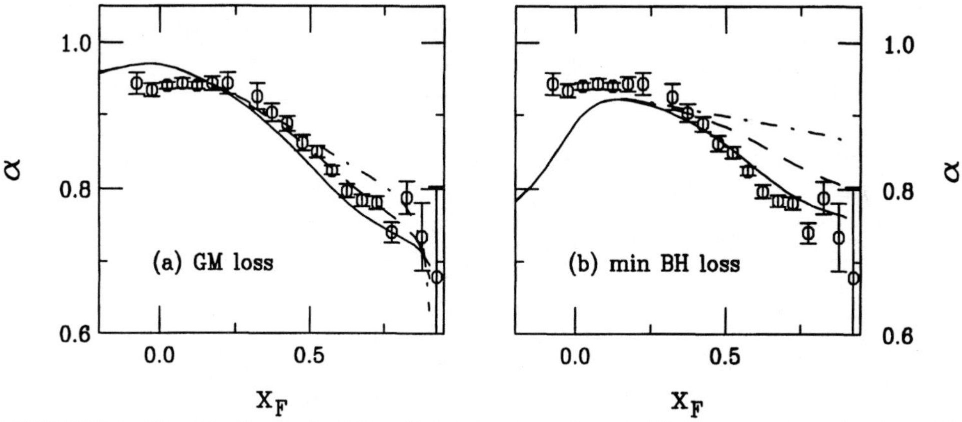

FIGURE 1. The effective probability of intrinsic charm is varied for pure octet production with (a) GM loss and (b) the minimum BH loss. The curves represent an effective intrinsic charm probability of 1% (solid), 0.31% (dashed) and 0% (dot-dashed). From Ref. [24]. Copyright 2000 by the American Physical Society.

The Drell-Yan data at 800 GeV do not show strong evidence for energy loss and, in fact, definitively rule out the maximum BH loss. However, since both the shadowing parameterizations and the GM loss model use the E772 data [19] data to fix model parameters, the importance of either mechanism for Drell-Yan production is difficult to ascertain. More data on J/ψ, ψ' and Drell-Yan production as a function of x_F at other energies is needed to determine the importance of all effects. Such measurements have been proposed at 120 GeV [55].

NUCLEUS-NUCLEUS COLLISIONS

It has been suggested that nucleus-nucleus collisions could create a quark-gluon plasma [56]. The prediction relies on the fact that the properties of quarkonium bound states can be described nonrelativistically. In particular, the J/ψ of the $c\overline{c}$ states and the Υ of the $b\overline{b}$ states are rather tightly bound with correspondingly small radii, smaller than normal hadrons which have $r \sim 1$ fm. These larger hadrons are broken into their constituent quarks at T_c, the critical temperature for the formation of a quark-gluon plasma. However, the smaller quarkonium states will remain bound above T_c, until their breakup temperatures T_D have been reached. This breakup occurs when the bound state energy,

$$E(r,T) = 2m_Q + \frac{\langle p^2\rangle\langle r^2\rangle}{m_Q r^2} + V(r,T) \; , \tag{25}$$

no longer has a minimum [28,56]. The actual values of T_D depend on the functional form of the screening mass, $\mu(T)$, which will be discussed later.

At the CERN SPS, only the charmonium states are produced frequently enough in AB collisions for them to be studied with meaningful statistics. The ψ' is also measured but because its event rate is only a few percent of the J/ψ, its measurement is not definitive. The J/ψ measurement is typically presented as a ratio with respect to Drell-Yan production to reduce systematic errors. The ratio has been shown as a function of global transverse energy, E_T, path length, L, and system size, AB. In this section, we discuss the presentation of the SPS data and describe several models used to confront the data. We then make predictions of J/ψ and Υ suppression patterns at RHIC, where experiments are just beginning, and the future LHC at CERN.

The Data

The only group currently measuring J/ψ production in nucleus-nucleus collisions is the NA38 (S+U) and NA50 (Pb+Pb) collaboration. The experimental setup consists of a dimuon spectrometer to measure the signal dimuons and up to three ways of measuring the centrality of the collision. These centrality detectors are a zero degree calorimeter (ZDC), an electromagnetic calorimeter, and a multiplicity counter. The ZDC measures the forward going energy of the projectile remaining after the collision while the electromagnetic calorimeter and multiplicity counter collect the energy and particles, respectively, blown to the sides by the collision. The energy measured by the calorimeter is referred to as transverse energy or E_T. The ZDC and E_T measurements are strongly correlated since large energy in the ZDC corresponds to low E_T, a peripheral collision with high impact parameter, b, while little energy in the ZDC corresponds to a central collision with high E_T.

The lead target itself has played an important role in the interpretation of the data, as we now describe. The lead beam runs began in 1995 at the CERN SPS. The 1995 and 1996 runs used seven subtargets amounting to 17% and 30% of an interaction length respectively. The longer interaction length in the 1996 run significantly increased the J/ψ database. However, the large number of subtargets and the relatively long interaction length increased the probability of reinteractions in the same target or another subtarget. Reinteractions can lead to a medium E_T event appearing to be at higher E_T. To remedy this situation, in the 1998 run the NA50 collaboration used only a single target with 7% of an interaction length. Fewer J/ψ's were obtained but the reinteraction problem was considerably reduced. Since only data with $E_T > 40$ GeV were obtained in 1998, the NA50 collaboration will take data with the target in vacuum to improve their low E_T data. As we will see, any misunderstanding of the high E_T data could lead to quite different physics interpretations.

Over the years of SPS heavy ion operations, J/ψ and Drell-Yan data have been collected at 158 GeV (Pb+Pb), 200 GeV (pW, pU, O+U, and S+U), and 450 GeV (pp, pd, pC, pAl, pCu, and pW). It is attractive to try to present all the data collectively. This requires a number of adjustments for rapidity and angular acceptance, isospin of the projectile and target, the Drell-Yan mass region, and projectile energy, as we outline below. Once these adjustments have been carried out, the E_T integrated data is plotted as a function of AB. Two facts are immediately evident from the data. The first is that the ratio of measured Drell-Yan to a leading order calculation is independent of AB, signifying a linear A dependence and a constant K factor. The second is that the J/ψ per nucleon cross section follows the $(AB)^\alpha$ dependence expected from pA interactions until the Pb+Pb point which lies 30% below the "normal" curve. This difference is the anomalous suppression in Pb+Pb interactions [3].

The differential J/ψ data has been presented two ways: as a function of E_T, and L [3]. Both are approximate measures of the centrality, or impact parameter, of the collision. Low impact parameter, high E_T and L, collisions are almost head on while high impact parameter, low E_T and L, are grazing collisions where the nuclei barely touch. When the data is presented as a function of L, all the data can be shown together after the adjustments mentioned above are made. All the data are reported in the rapidity interval $0 < y_{\rm cm} < 1$ and in the measured angular interval in the Collins-Soper frame $|\cos\theta_{\rm CS}| < 0.5$. Muon pairs from J/ψ decays are isotropic in phase space [57] so that a factor of two is needed to adjust calculations of the J/ψ cross section from $|\cos\theta_{\rm CS}| < 1$ to $|\cos\theta_{\rm CS}| < 0.5$. However, the Drell-Yan cross section is proportional to $1 + \cos^2\theta_{\rm CS}$ [17], resulting in a reduction by a factor of 2.46 between the full region and the measured angular interval.

The continuum/Drell-Yan cross section was originally adjusted for energy, isospin and mass interval by calculations with the GRV LO parton distributions [41]. More recently the MRS A' [58] next-to-leading order, low Q^2, parton distribution functions have been used. The isospin correction is used to adjust the nucleus-nucleus data to the pp value at the same energy and mass interval, see Eq. (4). The Drell-Yan K factor is obtained by comparing the measured cross section in the given mass interval to the calculated LO cross section. The isospin correction and K factor depend quite strongly on the parton distribution functions used in the calculation. The GRV LO set gives an isospin correction in Pb+Pb collisions of pp/PbPb = 1.30 at $\sqrt{s} = 19.4$ GeV and $2.9 < M < 4.5$ GeV while the MRS A' set indicates a correction of only 1.03 for the same calculation.

The Drell-Yan data are reported in the mass range $2.9 < M < 4.5$ GeV. The original S+U data reported in Ref. [1] was for the "continuum" in the range $1.7 < M < 2.7$ GeV and included not only low-mass Drell-Yan pairs but also contributions from open charm decays and an intermediate mass enhancement [59]. After the intermediate mass enhancement was identified, the K factor was fit to the continuum above $M = 4.2$ GeV, a region where only Drell-Yan pairs are expected to be important since dileptons from open charm decays are small. This Drell-Yan calculation was then extrapolated backwards to lower masses and the S+U data was reported for Drell-Yan pairs alone in the mass interval $1.5 < M < 5.5$ GeV [59]. The Pb+Pb data was reported in the more restricted interval $2.9 < M < 4.5$ GeV and the S+U data was consequently adjusted to the Pb+Pb mass range [3].

Most recently, the 1996 and 1998 Pb+Pb data have been compared to "minimum bias" Drell-Yan. This has been done because the accuracy of the J/ψ to Drell-Yan ratio is limited by Drell-Yan production. The minimum bias E_T distribution does not include the lepton pair trigger and thus has a many times greater statistical base. The minimum bias data reduced the error bars considerably [60]. The 1998 run with the thin target showed that reinteraction effets were significant, leading to a decrease of the J/ψ to Drell-Yan ratio at large E_T, as seen in Fig. 2. The reinteractions in the 1996 data caused the J/ψ to Drell-Yan ratio to appear to be constant at large E_T, as also shown in Fig. 2.

The energy adjustment of the J/ψ cross section has been performed with the "Schuler parameterization" [3]. No isospin correction is needed for the J/ψ since its production cross section is dominated by gluon fusion. The dilepton invariant mass range corresponding to the J/ψ peak is $2.7 < M < 3.5$ GeV.

Models of J/ψ Suppression

We now discuss several models of J/ψ suppression and compare these to the data. We first describe analytical models in some detail. With this foundation, we then briefly discuss suppression in microscopic transport models. Since the success of some of these calculations depends strongly on the comover cross section, we mention recent attempts to quantify this cross section. We then describe how

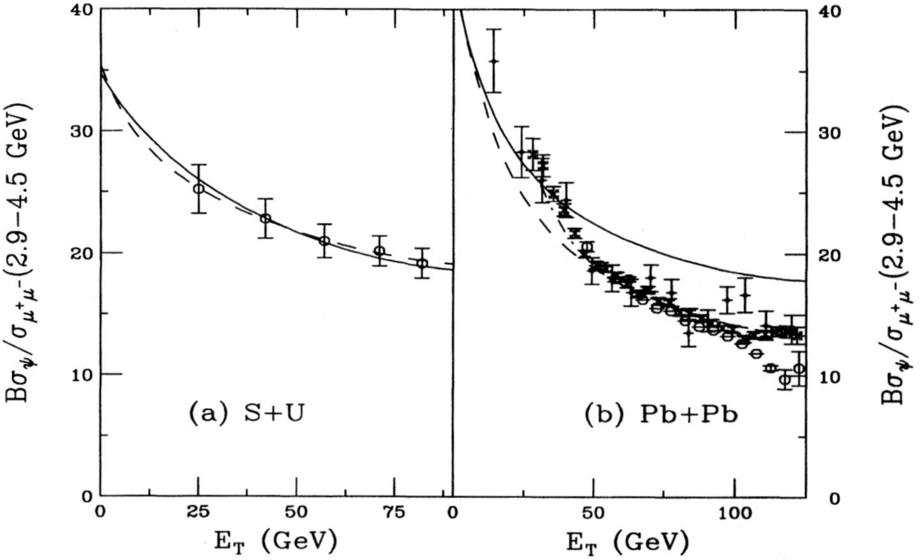

FIGURE 2. The calculated J/ψ to Drell-Yan ratios are compared with data from (a) S+U and (b) Pb+Pb interactions [9,60]. The 1996 Pb+Pb J/ψ to measured Drell-Yan data is given by the '+' symbols [60], the 1996 J/ψ to minimum bias data is given by the '×' symbols [60], and the 1998 J/ψ to minimum bias data [9] is given by the circles. The solid curves show $\sigma_{\psi N} = 4.8$ mb and $\sigma_{\psi co} = 0.67$ mb. The dashed curve in (a) shows nuclear absorption alone with $\sigma_{\psi N} = 7.3$ mb. Two additional calculations are shown in (b): increasing the participant density by a factor of two (dashed) and a quark-gluon plasma with $n_f = 3$ (dot-dashed).

nuclear shadowing effects could alter this picture. Finally, we discuss how the assumption of a phase transition affects the description of the data.

Analytical Models

To quantitatively compare suppression calculations to the data requires a model of the global E_T distribution. Early results from the CERN SPS heavy-ion measurements showed that the global transverse energy of the collision, E_T, is proportional to the multiplicity of produced particles [61]. The mean multiplicity is directly proportional to the number of nucleon participants in the initial collisions [62],

$$\begin{aligned} N_{AB}(b) &\approx \int d^2s \left[T_A(\vec{s})\left(1 - e^{-\sigma_{\rm in}^{pp} T_B(\vec{b}-\vec{s})}\right) + T_B(\vec{b}-\vec{s})\left(1 - e^{-\sigma_{\rm in}^{pp} T_A(\vec{s})}\right)\right] \\ &= \int d^2s \left[N_A(b,s) + N_B(b, b-s)\right] \ , \end{aligned} \tag{26}$$

where $\sigma_{\rm in}^{pp}$ is the inelastic pp interaction cross section. The actual number of particles produced at a given impact parameter fluctuates. Thus the probability to produce E_T at b is

$$p(E_T; b) = \frac{1}{\sqrt{2\pi\sigma^2(b)}} \exp\left(- \frac{(E_T - \overline{E}_T^{AB}(b))^2}{2\sigma^2(b)}\right) \ , \tag{27}$$

where the mean E_T, $\overline{E}_T^{AB}(b)$, and standard deviation, $\sigma(b)$, are [63]

$$\overline{E}_T^{AB}(b) = \epsilon_P N_{AB}(b) \qquad \sigma^2(b) = \omega \epsilon_P \overline{E}_T^{AB}(b) \ . \tag{28}$$

The energy per participant is ϵ_P and ω governs the fluctuations in $N_{AB}(b)$. The parameters ϵ_P and ω are chosen to agree with the NA38/NA50 neutral E_T distributions. In S+U interactions $\epsilon_P = 0.74$ GeV. The 1996 Pb+Pb data were originally analyzed with $\epsilon_P = 0.4$ GeV. However, after the E_T scale was adjusted to agree with that obtained using the RQMD event generator [64], the E_T scale was reduced [65] so that $\epsilon_P = 0.27$ [66]. A smaller ϵ_P is needed for Pb+Pb because the pseudorapidity acceptance of the electromagnetic calorimeter in the laboratory frame was reduced to $1.1 < \eta < 2.3$ from $1.7 < \eta < 4.1$ for S+U. The same value, $\omega = 3.2$, is used to set the width of the fluctuations in both cases.

The minimum bias cross section as a function of E_T is then

$$\frac{d\sigma}{dE_T} = \int d^2b \, p(E_T; b) \ . \tag{29}$$

The number of hard processes produced in the collision is proportional to the nuclear overlap per area,

$$T_{AB}(b) = \int d^2s T_A(\vec{s}) T_B(\vec{b} - \vec{s}) \ . \tag{30}$$

The Drell-Yan production cross section, adjusted for isospin, in nuclear collisions is then

$$\frac{d\sigma_{AB\to\mu^+\mu^-}}{dE_T} = \sigma_{NN\to\mu^+\mu^-} \int d^2b \int d^2s\, T_A(\vec{s})T_B(\vec{b}-\vec{s})p(E_T;b)\ . \tag{31}$$

Note that the minimum bias analysis obtains the adjusted Drell-Yan cross section by multiplying the measured minimum bias E_T by a calculation of the ratio of Eq. (31) to Eq. (29) with $\sigma_{NN\to\mu^+\mu^-}$ replaced by $\sigma_{\rm in}^{pp}$. This ratio is the number of nucleon-nucleon collisions [60].

In Fig. 3 the Drell-Yan E_T distributions for S+U and Pb+Pb interactions calculated with Eq. (31) are compared with data from Refs. [67,68] respectively. The continuum S+U data for $M > 4$ GeV is from Ref. [67]. The Pb+Pb Drell-Yan calculation is compared to the number of pairs with $M > 4.2$ GeV [68] using the 1995 E_T scale since the Drell-Yan E_T distribution with the new E_T scale has not been published. Note that the integrated Drell-Yan cross section [3] agrees with the calculated value.

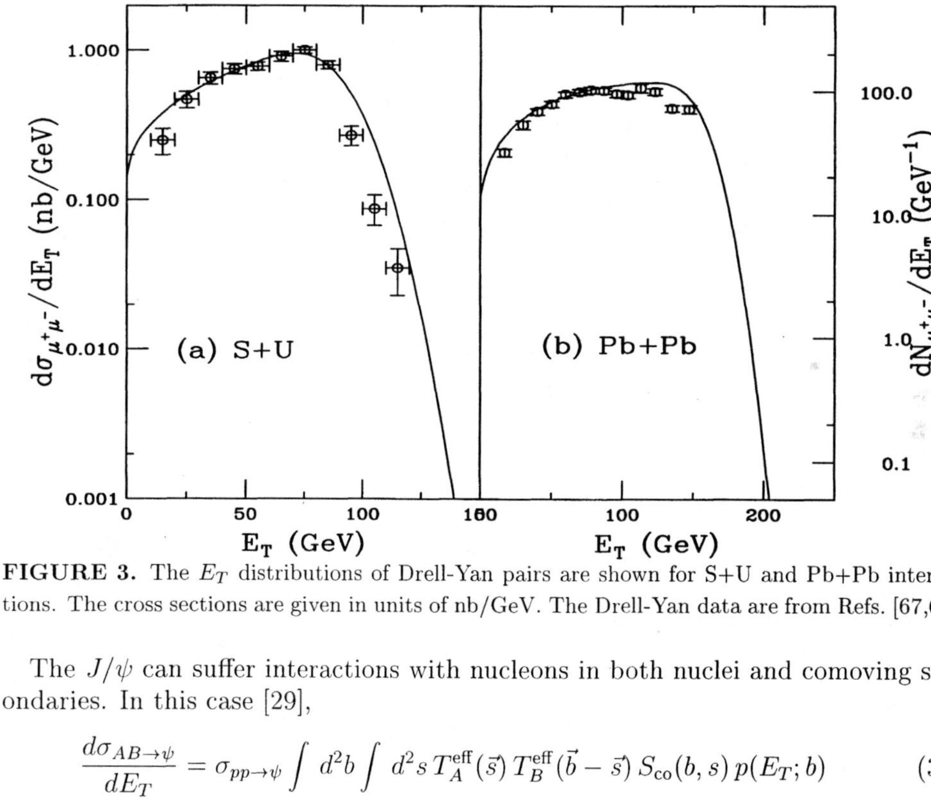

FIGURE 3. The E_T distributions of Drell-Yan pairs are shown for S+U and Pb+Pb interactions. The cross sections are given in units of nb/GeV. The Drell-Yan data are from Refs. [67,68].

The J/ψ can suffer interactions with nucleons in both nuclei and comoving secondaries. In this case [29],

$$\frac{d\sigma_{AB\to\psi}}{dE_T} = \sigma_{pp\to\psi} \int d^2b \int d^2s\, T_A^{\rm eff}(\vec{s})\, T_B^{\rm eff}(\vec{b}-\vec{s})\, S_{\rm co}(b,s)\, p(E_T;b) \tag{32}$$

where *e.g.* $T_A^{\rm eff}(\vec{s}) = T_A(\vec{s})S_A^{\rm abs}(\vec{s})$. The comover survival probability, $S_{\rm co}(b,s)$, is modified for AB collisions as [11]

$$S_{\rm co}(b,s) = \exp\left\{-\langle\sigma_{\psi{\rm co}}v\rangle n_{AB}(b,s)\ln\left(\frac{\tau_I(b)}{\tau_0(b,s)}\right)\right\} , \qquad (33)$$

where $n_{AB}(b,s)$ is the participant density, $n_{AB}(b,s) = N_A(b,s) + N_B(b,b-s)$, as in Eq. (26). The comover formation time depends on the path length $L_A(b) = N_A(b)/2\sigma_{\rm in}^{pp}\rho_A(0)$ [69,70] so that

$$\tau_0(b,s) = 1 + \frac{L_A(b,s)}{\gamma_A(b,s)} + \frac{L_B(b,s)}{\gamma_B(b,s)} \qquad (34)$$

is ~ 2 fm in central collisions and ~ 1 fm in the most peripheral collisions. The comovers interact with the J/ψ only if $\tau_I(b) > \tau_0(b,s)$ where

$$\tau_I(b) = \begin{cases} R_A/v & , \quad b < R_B - R_A \\ (R_A + R_B - b)/(2v) & , \quad R_B - R_A < b < R_B + R_A \end{cases} . \qquad (35)$$

The calculations are compared to the NA38 and NA50 data in Fig. 2. While the E_T dependence of the S+U data can be described by either absorption alone or by absorption and comover interactions, the extrapolation to Pb+Pb interactions fails. In this picture, there are two possible explanations for this. Either the comover density increases more rapidly than the number of participants in Pb+Pb interactions [71] or a threshold has been crossed, possibly resulting in the production of a new state of matter. The dashed curve in Fig. 2(b) shows the effect of doubling the participant density in Pb+Pb collisions. While the agreement with the 1996 data is reasonable above 50 GeV, there is no way to significantly increase the suppression above 100 GeV, as observed in the 1998 data, without introducing important elements that do not appear in light ion collisions.

Capella and collaborators [71] have refined the comover density to include the fluctuations beyond the turnover of the E_T distribution at high E_T, similar to the early calculations of Gavin and Vogt [29]. Their comover density, $N_{\rm co}(b,s)$, which includes contributions proportional to the number of collisions as well as the number of participants, is thus multiplied by the ratio $E_T/E_T^{NF}(E_T)$. The non-fluctuating transverse energy, $E_T^{NF}(b)$, replaces $\overline{E}_T^{AB}(b)$ in Eq. (27) with [71]

$$E_T^{NF}(b) = k[A - T_A(b)]E_{\rm in} + q\int d^2s\,dy\,N_{\rm co}(b,s) , \qquad (36)$$

where $q = 0.56$ GeV and $E_{\rm in} = 158$ GeV. The integral over s and y gives the comover density in the rapidity interval covered by NA50 without the multiplicative factor proportional to E_T. The first term of E_T^{NF} is used to fix agreement of the transverse and zero degree energies with $k = 1/4000$ from the shape of the E_T-ZDC correlation at low E_T. The width of the Gaussian is $qaE_T^{NF}(b)$ with $a = 0.94$. In this calculation, the number comovers changes faster above the knee of the E_T distribution due to fluctuations, causing the J/ψ to Drell-Yan ratio to decrease faster with E_T than previous comover models. However, the inflection is small and does not significantly improve the agreement with the data.

One of the most important considerations regarding the NA50 analysis is their comparison curve which assumes nuclear absorption alone with $\sigma_{\psi N} = 6.4$ mb. The absorption cross section is obtained by fitting α with Eq. (9) when A is large. This simple absorption curve can explain the trend of all the NA38/NA50 data except the Pb+Pb data. The departure of the 1996 data from this curve at $E_T \approx 40$ GeV has been referred to as a "step". However, there is no significant change in slope in the data at this E_T [72]. The "anomalous" J/ψ suppression is the deviation from this curve. The only change in slope appears in the 1998 data which decrease above $E_T = 100$ GeV. Most simple models of comover absorption cannot explain any decrease at high E_T. Capella's calculation includes a mechanism, the fluctuations, by which this decrease might occur [71] but it is too weak to explain the data.

Microscopic Models

A way to move beyond the simple analytical models described in the previous section is to go to a microscopic transport approach. A number of microscopic hadronic transport models have been used to study J/ψ suppression [73–79]. To simulate J/ψ suppression in microscopic models, energy loss in soft particle interactions with nucleons must be neglected for the A dependence of charmonium to be consistent with that observed in pA interactions [8]. In the UrQMD model [80], perturbative production and hadronic cascades are treated separately to avoid strong correlations betwen hard and soft processes [79]. The charmonium-nucleon absorption cross section is assumed to grow in size [81], similar to the color singlet model of absorption described earlier. The transport calculation includes all states of eight different meson multiplets which are populated through string fragmentation and which interact with the J/ψ with a constant cross section equal to 2/3 the asymptotic J/ψ-nucleon cross section. Contributions from χ_{cJ} and ψ' interactions and decays are included in the total J/ψ suppression. In the model, only $\approx 37\%$ of the total J/ψ dissociation by comovers is due to π and ρ interactions.

The E_T dependence in transport models is obtained by simply adding up the transverse energies of all particles falling into the acceptance. In the lefthand side of Fig. 4, we show the Drell-Yan E_T distribution obtained in the model for Pb+Pb interactions, compared to the 1995 Drell-Yan data. The E_T scale of the data has been reduced 20% to simulate the NA50 adjusted E_T scale [79]. The results are encouraging, especially since the UrQMD model successfully reproduced the NA49 S+Au and Pb+Pb minimum bias data [82] without adjustment.

The J/ψ to Drell-Yan ratio is compared to the Pb+Pb data in the righthand side of Fig. 4. The gross features of the 1996 data are relatively well described by the model except at low E_T. The model calculation is smooth and saturates with increasing E_T. The high E_T decrease in the 1998 data is inconsistent with the simulation, as shown in the triangular points. These results are in agreement with those found by other transport models [73–77]. The major difference between the analytic calculations described previously and the transport models is the greater

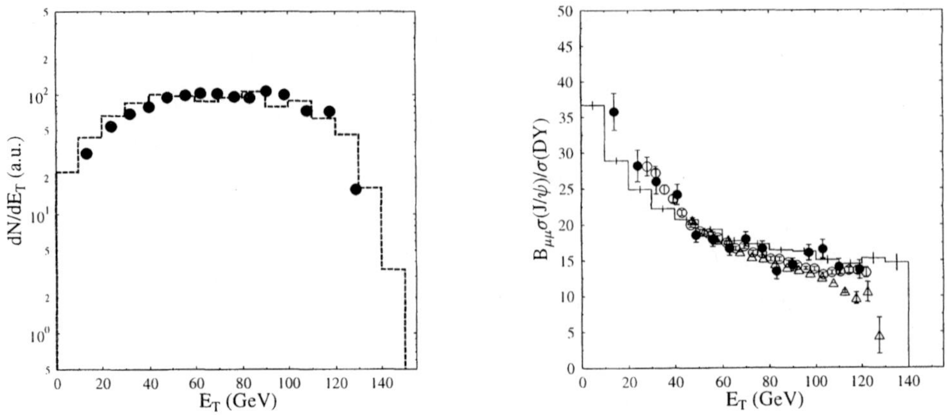

FIGURE 4. The lefthand plot shows the Drell-Yan E_T distribution with the E_T axis of the data rescaled by 0.8 to agree with the new E_T scale. The righthand plot shows the J/ψ to Drell-Yan ratio as a function of E_T for Pb+Pb collisions at 158 GeV [9,60] compared to the 1996 Drell-Yan data (closed circles), 1996 minimum bias data (open circles) and 1998 minimum bias data (triangles). Modified from Ref. [79]. Copyright 1999 by the American Physical Society.

increase in the number of interacting particles. In the transport model, the meson yield increases proportional to $N_{AB}^{1.25}$ in central collisions.

The Comover Cross Section

The largest uncertainty in all of these models is the J/ψ-comover cross section. All the hadronic models discussed here have typically assumed $\langle \sigma_{\psi\mathrm{co}} v \rangle \approx$ 1-3 mb. However, calculations within the framework of short distance QCD suggest a strong suppression of such J/ψ dissociation due to the relative abscence of hard gluons in mesons [83]. If true, J/ψ-comover interactions would play a negligible role in the observed suppression. Of course, indirect contributions from χ_{cJ} and ψ' decays could still be suppressed in this scenario because the mass threshold is smaller or negligible for pion interactions with χ_{cJ} and ψ'. Interactions of these more massive charmonium states with mesons have not yet been calculated.

More recently, a number of groups have modeled J/ψ interactions with π and ρ mesons in the context of meson exchange models [84–87]. These calculations show that while threshold suppression is important, even for $\rho J/\psi \to D\overline{D}$, the calculated dissociation rate $\langle \sigma_{\psi\mathrm{co}} v \rangle$ is compatible with a 1-2 mb cross section [87].

Shadowing of the nuclear parton distribution functions was described in the discussion of pA interactions. However, the effect of shadowing on the E_T dependence of the NA50 data has only recently been considered. There is some evidence [88] that shadowing depends on the parton location in the nucleus. This spatial dependence could affect the interpretation of the J/ψ to Drell-Yan ratio in heavy ion collisions because quarks and gluons are affected differently by shadowing.

Shadowing is assumed to be proportional to the local nuclear density [89]. The spatial dependence is parameterized as

$$S^i_{\rm WS} = S^i(A,x,Q^2,\vec{r},z) = 1 + N_{\rm WS}[S^i(A,x,Q^2) - 1]\frac{\rho(s)}{\rho_0} \tag{37}$$

where $N_{\rm WS}$ is chosen so that $(1/A)\int d^3s\rho(s)S^i_{\rm WS} = S^i$. For lead, $N_{WS} = 1.32$. At large radii, $s \gg R_A$, $S^i_{\rm WS} \rightarrow 1$ while at the nuclear center, the modifications are larger than the average S^i. An alternative parameterization, proportional to the thickness of a spherical nucleus [35], leads to a slightly larger modification in the nuclear core. The spatially averaged shadowing parameterizations are the same as those described earlier.

Shadowing affects different Drell-Yan mass regions differently because low M corresponds to lower x than high M. Thus the experimental K factor could be underestimated in the mass region $2.9 < M < 4.5$ GeV since the Drell-Yan normalization is obtained by extrapolating from the region $M > 4.2$ GeV. This discrepancy would increase in peripheral collisions showing that using a calculation to extrapolate to an unmeasured region could be problematic [89].

In Fig. 5, the J/ψ to Drell-Yan ratio obtained from spatially-dependent shadowing alone is presented as a function of E_T. After adjusting for the NA50 acceptance, with no shadowing, the J/ψ to Drell-Yan ratio ~ 40.3, in agreement with the NA50 data [3]. The combined Drell-Yan shadowing and J/ψ antishadowing in the x region of NA50 increases the ratio at low E_T to 40.6 with S_1, 44.5 for S_2 and 54.4 using S_3. The S_1 ratio is independent of E_T because S_1 is identical for quarks and gluons. However, S_2 and S_3 vary with E_T. The S_2 ratio rises about 7% while the S_3 ratio increases $\approx 11\%$ as $\langle E_T \rangle$ grows from 14 GeV to 120 GeV. These enhancements are opposite to the decrease with E_T [3], neglecting shadowing. Because of the uncertainties in the nuclear gluon distribution, it is difficult to draw detailed conclusions. However, S_1 should represent a lower limit and S_3 an upper limit. A stronger spatial dependence would slightly increase the effect with E_T while other parameterizations might predict a smaller effect. The calculation suggests that if the S_3 parameterization is correct, the suppression effect could be underestimated. This should be investigated further.

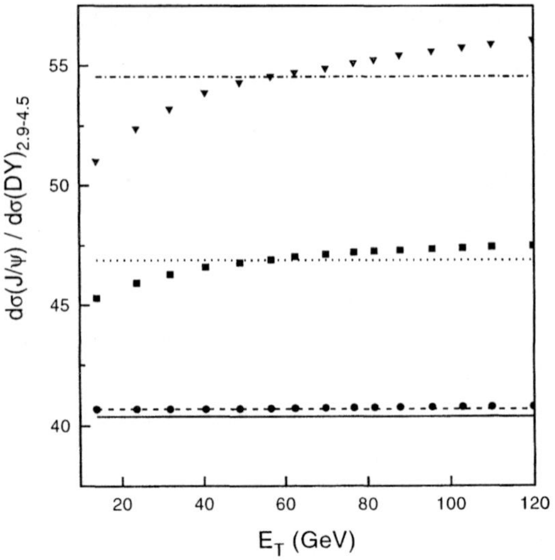

FIGURE 5. The ratio of J/ψ to Drell-Yan production, as a function of transverse energy, E_T. The curves correspond to b-averaged results with S_1 (dashed), S_2 (dotted) and S_3 (dot-dashed). The spatial dependence is illustrated for $S_{1,\rm WS}$ (circles) $S_{2,\rm WS}$ (squares) and $S_{3,\rm WS}$ (triangles). From Ref. [89]. Copyright 1999 by the American Physical Society.

Plasma Screening?

In this section, we discuss two calculations that consider a quark-gluon plasma origin of the J/ψ suppression seen in Pb+Pb collisions. The first describes a simple model of the phase transition based on energy density and discusses possible initial conditions leading to the transition [32]. The second, models the phase transition by percolation and thus does not rely on the model-dependent energy density [90].

As discussed previously, the quarkonium states break up in a quark-gluon plasma if their bound state energy can no longer be minimized at some finite temperature T_D. The temperature at which breakup occurs depends strongly on $\mu(T)$, the screening mass which modifies the quarkonium potential at finite temperature. We describe two estimates of the screening mass which lead to different predictions of the quarkonium suppression patterns, referred to hereafter as Case I and Case II. Case I is based on perturbative estimates of the screening mass [91],

$$\frac{\mu(T)}{T_c} = \sqrt{1 + \frac{n_f}{6}\, g\left(\frac{T}{T_c}\right)}\, \frac{T}{T_c} \, . \qquad (38)$$

We use $n_f = 3$ and $T_c = 170$ MeV [92]. In their prediction of J/ψ suppression, Matsui and Satz [56] used a parameterization based on SU(N) lattice simulations [93],

$$\frac{\mu(T)}{T_c} \simeq 4\frac{T}{T_c} \qquad (39)$$

where $T_c = 260$ MeV for SU(3) [94]. This is our Case II. The resulting values of T_D in both scenarios are given in Table 1.

		T_D (MeV)			
$c\bar{c}$	Case I ($n_f = 3$)	Case II (SU(3))	$b\bar{b}$	Case I ($n_f = 3$)	Case II (SU(3))
J/ψ	406	260	Υ	994	391
ψ'	189	260	Υ'	386	260
χ_{cJ}	178	260	χ_{bJ}	314	260

TABLE 1. The values of T_D for the two choices of $\mu(T)$, Eq. (38) from perturbative estimates assuming the high-temperature limit and the pure gluon SU(3) case, Eq. (39).

The importance of any possible plasma effects depends on the energy density, ϵ, of the system in this idealized model. The energy density is not directly measurable but must be inferred from the relationship between E_T and the average number of collisions per area [70,95,96]. This correlation cannot fix $\epsilon(E_T)$ precisely—fluctuations in E_T cause the energy density bins to overlap, especially in the most central collisions. Comparing the ranges of ϵ reached in S+U and Pb+Pb collisions suggests that the maximum ϵ obtained in S+U collisions is less than that at which the anomalous suppression sets in in the new Pb+Pb data [97].

The plasma predictions build upon the naive hadronic suppression model just discussed. The survival probability, Eq. (12), is modified to include plasma screening effects on each state. The calculated J/ψ to Drell-Yan ratio including quark-gluon plasma production with radius $R = R_{\rm Pb}$ and $p_{T\psi} \approx 0$ are compared with the Pb+Pb data in Fig. 2(b). The calculation is in better agreement with the low E_T data than simply increasing the participant density and, in fact, does very well in describing all the data below $E_T \approx 100$ GeV. However, it is obvious that an additional suppression mechanism is needed at large E_T to describe the new data. Capella *et al.* [71] have assumed this is due to the E_T fluctuations at small impact parameter. However, as explained earlier, this effect is too small to explain the drop. Blaizot *et al.* have obtained a better fit by assuming the energy density increases due to fluctuations in the number of participants [96,66]. They were able to explain the original anomalous suppression [60] assuming a threshold for additional suppression when $n_{\rm PbPb} \geq n_{\rm SU}^{\rm max}$ [96]. Including two thresholds gives excellent agreement with the 1998 data [66].

Satz and collaborators have proposed modeling the quark-gluon plasma phase transition by percolation [98]. This method of studying properties of phase transitions has been applied to many systems [99]. Percolation may be described as the populating of an area πR^2 with thin discs of area πr^2. Discs which, when randomly distributed over the surface area πR^2, overlap each other, begin to form clusters. When the overlapping clusters span the entire system, the percolation point has been reached. If the discs have color, like partons, the percolation point corresponds to the onset of deconfinement.

In nucleus-nucleus collisions, the percolation point is reached when $\approx 50\%$ of the surface is covered with discs, corresponding to a cluster density of $n_{\rm cl} = 6\,{\rm fm}^{-2}$, reached at $b \approx 8$ fm [90]. This model avoids the uncertainties in the determination of the energy density that the calculation of the dot-dashed curve in Fig. 2(b) is subject to since the $E_T - b$ correlation is better defined and less model dependent.

Note that in both calculations, the χ_{cJ} and ψ' states are both essentially suppressed at the transition point. The J/ψ must be dissociated at a higher temperature. This occurs at a smaller impact parameter than the percolation point, assumed to occur at an energy density a factor of 1.6 greater than the energy density at the critical point. In this case, the model dependence of the energy density is again introduced. Thus, in this picture, two "steps" are then obtained which are smeared out by fluctuations in E_T with impact parameter. If the number of gluons increases above the nonfluctuating E_T, then more J/ψ suppression can occur at large E_T, in accord with the 1998 data [100].

Quarkonium Suppression at RHIC and LHC

The colliders at RHIC and LHC will collide nuclei at energies factors of 10 and 275 greater than the energy of the CERN SPS, opening up a new frontier in quarkonium physics.

Not only will J/ψ's be produced in the normal channels but new production channels will open up. The $b\bar{b}$ production cross section is large enough for $B \to J/\psi X$ decays to contribute to the observed J/ψ yield, particularly if the normal channels are suppressed by plasma production. This late-time contamination could probably be separated from initially produced and suppressed charmonium by detecting the decay vertex. Additionally, such J/ψ's will have lower transverse momentum, p_T, relative to the maximum p_T at which initial J/ψ's can be produced.

At high energies, J/ψ's can also be produced at later times than those produced in the initial nucleon-nucleon collisions. At RHIC, up to 10 $c\bar{c}$ pairs might be produced while at LHC, up to 400 $c\bar{c}$ pairs could be produced [13]. These $c\bar{c}$ pairs could propagate through the system, find each other, and recombine to form a J/ψ [101,102]. The calculations of recombination processes depend strongly on the initial $c\bar{c}$ production rate and the comover absorption cross section, needed to construct the reverse reaction $D\overline{D} \to J/\psi\pi$. A wide range of assumptions show that either this contribution is negligible or leads to J/ψ enhancement. However, these secondary J/ψ's would also have lower p_T than the initial production. Therefore, the suppression signal of low p_T J/ψ's could be contaminated from these additional sources, especially at the LHC where all these cross sections will be larger. Large p_T J/ψ's and the Υ family could be more useful plasma probes than low p_T J/ψ's. One would then want to look at the quarkonium suppression as a function of p_T.

Unfortunately the dilepton continuum will not be a useful reference process since $c\bar{c}$ and $b\bar{b}$ decays to dileptons are at least as large or larger than Drell-Yan production for high masses [103] and these decays could be influenced by energy loss and modify the spectrum as a function of p_T [104,105].

A viable alternative would be to compare yields of quarkonium states, $\psi'/J/\psi$ [27,106–109] and Υ'/Υ [110–113] as a function of p_T. These ratios have been found to be independent of p_T in pp, $p\bar{p}$ and pA interactions. Nuclear effects due to absorption, shadowing, and energy loss should affect the similar mass quarkonium states in nearly identical ways, canceling in the ratio [114]. Any p_T dependence due to comover scattering has been shown to be weak [115–117].

We note, however, that in the secondary production model of Ref. [101], while secondary J/ψ production is negligible, the low p_T ψ' suppression signal could be wiped out by the large secondary ψ' production. The possible Υ suppression might then be the clearer quark-gluon plasma signal. In this section, we show some predictions for the $\psi'/J/\psi$ and Υ'/Υ ratios at RHIC and LHC for initial production and suppression alone. If the ratios exhibit a significant p_T-dependence at large p_T in nucleus-nucleus collisions, it will be virtually certain that a quark-gluon plasma was formed.

The initial conditions strongly influence the p_T dependence of the suppression. We have made two different assumption about the initial conditions. The first, a minijet plasma, assumes fast equilibration at $\tau \sim 0.1$ fm with high initial temperature. This scenario depends on the behavior of the nuclear parton distribution functions, especially at the low momentum fractions probed in the initial nucleon-nucleon collisions at the LHC. Equilibration at these high temperatures is actually unlikely, even at the LHC, and less likely at RHIC [118]. Therefore we show results for minijets only at the LHC, with and without nuclear shadowing.

Alternatively, the initial conditions could be dominated by kinetic equilibration processes [119] with a correspondingly longer equilibration time, $t_0 \sim 0.5 - 0.7$ fm. This time is reached when the momentum distributions are locally isotropic due to elastic scatterings and the expansion of the system. Chemical equilibrium is generally not assumed but the system moves toward equilibrium as a function of time. Then the cooling of the plasma is more rapid than the Bjorken scaling [120] adopted here, producing incomplete suppression at low p_T. Because the equilibration time of the parton gas is longer than that obtained from the minijet initial conditions, the time the system spends above the breakup temperature is also longer, leading to stronger suppression even though T_0 is lower.

The time at which the temperature drops below T_D and the state can no longer be suppressed, $t_D = t_0(T_0/T_D)^3$, and the maximum quarkonium p_T for which the resonance is suppressed, $p_{T,\rm m} = M\sqrt{(t_D/\tau_F)^2 - 1}$, are given in Table 2 for Cases I and II with both the parton gas and minijet initial conditions. Unfortunately the short equilibration time of the minijet system correspondingly reduces the plasma lifetime in the scaling expansion, causing the minijet plasma to be too short-lived to produce quarkonium suppression in some cases. Results for the minijet initial conditions are given for the GRV 94 LO parton densities for both $S = 1$ (no shadowing) and the lowest temperatures obtained with shadowing when $S = S_1$. Note that the reduction of the initial temperature due to shadowing significantly reduces the p_T range of the suppression. However, this result can be distinguished

	Case I ($n_f = 3$)		Case II (SU(3))	
	parton gas (RHIC)			
	$T_0 = 550$ MeV, $t_0 = 0.7$ fm			
	t_D (fm)	$p_{T,\max}$ (GeV)	t_D (fm)	$p_{T,\max}$ (GeV)
J/ψ	1.2	2.8	6.63	22.6
ψ'	12.3	30.12	6.63	15.9
χ_{cJ}	13.7	23.75	6.63	11.1
Υ	-	0	1.95	22.3
Υ'	-	0	6.63	33.4
χ_{bJ}	-	0	6.63	23.2
	parton gas (LHC)			
	$T_0 = 820$ MeV, $t_0 = 0.5$ fm			
	t_D (fm)	$p_{T\text{m}}$ (GeV)	t_D (fm)	$p_{T\text{m}}$ (GeV)
J/ψ	4.12	13.96	15.69	54.0
ψ'	40.8	100.6	15.69	38.5
χ_{cJ}	48.9	85.47	15.69	27.2
Υ	-	0	4.6	56.53
Υ'	4.79	23.16	15.69	81.98
χ_{bJ}	8.90	32.42	15.69	58.9
	minijet plasma, $S = 1$ (LHC)			
	$T_0 = 820$ MeV, $t_0 = 0.1$ fm		$T_0 = 1.05$ GeV, $t_0 = 0.1$ fm	
	t_D (fm)	$p_{T\text{m}}$ (GeV)	t_D (fm)	$p_{T\text{m}}$ (GeV)
J/ψ	-	0	6.59	22.7
ψ'	8.17	19.8	6.59	15.8
χ_{cJ}	9.78	16.75	6.59	11.0
Υ	-	0	1.94	22.2
Υ'	-	0	6.59	33.2
χ_{bJ}	-	0	6.59	23.05
	minijet plasma, $S = S_1$ (LHC)			
	$T_0 = 699$ MeV, $t_0 = 0.1$ fm		$T_0 = 897$ MeV, $t_0 = 0.1$ fm	
	t_D (fm)	$p_{T\text{m}}$ (GeV)	t_D (fm)	$p_{T\text{m}}$ (GeV)
J/ψ	-	0	4.11	13.8
ψ'	5.06	11.9	4.11	9.4
χ_{cJ}	6.06	10.0	4.11	6.3
Υ	-	0	1.21	11.7
Υ'	-	0	4.11	19.2
χ_{bJ}	-	0	4.11	12.1

TABLE 2. Values of t_D, and $p_{T\text{m}}$ with Cases I and II for $\mu(T)$ with a parton gas and a minijet plasma with $S = 1$ and S_1 from the GRV 94 LO calculation.

from a case with no significant shadowing and a plasma with a smaller spatial extent [114].

Since it has been demonstrated that the χ_{cJ} and ψ' contributions to large p_T J/ψ production can be subtracted at $p\overline{p}$ colliders [27], the direct or 'prompt' ratio is displayed. However, it is doubtful that the prompt Υ rate can be successfully extracted because the feeding from χ_{bJ} states will be difficult to disentangle [121]. The Υ family is also more complex, including feeddown to the Υ from Υ', Υ'' and two sets of χ_{bJ} states and feeddown to the Υ' from the Υ'' and $\chi_{bJ}(2P)$ states. Thus in the Υ'/Υ ratio, all sources of Υ' and Υ, each associated with a different suppression factor, must be considered [114]:

$$\frac{\Upsilon'}{\Upsilon}\Big|_{\rm indirect} \equiv \frac{\Upsilon' + \chi_{bJ}(2P)(\to \Upsilon') + \Upsilon''(\to \Upsilon')}{\Upsilon + \chi_{bJ}(1P, 2P)(\to \Upsilon) + \Upsilon'(\to \Upsilon) + \Upsilon''(\to \Upsilon)} \,. \tag{40}$$

In computing this 'indirect' Υ'/Υ ratio it is assumed that the suppression factor is the same for the $\chi_{bJJ}(2P)$ and $\chi_{bJ}(1P)$ states and that identical suppression factors can be used for the Υ' and Υ''. The relative production and suppression rates in the color evaporation model, including the χ_{bJ} states, can be found in Ref. [114].

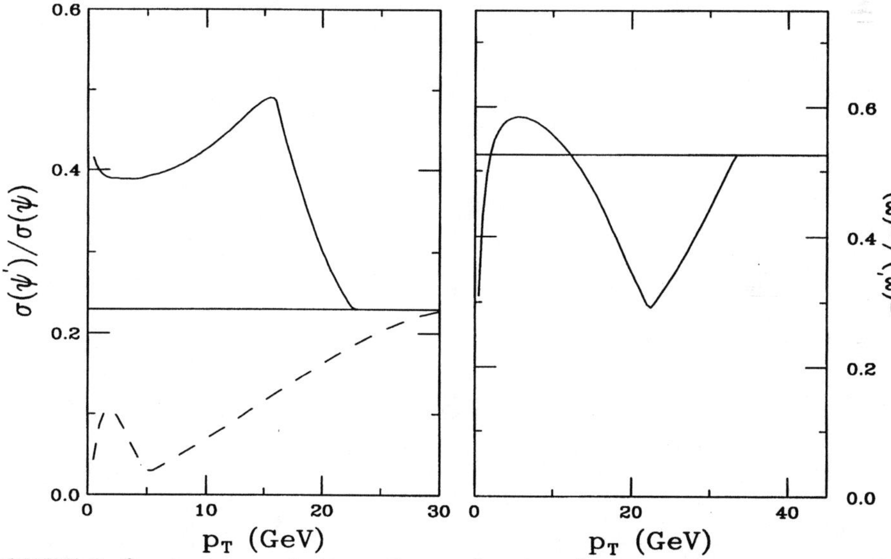

FIGURE 6. Quarkonium production ratios as a function of p_T assuming a parton gas at RHIC with plasma radius $R = R_{\rm Pb}$. The lefthand plot gives the direct or prompt ψ'/ψ ratio assuming Case I (dashed) and Case II (solid). The righthand plot shows the indirect Υ'/Υ ratio assuming case I. The horizontal lines represent the pp ratios.

Because the minijet plasma cannot effectively suppress the quarkonium states at RHIC, only the parton gas results are shown in Fig. 6. The lefthand plot shows the $\psi'/J/\psi$ ratio. Plasma screening in Case II indicates that the J/ψ is suppressed over a larger p_T range than the ψ' due to the difference in formation times. The ratio of cross sections is thus larger than that found in previous experiments. For a plasma with $R = R_{\rm Pb}$, the ratio continues to grow until the ψ' is no longer suppressed at $p_T = 16$ GeV and then drops smoothly to the pp value at $p_T = 23$ GeV. In Case I the ψ' is more strongly suppressed than the J/ψ, leading to a smaller ratio than in pp collisions. The J/ψ is no longer suppressed at $p_T = 3$ GeV, producing a small kink in the ratio which then proceeds to increase to the pp value when the ψ' is no longer suppressed.

The results for the indirect Υ'/Υ ratio at RHIC are shown on the righthand side of Fig. 6 for a parton gas with radius $R = R_{\rm Pb}$. At low p_T and also at $p_T > 10$ GeV, the total Υ' rate is suppressed more than the total Υ rate. The kink appears at $p_T \approx 23$ GeV when both the Υ and χ_{bJ} are no longer suppressed. If Υ suppression is observed at RHIC, the equilibration time must be relatively long. A significant deviation from the previous results for the ratio in pp and $p\bar{p}$ collisions would signal plasma formation.

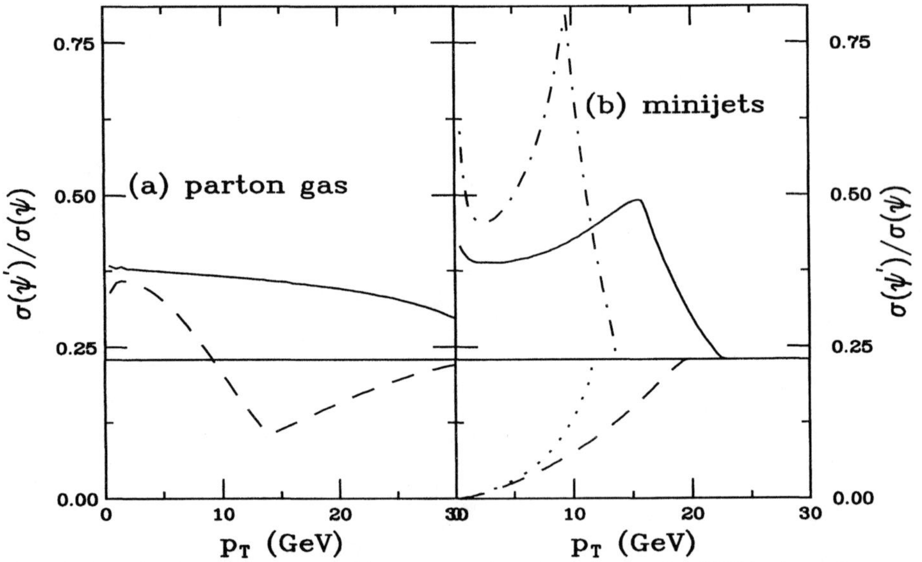

FIGURE 7. The direct or prompt ψ'/ψ ratio as a function of p_T at the LHC is shown for several choices of initial conditions and $R = R_{\rm Pb}$. In (a), parton gas results are shown for Case I (dashed) and Case II (solid). In (b) minijet results are given for both cases without shadowing, Case I (dashed) and Case II (solid), and with $S = S_1$, Case I (dotted) and Case II (dot-dashed). The horizontal curve represents the pp ratio.

In Fig. 7, the ratio of the directly produced ψ' and J/ψ cross sections are shown for several sets of initial conditions at the LHC. The parton gas produces suppression over nearly twice the p_T range as the minijet initial conditions, as shown in Table 2. In Case II, the J/ψ is more suppressed than the ψ' for a large range of p_T, up to 54 GeV for the parton gas. In Case I, the ψ' is more suppressed than the J/ψ except when $p_T < 9$ GeV in the parton gas. The kink in the dashed curve appears when the J/ψ is no longer suppressed. In each case, the p_T signature obvious in the $\psi'/J/\psi$ production ratios is unique if the full p_T range can be measured. Otherwise it may be difficult to distinguish between the parton gas and minijet plasma initial conditions for $p_T < 20$ GeV at the LHC unless the measurement is made with sufficiently high statistics. Note that even though the decreased initial temperature of the minijet gas when shadowing is included reduces the p_T range of the suppression, the shape of the ratio remains similar.

Figure 8 gives the indirect results for the Υ'/Υ ratio at the LHC. In a parton gas assuming a plasma like Case II, all the Υ states can be suppressed for $p_T > 50$ GeV, producing the rather flat ratio given in the solid curve. A measurement at the 20% level is thus needed to distinguish between the pp value of the ratio and the QGP prediction. Substantial systematic errors in the ratio could make the detection of a deviation quite difficult due to the slow variation with p_T. This is a disadvantage of the indirect ratio: the prompt $\psi'/J/\psi$ ratio is enhanced by nearly a factor of two over the pp value, making detection easier. With the slowly growing screening mass of Case I, the direct Υ rate is not suppressed while the Υ' and χ_b states are suppressed. Under these conditions, the indirect ratio is less than the pp value until the Υ' is no longer suppressed and then is slightly enhanced by the χ_{bJ} decays until they also no longer suffer from plasma effects. Thus although the indirect ratio is less sensitive to the plasma, the Υ'/Υ and $\psi'/J/\psi$ ratios together can significantly constrain plasma models, especially if the quarkonium states can be measured with sufficient accuracy up to high p_T. Again, the shape of the ratio is similar when the effect of shadowing on the initial conditions is included although the range of the suppression is reduced.

SUMMARY

A wealth of interesting data on charmonium production exists in fixed-target interactions with both proton and nuclear beams. The strong kinematics dependence of the pA data indicates that more than simple absorption by nucleons is taking place. However, the difficulties involved in interpreting the results as a function of x_F arises in part because, so far, shadowing and energy loss effects are not separable due to the way the Drell-Yan data has been incorporated into the shadowing parameterizations. More direct measures of the gluon distribution in the nucleus would help pin down the nuclear parton distributions more precisely without having to rely too strongly on the Drell-Yan data. Since Gavin and Milana fit their energy loss to the A dependence of the Drell-Yan data without including shadowing and the shadowing models include this same data in their fits without assuming any energy loss, the results of a full calculation remain somewhat ambiguous. Until this is sorted out it is difficult to discern what role, if any, energy loss may play in the pA data. The importance of intrinsic charm is larger for models where the energy loss is not too high but a small intrinsic charm component of J/ψ production seems

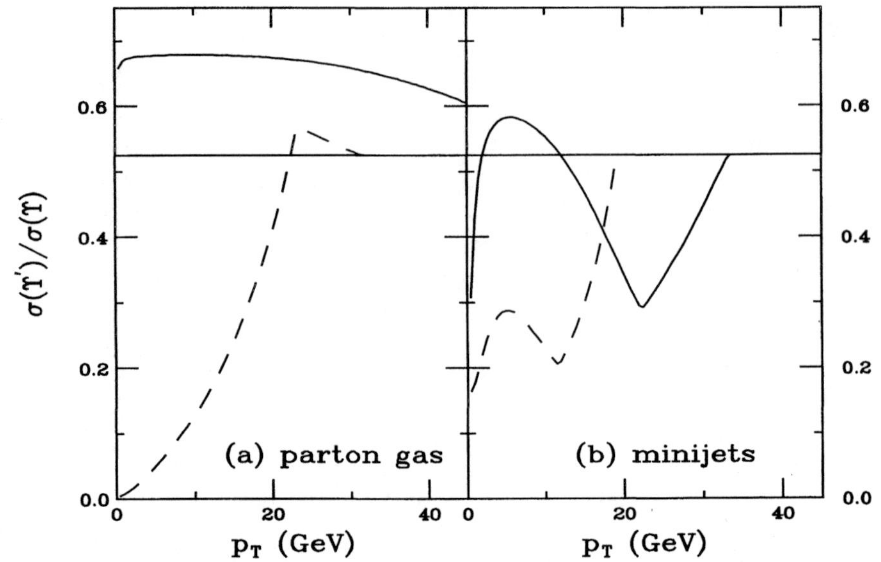

FIGURE 8. The Υ'/Υ LHC ratio computed from Eq. (40) is shown for several initial conditions and $R = R_{\rm Pb}$. In (a), parton gas results are shown for Case I (dashed) and Case II (solid). In (b) minijet results are given for Case II without shadowing (solid) and with $S = S_1$ (dashed). The horizontal curve represents the *pp* ratio.

to improve the agreement with the data. More data at other energies could provide a means of distinguishing between models [24].

The interpretation of the present nucleus-nucleus data on J/ψ suppression is also somewhat ambiguous even though most practitioners agree that some novel effect appears in the Pb+Pb data. Whether this is interpreted as increased comover density or as crossing a threshold (like a phase transition would account for), everyone can agree that some effect of high density appears in the Pb+Pb data that is not present in the S+U data. The upcoming NA50 run with target in vacuum will provide better J/ψ data at low E_T than is presently available. This new data will clarify whether or not the peripheral Pb+Pb collisions behave like central and semi-central S+U interactions—a prerequisite for determining the onset of the anomalous suppression. The minimum bias comparison [9,60] has considerably reduced the statistical uncertainites of the J/ψ suppression data but since the minimum bias analysis is rather model dependent, high statistics Drell-Yan data at large masses should ideally be used. Unfortunately, a very precise Drell-Yan measurement is not likely to be forthcoming from Pb+Pb collisions.

Perhaps definitive data will not be obtained until quarkonium data is available from heavy ion colliders. In these lectures, the p_T dependence of the quarkonium ratios $\psi'/J/\psi$ and Υ'/Υ has been suggested as a measurement through which quark-gluon plasma production might clearly manifest itself [11,114]. It will be particularly interesting to see if Υ production can be suppressed since screening has been predicted to be much less effective for the Υ than the other quarkonium states [28]. Clear evidence for Υ suppression will most likely have to wait until data is available from the LHC.

ACKNOWLEDGMENTS

I thank N. Armesto, M. Bedjidian, J.-P. Blaizot, B. Chaurand, D. Denegri, L. Gerland, M. Gonin, L. Kluberg, M.J. Leitch, H. Satz and C. Spieles for enjoyable discussions. I thank the organizers for their hospitality.

REFERENCES

1. Baglin, C., *et al.* (NA38 Collab.), *Phys. Lett.* **B251**, 472 (1990); Phys. Lett. **B270**, 105 (1991); *Phys. Lett.* **B345**, 617 (1995).
2. Gonin, M., *et al.* (NA50 Collab.), *Nucl. Phys.* **A610**, 404c (1996).
3. Abreu, M.C., *et al.* (NA50 Collab.), *Phys. Lett.* **B410**, 327 (1997); 337 (1997).
4. Badier, J., *et al.* (NA3 Collab.), *Z. Phys.* **C20**, 101 (1983).
5. Katsanevas, S., *et al.* (E537 Collab.), *Phys. Rev. Lett.* **60**, 2121 (1988).
6. Alde, D.M., *et al.* (E772 Collab.), *Phys. Rev. Lett.* **66**, 133 (1991).
7. Leitch, M.J., *et al.* (E789 Collab.), *Nucl. Phys.* **A544**, 197c (1992).
8. Leitch, M.J., (E866 Collab.), *Nucl. Phys.* **A661**, 554c (1999); Leitch, M.J., *et al.* (E866 Collab.), *Phys. Rev. Lett.* **84**, 3256 (2000).
9. Abreu, M.C., *et al.*, (NA50 Collab.), *Phys. Lett.* **B477**, 28 (2000).
10. Barger, V., Keung, W.Y., and Philips, R.N., *Z. Phys.* **C6**, 169 (1980); *Phys. Lett.* **91B**, 253 (1980).
11. Vogt, R., *Phys. Rept.* **310**, 197 (1999).
12. Martin, A.D., Roberts, R.G., Stirling, W.J., and Thorne, R.S., *Eur. Phys. J.* **C4**, 463 (1998); *Phys. Lett.* **B443**, 301 (1998).
13. Gavai, R.V., *et al.*, *Int. J. Mod. Phys.* **A10**, 3043 (1995).
14. Bodwin, G.T., Braaten, E., and Lepage, G.P., *Phys. Rev.* **D51**, 1125 (1995).
15. Beneke, M. and Rothstein, I.Z., *Phys. Rev.* **D54**, 2005 (1996).
16. Lai, H.L. *et al.*, *Phys. Rev.* **D51**, 4763 (1995).
17. Gavin, S. *et al.*, *Int. J. Mod. Phys.* **A10**, 2961 (1995).
18. Collins, J.C., Soper, D.E., and Sterman, G., in *Perturbative Quantum Chromodynamics*, edited by A.H. Mueller, (World Scientific, Singapore, 1989), p. 1.
19. Alde, D.M., *et al.* (E772 Collab.), *Phys. Rev. Lett.* **64**, 2479 (1991).
20. Alde, D.M. *et al.* (E772 Collab.), *Phys. Rev. Lett.* **66**, 2285 (1991).
21. Vogt, R., Brodsky, S.J., and Hoyer, P., *Nucl. Phys.* **B360**, 67 (1991).
22. Brodsky, S.J., Hoyer, P., Peterson, C., and Sakai, N., *Phys. Lett.* **B93**, 451 (1980); Brodsky, S.J., Peterson, C., and Sakai, N., *Phys. Rev.* **D23**, 2745 (1981).
23. Brodsky, S.J. and Hoyer P., *Phys. Rev. Lett.* **63**, 1566 (1989).
24. Vogt, R., *Phys. Rev.* **C61**, 035203 (2000).
25. deJager, C.W., deVries, H., and deVries, C., *Atomic Data and Nuclear Data Tables* **14**, 485 (1974).

26. Kharzeev, D. and Satz, H. *Phys. Lett.* **B366**, 316 (1996).
27. Sansoni, A. (CDF Collab.), *Nucl. Phys.* **A610**, 373c (1996).
28. Karsch, F., Mehr, M.T., and Satz, H., *Z. Phys.* **C37**, 617 (1988).
29. Gavin, S. and Vogt, R., *Nucl. Phys.* **B345**, 104 (1990).
30. Blaizot, J.-P. and Ollitrault, J.-Y., *Phys. Lett.* **217B**, 386 (1989).
31. Zhang, X.-F., Qiao, C.-F., Yao, X.-A., and Chao, W.-Q., hep-ph/9711237; in *Quarkonium Production in Relativistic Nuclear Collisions*, edited by X.-N. Wang and B. Jacak, (Singapore, World Scientific, 1999), p. 111.
32. Vogt, R., *Phys. Lett.* **B430**, 15 (1998).
33. Hüfner, J. and Povh, B., *Phys. Rev. Lett.* **58**, 1612 (1987).
34. Arneodo, M., *Phys. Rep.* **240**, 301 (1994).
35. Emel'yanov, V., Khodinov, A., Klein, S.R., and Vogt, R., *Phys. Rev. Lett.* **81**, 1801 (1998); *Phys. Rev.* **C56**, 2726 (1997).
36. Eskola, K.J., Qiu, J., and Czyzewski, J., private communication.
37. Eskola, K.J., *Nucl. Phys.* **B400**, 240 (1993).
38. Duke, D.W. and Owens, J.F., *Phys. Rev.* **D30**, 49 (1984).
39. Eskola, K.J., Kolhinen, V.J., and Ruuskanen, P.V., *Nucl. Phys.* **B535**, 351 (1998).
40. Eskola, K.J., Kolhinin, V.J., and Salgado, C.A., *Eur. Phys. J.* **C9**, 61 (1999).
41. Glück, M., Reya, E., and Vogt, A., *Z. Phys.* **C53**, 127 (1993).
42. Wang, X.-N. and Gyulassy, M., *Phys. Rev. Lett.* **68**, 1480 (1992); Baier, R., Dokshitser, Yu.L., Peigne, S., and Schiff, D., *Phys. Lett.* **B345**, 277 (1995); Baier, R., Dokshitser, Yu.L., Mueller, A.H., Peigne, S., and Schiff, D., *Nucl. Phys.* **B478**, 577 (1996); *Nucl. Phys.* **B483**, 291 (1997); Mustafa, M.G., Pal, D., Srivastava, D.K., and Thoma, M.H., *Phys. Lett.* **B428**, 234 (1998).
43. Baier, R., Dokshitser, Yu.L., Mueller, A.H., Peigne, S., and Schiff, D., *Nucl. Phys.* **B484**, 265 (1997).
44. Gavin, S. and Milana, J., *Phys. Rev. Lett.* **68**, 1834 (1992).
45. Brodsky, S.J. and Hoyer, P., *Phys. Lett.* **B298**, 165 (1993).
46. Kharzeev, D. and Satz, H., *Z. Phys.* **C60**, 389 (1993).
47. Gunion, J.F., *Phys. Lett.* **88B**, 150 (1979).
48. Landau, L.D. and Pomeranchuk, I.Ya., *Dokl. Akad. Nauk SSSR* **92**, 535 (1953); 735 (1953); Migdal, A.B., *Phys. Rev.* **103**, 1811 (1956).
49. Baier, R., Dokshitser, Yu.L., Mueller, A.H., and Schiff, D., *Nucl. Phys.* **B531**, 403 (1998).
50. Luo, M., Qiu, J., and Sterman, G., *Phys. Rev.* **D49**, 4493 (1994).
51. Brodsky, S.J., Hoyer, P., Mueller, A.H., and Tang, W.K., *Nucl. Phys.* **B369**, 519 (1992).
52. Vogt, R. and Brodsky, S.J., *Phys. Lett.* **B349**, 569 (1995).
53. Aubert, J.J. *et al.* (EMC Collab.), *Phys. Lett.* **110B**, 73 (1982); Hoffmann, E. and Moore, R., *Z. Phys.* **C20**, 71 (1983).
54. Harris, B.W., Smith, J., and Vogt, R., *Nucl. Phys.* **B461**, 181 (1996).
55. Proposal for Drell-Yan measurements of nucleon and nuclear structure with the FNAL main injector, FNAL proposal P906 (D.F. Geesaman, spokesperson).
56. Matsui, T. and Satz, H., *Phys. Lett.* **B178**, 416 (1986).
57. Romana, A., Ph.D. thesis, Universite de Paris Sud, 1980; Gribushin, A., *et al.*

(E672 Collab.), *Phys. Rev.* **D53**, 4123 (1996); Akerlof, C., *et al.* (E537 Collab.), *Phys. Rev.* **D48**, 5067 (1993).
58. Martin, A.D., Roberts, R.G., and Stirling, W.J., *Phys. Lett.* **B354**, 155 (1995).
59. Lourenço, C. *et al.* (NA38 Collab.), *Nucl. Phys.* **A566**, 77c (1994).
60. Abreu, M.C., *et al.*, (NA50 Collab.), *Phys. Lett.* **B450**, 456 (1999).
61. Albrecht, R., *et al.*, *Z. Phys.* **C38**, 3 (1988); Schukraft, J., *et al.*, *Z. Phys.* **C38**, 59 (1988); Bamberger, A., *et al.*, *Z. Phys.* **C38**, 89 (1988).
62. Margetis, S., *et al.* (NA49 Collab.), *Nucl. Phys.* **A590**, 355c (1995).
63. Baym, G., Friedman, G., and Sarcevic, I., *Phys. Lett.* **B219**, 205 (1989).
64. Sorge, H., Stöcker, H., and Greiner, W., *Ann. Phys.* **192**, 266 (1989).
65. Romana, A., *et al.* (NA50 Collab.), in Proceedings of the 33rd Rencontres de Moriond, *QCD and High Energy Hadronic Interactions*, Les Arcs, France, 1998.
66. Blaizot, J.-P., Dinh, M., and Ollitrault, J.-Y., nucl-th/0007020.
67. Borhani, A., (NA38 Collab.), Ph.D. thesis, Ecole Polytechnique, Palaiseau, (1996).
68. Ramello, L., *et al.* (NA50 Collab.), *Nucl. Phys.* **A638**, 261c (1998).
69. Gerschel, C. and Hüfner, J., *Z. Phys.* **C56**, 171 (1992).
70. Gavin, S. and Vogt, R., hep-ph/9610432.
71. Capella, A., Ferreiro, E.G., and Kaidalov, A.B., hep-ph/0002300.
72. Nagle, J., private communication.
73. Cassing, W. and Ko, C.M., *Phys. Lett.* **B396**, 39 (1997).
74. Cassing, W. and Bratkovskaya, E.L., *Nucl. Phys.* **A623**, 570 (1997).
75. Sorge, H., Shuryak, E., and Zahed, I., *Phys. Rev. Lett.* **79**, 2775 (1997).
76. Geiss, J., Greiner, C., Bratkovskaya, E.L., Cassing, W., and Mosel, U., *Phys. Lett.* **B447**, 31 (1999).
77. Kahana, D.E. and Kahana, S.H., nucl-th/9808025.
78. Spieles, C., Vogt, R., Gerland, L., Bass, S.A., Bleicher, M., Frankfurt, L., Strikman, M., Stöcker, H., Greiner, W., *Phys. Lett.* **B458**, 137 (1999).
79. Spieles, C., Vogt, R., Gerland, L., Bass, S.A., Bleicher, M., Stöcker, H., and Greiner, W., *Phys. Rev.* **C60**, 054901 (1999).
80. Bass, S.A. *et al.*, *Prog. Part. Nucl. Phys.* **41**, 225 (1998); version 1.1 available from http://www.th.physik.uni-frankfurt.de/~urqmd/urqmd.html.
81. Gerland, L., Frankfurt, L., Strikman, M., Stöcker, H., and Greiner, W., *Phys. Rev. Lett.* **81**, 762 (1998).
82. Alber, T., *et al.* (NA49 Collab.), *Phys. Rev. Lett.* **75**, 3814 (1995).
83. Kharzeev, D. and Satz, H., *Phys. Lett.* **B334**, 155 (1994).
84. Martins, K., Blaschke, D., and Quack, E., *Phys. Rev.* **C51**, 2723 (1995).
85. Matinian, S. and Muller, B., *Phys. Rev.* **C58**, 2994 (1998).
86. Haglin, K., *Phys. Rev.* **C61**, 031902 (2000).
87. Lin, Z. and Ko, C.M., nucl-th/9912046.
88. Kitagaki, T., *et al.* (E745 Collab.), *Phys. Lett.* **B214**, 281 (1988).
89. Emel'yanov, V., Khodinov, A., Klein, S.R., and Vogt, R., *Phys. Rev.* **C59**, 1860 (1999).
90. Nardi, M. and Satz, H., *Phys. Lett.* **B442**, 14 (1998).
91. Gross, D.J., Pisarski, R.G., and Yaffe, L.G., *Rev. Mod. Phys.* **53**, 43 (1981).
92. Fingberg, J., Heller, U.M., and Karsch, F., *Nucl. Phys.* **B392**, 493 (1993).

93. Petersson, B., *Nucl. Phys.* **A525**, 237c (1991).
94. Boyd, G., *et al.*, *Nucl. Phys.* **B469**, 419 (1996).
95. Kharzeev, D., Lourenço, C., Nardi, M., and Satz, H., *Z. Phys.* **C74**, 307 (1997).
96. Blaizot, J.-P. and Ollitrault, J.-Y., *Phys. Rev. Lett.* **77**, 1703 (1996).
97. Chaurand, B., private communication.
98. Satz, H., *Reports on Progress in Physics*, in press, hep-ph/0007069.
99. Isichenko, M.B., *Rev. Mod. Phys.* **64**, 961 (1992).
100. Satz, H. and Srivastava, D.K., *Phys. Lett.* **B475**, 225 (2000).
101. Redlich, K. and Braun-Munzinger, P., hep-ph/0001008.
102. Ko, C.M., Zhang, B., Zhang, X.-F., and Wang, X.-N., *Phys. Lett.* **B444**, 237 (1998).
103. Gavin, S., McGaughey, P.L., Ruuskanen, P.V., and Vogt, R., *Phys. Rev.* **C54**, 2606 (1996).
104. Lin, Z., Vogt, R., and Wang, X.-N., *Phys. Rev.* **C57**, 899 (1998).
105. Lin, Z. and Vogt, R., *Nucl. Phys.* **B544**, 339 (1999).
106. Lourenço, C., *et al.* (NA38/NA50 Collab.), in Proceedings of EPS Int. Conf. on High Energy Physics, Brussels, Belgium, 1995, EPS HEP Conf. 1995:363, CERN-PRE-95-001.
107. Antoniazzi, L., *et al.*, *Phys. Rev. Lett.* **70**, 383 (1993).
108. Antoniazzi, L., *et al.*, *Phys. Rev.* **D46**, 4828 (1992).
109. Ronceux, B., *et al.*, (NA38 Collab.), *Nucl. Phys.* **A566**, 371c (1994).
110. Yoh, J.K., *et al.*, *Phys. Rev. Lett.* **41**, 684 (1978); Ueno, K., *et al.*, *Phys. Rev. Lett.* **42**, 486 (1979).
111. Yoshida, T., *et al.*, *Phys. Rev.* **D39**, 3516 (1989).
112. Moreno, G., *et al.*, *Phys. Rev.* **D43**, 2815 (1991).
113. Abe, F., *et al.* (CDF Collab.), *Phys. Rev. Lett.* **75**, 4358 (1995).
114. Gunion, J.F. and Vogt, R., *Nucl. Phys.* **B492**, 301 (1997).
115. Ftáčnik, J., Lichard, P., and Pišút, J., *Phys. Lett.* **B207**, 194 (1988).
116. Gavin, S., Gyulassy, M., and Jackson, A., *Phys. Lett.* **B207**, 257 (1988).
117. Vogt, R., Prakash, M., Koch, P., and Hansson, T.H., *Phys. Lett.* **B207**, 263 (1988).
118. Emel'yanov, V., Khodinov, A., Klein, S.R., and Vogt, R., *Phys. Rev.* **C61**, 044904 (2000).
119. Xu, X.-M., Kharzeev, D., Satz, H., and Wang, X.-N., *Phys. Rev.* **C53**, 3051 (1996).
120. Bjorken, J.D., *Phys. Rev.* **D27**, 140 (1983).
121. CMS Technical Proposal, CERN/LHCC 94-38 (1994).

Electroweak Interactions at LEP

John Swain*

**Department of Physics, Northeastern University, Boston, MA 02115* [1]

Abstract. The electroweak interactions are based on an extension of the electromagnetic (Maxwell) interactions, realized in a rather odd way so that the symmetries of the theory are not immediately obvious. This "broken" theory has been the subject of intense investigation at LEP, and has passed all tests with flying colours. These lectures are meant to complement the many excellent presentations of the Standard $SU(2)_L \times U(1)_Y$ electroweak interactions in three main ways: first to clarify the physical meaning of symmetries in particle physics, second, to summarize the recent tests of the standard model using LEP data, and finally to look at possible roles of gravity in understanding mass.

INTRODUCTION: WHAT IS A PARTICLE?

The first thing to get clear is the notion of a particle. This is often confusing to students as it tends to be taken for granted as something that's rather obvious. Ryder has written one of the very few recent books [1] that really tries to deal with this, and I strongly recommend it to anyone interested in learning some quantum field theory.

The physical idea is simple: a particle is supposed to be some sort of tiny bit of stuff with well-defined properties and out of which more complicated things are built. This simple notion, that the world is made of little pieces, has been called by Feynman the most important single piece of information which we might want to pass on to future generations in the event of some great catastrophe.

Incidentally, you should free yourself of ideas like "wave-particle duality" as quickly as possible. The simple experimental fact is that the only things that ever get measured are particles. True, they often display wave-like behaviour, but nobody looking up close has ever seen a wave – it's just particles. The "wave-particle duality" is really more of a philosophical topic than a physical one, arising as it did from the early versions of quantum mechanics which took classical particles and made wavefunctions out of them. Such a "first quantized" picture always needs to be upgraded to a "second quantized" one where the fundamental objects

1) Supported by the National Science Foundation.

CP562, *Particles and Fields: Ninth Mexican School,* edited by G. Herrera Corral and L. Nellen

are particles again, and in fact the number of particles need not remain constant (a very useful thing in particle physics!).

Symmetry and Particles

Thinking about how to more rigorously define a particle immediately leads us to relate the notion of particles with a notion of symmetry. For example, if we have two electrons at two different points in space we might say that they're different – after all, one is *here* and the other one is *there*. Of course this somehow seems to be a bit much, and we usually decide that the notion of *here* versus *there* ought not to be relevant. In other words, we only ask that the particles be the same *modulo position* – or, equivalently, we use the assumed homogeneity of space to mean that there's nothing special about two different points.

Extending this idea, we also want to be sure that we don't regard two electrons at different times as being different particles. Similarly, two electrons rotated with respect to each other should be the same, as should two electrons which are boosted with respect to each other.

The sum total of all these transformations which we take to leave the identity of a particle unchanged is the Poincaré group, and contains translations in space and time, rotations, and boosts. That these transformations form a group is just a fancy way of saying that for any transformation there's one that undoes it, and there's a special transformation called the identity that does nothing.

Now we can look for ways to represent the elements of the Poincaré group with linear operators (matrices). Why linear operators? Well, first of all they're simple to work with. Second, their theory is well-developed. Third, linearity carries with it the useful notion that particles can be superposed – in some sense they interact weakly. These are all assumptions which must be justified *a posteriori* in light of their usefulness, but you should be aware that nonlinear realizations of groups certainly exist, as do nonlinear particle-like objects such as solitons.

We will also ask that the representations corresponding to elementary particles be *irreducible*. This just means that under a general transformation in the Poincaré group, all the components of the vector that gets acted on are mixed up with each other. If a representation has invariant subspaces, then we may as well identify each separate subspace with a different particle, since they're completely decoupled.

Wigner [2] studied the finite dimensional irreducible representations (there goes another assumption!) and found that they are labelled in terms of two Casimir operators : $P^\mu P_\mu = m^2$ and $m^2 s(s+1)$ where P^μ is the 4-momentum that generates translations in space and time, and s is the spin which tells you how much a particle has to be rotated to return to its original state (one full turn for a vector; any angle for a scalar, *etc.*).

Casimir operators are matrices which commute with all the transformations in the group and can thus be thought of as intrinsic properties of a particle which are the same regardless of whether they are measured before or after a group transformation is applied.

The list one gets is the following:

- $m^2 < 0$: tachyons – in classical mechanics these states correspond to speeds greater than the speed of light, but in quantum mechanics are usually take as a sign of instability. This is easy to see if you think of the phase associated to a particle at rest as $\exp(iEt/\hbar) = \exp(imc^2t/\hbar)$ which decays to zero or blows up for m imaginary.
- $m^2 = 0$: massless particles with only two possible states: spin angular momentum $\vec{\sigma}$ and linear momentum $\vec{p}$ parallel or antiparallel
- $m^2 > 0$: massive particles with $2s+1$ states where s is a half or whole positive integer, greater than or equal to zero.

Following Wigner then, a particle *is* an irreducible representation of the Poincaré group.

Note what we have now is a rather simple and elegant definition of what a particle is: two numbers, spin and mass. One of them, the spin, is drawn from a very easy to understand spectrum of values as is shown in elementary treatments of the quantum theory of angular momentum. The other, mass, is drawn from a very weird pattern of observed masses of elementary particles about which identically zero is understood.

Note also that the very concept of a particle is inextricably linked with the idea of what we think good symmetries of the universe should be!

Internal Symmetries

So far we've looked at symmetries that move particles around in space and time. We can also consider internal symmetries which do not act in space and time. For example, the wavefunction Ψ of a particle in quantum mechanics carries with it a phase, a sort of complex number that lies on a circle in some sort of abstract internal space. Differences in phases are important, but the overall definition of the zero for these phases is completely arbitrary, and we can redefine it without changing a particle:

$$\Psi(\vec{x}, t) \to e^{i\theta}\Psi(\vec{x}, t) \tag{1}$$

which corresponds to a shift in the phase of Ψ by a constant additive term of θ.

This is illustrated in figure 1. You might find it helpful to think of the phases as little clocks with just the second hand shown. Physics is all in the differences between the phases (interference effects in quantum mechanics) and redefining the phase globally by some constant amount has no meaning – just like the shifts on and off daylight savings time.

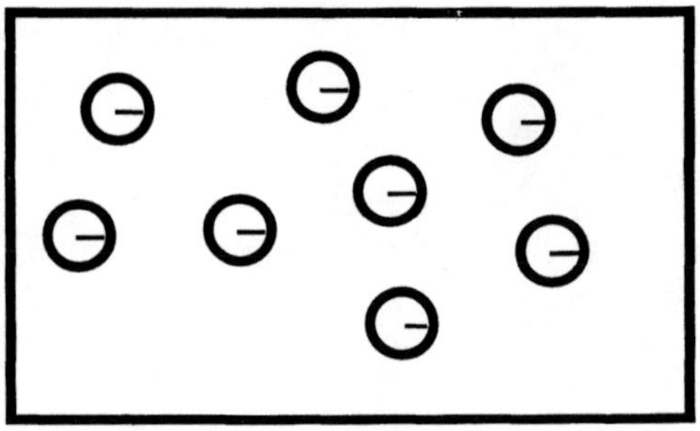

FIGURE 1. Redefining phases by a constant shift everywhere in space and time.

Global and Local Symmetries

While the symmetry we just looked at is perfectly well-defined, it does seem to violate the spirit of the principle of relativity. Shouldn't we be allowed to change the phases differently in different places? This would be something like a generalized daylight savings time where the phases are changed differently in different places by all sorts of amounts:

$$\Psi(\vec{x},t) \rightarrow e^{i\theta(\vec{x},t)}\Psi(\vec{x},t) \tag{2}$$

as shown schematically in figure 2.

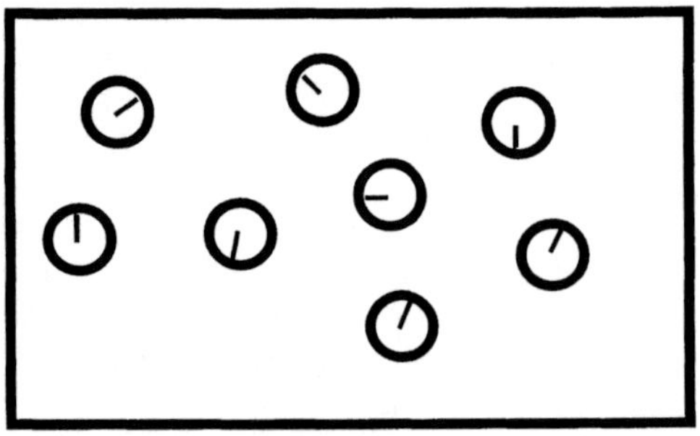

FIGURE 2. Redefining phases locally by different amounts.

The problem now is that there is no really sensible way to compare phases in one place and another – how is one to know if the difference in phase is a physical one, or is merely due to change in the definition of where a phase of zero is? The only way out is to introduce a new physical field which tells how the phase redefinitions are done! This is sketched in figure 3.

The idea now is that one can replace the usual derivatives ∂_μ with new "covariant" ones D_μ which contain two pieces: one, the usual ∂_μ which describes how a physical

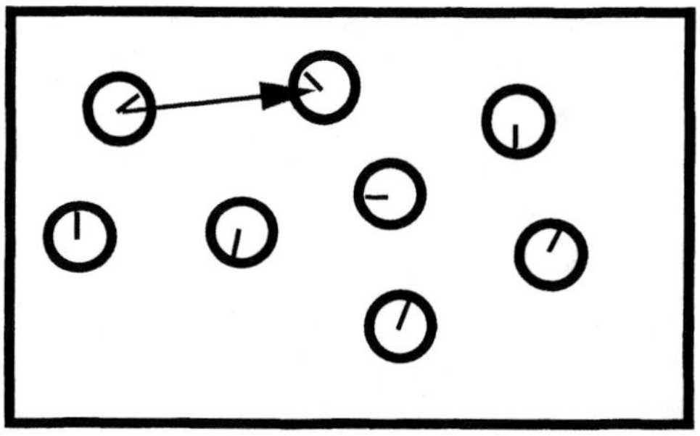

FIGURE 3. A gauge field is needed to make sensible comparisons.

quantity changes from one place to another, and another ieA^{μ} which describes how the coordinates (here in the internal space – the little circle) themselves have been changed in going from one place to another. (The factor i is conventional, and e can be associated with the electric charge of the particle in question. Particles with no electric charge can be described by real fields.)

$$\partial_{\mu} \rightarrow D_{\mu} = \partial_{\mu} + ieA^{\mu} \tag{3}$$

We have now found a way to get local invariance (*i.e.* the freedom to redefine phases from place to place as we please) at the price of introducing a new field A_{μ} which must be defined everywhere and will tell us how to compensate for these changes when we want to take derivatives. This symmetry under local changes of phase is called a "gauge symmetry" and the field A_{μ} is called a "gauge field", or a "connection". The extra symmetry in this sort of theory is called "gauge invariance".

This may all seem like some sort of crazy mathematical game: demand local gauge invariance, and then invent a new field A_{μ} in order to ensure that derivatives can still be taken sensibly! The amazing thing, however, is that if one makes some simple assumptions about the dynamics that should be attached to A_{μ}, we find the following amazing result: A_{μ} obeys equations which turn out to be Maxwell's equations, and the replacement of ordinary derivatives with the covariant ones gives the correct coupling of electromagnetism to matter!

The Lagrangian kinetic term for A_{μ}, up to constant factors, is just the simplest invariant thing that one can write down: $F_{\mu\nu} = [D_{\mu}, D_{\nu}]$ which has a nice geometric interpretation in terms of the concept of "curvature". D_{μ} generates displacements much as ∂_{μ} does, but now moves one about not just in spacetime, but in the internal space as well. $F_{\mu\nu}$ measures the possible failure to reach the same point in internal space by moving along a little displacement dx^{μ} and then a little displacement dx^{ν}, or doing them in the opposite order. If there's a difference, one says that connection is "curved" and $F_{\mu\nu}$ is the "curvature", which we identify with the field strength of the electromagnetic field.

Incidentally, if one allows the spacetime coordinates to be changed locally, then the covariant derivative will need another connection in spacetime (as opposed to the internal space) and this will lead to possible curvature in spacetime. A sensible kinetic term for this connection will then give rise to general relativity: Einstein's theory of gravity!

Feynman Diagrams

There's a nice diagrammatic way to think about what's happening which uses diagrams called Feynman diagrams. One is shown in figure 4.

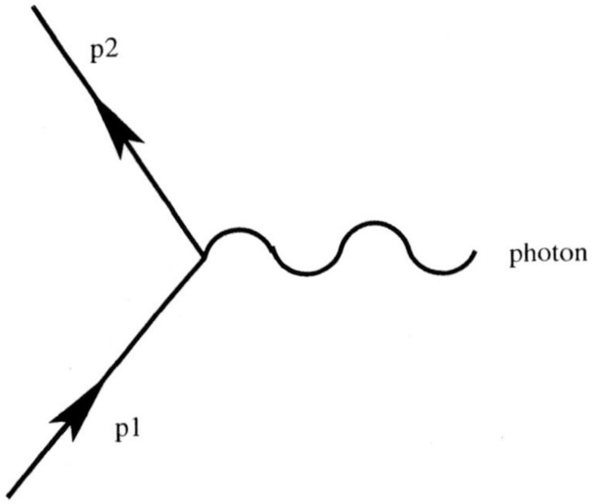

FIGURE 4. A piece of a Feynman diagram.

The two straight line segments represent a particle with two well-defined momenta, and different phases. The difference in phases is represented by a wiggly line which represents the connection, which itself is carried by a particle (remember, everything is a particle!) which in this case is a quantum of the connection field: a photon. Remember that despite a change in phase, we want to consider the identity of the particle to be unchanged – the photon is what we need in order that changes in phase can be made locally. The photon is an example of what is called a "gauge boson". It is the quantum mechanical manifestation of the field which was introduced to allow local changes in phase.

Such diagrams can be assigned mathematical expressions which correspond to quantum mechanical amplitudes for the process to take place. Sticking together lots of these diagrams and summing up over the amplitudes for various unobserved sub-processes gives the quantum mechanical description of electrodynamics.

A Ridiculous Symmetry?

Now how about a diagram that looks like the one shown in figure 5? Could this possibly make any sense?

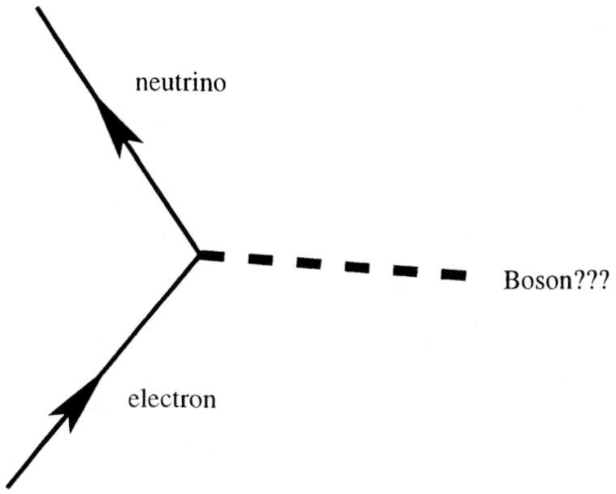

FIGURE 5. A symmetry?

Such transformations are certainly seen in nature and correspond to what we call the "weak interaction". The idea would be that, in some sense, an electron and a neutrino were different aspects of the same thing, just as an electron with one phase and another with another phase are different aspects of the same thing. Even more bizarre, we're asking that this redefinition can be done differently in different points of space and time! In other words, we're thinking of a group of transformations like

$$\left(\begin{array}{c} \nu_e' \\ e' \end{array} \right)_L = \left(\begin{array}{cc} \mathrm{U}_{11} & \mathrm{U}_{12} \\ \mathrm{U}_{21} & \mathrm{U}_{22} \end{array} \right) \left(\begin{array}{c} \nu_e \\ e \end{array} \right)_L ; \tag{4}$$

where the U_{ij} can vary in space and time. We also ask that $\mathrm{UU}^\dagger = 1$ (*i.e.* U is unitary) in order that we have a conservation of particles – a neutrino might turn into an electron or vice versa, but it shouldn't just disappear!).

How can this be? In what sense could a very light and chargeless neutrino be the same thing as a relatively massive and charged electron?

Spontaneously Broken Gauge Symmetry

There is obviously something physically wrong with a symmetry that says that an electron and a neutrino are different aspects of the same thing. It turns out there is a way to make sense of this, but we have to realize the symmetry in a slightly non-obvious way. Usually the symmetry is said to be "broken", but this is really a misnomer – the symmetry is present, though a little hard to see.

The way to do it is to assume that there is a new field ϕ which is a complex scalar doublet under $SU(2)_L$ which, for some dynamical reason acquires a vacuum expectation value[2]. We need it to be constant so that physics is the same everywhere, and we need it to be a scalar so that there are no preferred directions in space.

$$\langle\phi\rangle = \begin{pmatrix} 0 \\ v \end{pmatrix} \neq 0 \tag{5}$$

This is commonly referred to as "spontaneous symmetry breaking" and implies that while the theory respects the full $SU(2)_L \times U(1)_Y$ symmetry, the vacuum, for dynamical reasons does not – it only sees a leftover $U(1)$ symmetry which we identify with electromagnetism.

There is a simple analogy which may be helpful to think about. Suppose you lived in a big magnet. While the underlying laws of physics would be the spherically symmetric Maxwell equations, in fact you would notice a preferred direction along which the magnetic field pointed. This would come about not due to some defect in the spherical symmetry of the Maxwell theory, but rather due to the fact that the dynamics made the lowest energy state one in which the microscopic spins that make up the matter tended to line up. Any direction would be as likely as any other, so if you considered the full range of possible directions of the magnetic field you would find the full rotational symmetry of Maxwell's equations. As it happens, nature picked out *one* direction *spontaneously* despite the fact that a whole $SO(3)$ rotation group's worth of them would have been equally good.

If you heat up the magnet (*i.e.* your world) past the Curie temperature, the magnetization would disappear and you'd see the full $SO(3)$ symmetry. If you cooled it again, a new direction would be selected at random for magnetization to appear and the full $SO(3)$ symmetry would again be replaced by the smaller $SO(2)$ symmetry of rotation about the direction of the magnetic field.

In the Standard Model, if one does perturbation theory around a vacuum which does not respect $SU(2)_L \times U(1)_Y$, one finds that of the four gauge bosons corresponding to the four degrees of freedom associated with $SU(2)_L \times U(1)_Y$ transformations, three become massive and one is left massless. The massive degrees of freedom correspond to the longitudinal degrees of freedom of three (now massive and spin-1, and thus in need of an extra component) gauge bosons which are identified with the $W^{\pm}$ which convert left-handed electrons into neutrinos and vice versa and the Z^0 boson which is a sort of heavy photon. Those three degrees of freedom correspond to three degrees of freedom of the four in the complex scalar doublet, the last one being identified with a neutral scalar field called the "Higgs field". One direction in the $SU(2)_L \times U(1)_Y$ space corresponds to a degree of freedom which is still massless and is called the "photon" by definition. This direction

[2] What is usually done is to introduce a potential for the field and a negative mass-squared (which you will recall indicates an instability) so that the lowest energy configuration is not one with the field identically zero everywhere, but one where the field takes on a uniform constant value.

may lie partly within the $SU(2)_L$ internal space and partly within the $U(1)_Y$ internal space. The mixing of the gauge bosons associated with the two groups is parametrized in terms of a mixing angle called θ_W which must be determined from experiment, but is related to the $W^{\pm}$ and Z^0 boson masses.

While the $W^{\pm}$ bosons couple only to left-handed fields, the Z^0 and photon couple to both right and left handed fields. In the case of the photon, the coupling is to electric charge and corresponds to what one would expect from the simple $U(1)$ theory. The Z^0 couples to a mixture of right and left-handed components of each fermion which depend on their quantum numbers under $U(1)_Y$, and these couplings are testable predictions of the Standard Model in terms of θ_W.

The fact that there is a vacuum expectation value of a scalar field in otherwise empty space also gives us a way to introduce masses for fermions without spoiling the $SU(2)_L \times U(1)_Y$ symmetry. First let's see why fermion masses might be a problem. The left-handed electron and electron neutrino go into the same $SU(2)_L$ doublet. Right-handed electrons certainly exist (just run fast enough to overtake a left-handed electron and it will be right-handed for you!) but there don't seem to be any right-handed neutrinos – in other words, the right-handed field that corresponds to the "electron/neutrino field" (recall these are supposed to be different aspects of the same thing) is just a right-handed electron which is an $SU(2)_L$ singlet.

Mass, as we just hinted at in the last paragraph, can be seem as something that couples the left and right-handed pieces of a fermion together[3]. In the absence of mass, a left-handed neutrino, say, is always left-handed – there is no way to run past it and flip its helicity since massless particles always go at the speed of light. The only way we can write a term coupling a left-handed $SU(2)$ doublet and right-handed $SU(2)$ singlet together is to have another $SU(2)$ doublet field whose group index can be contracted with the index on the doublet. Otherwise the product of a singlet and a doublet is not $SU(2)$ invariant[4].

In the Standard Model, the scalar field which is introduced to break $SU(2)_L \times U(1)_Y$ symmetry is conveniently used to give masses to all the (chiral) fermions. These masses are never *predicted*, but can be set to any arbitrary value, which is one of the rather unsatisfactory features of the Higgs mechanism as an explanation for mass – for each mass one has just writes a corresponding coupling to the scalar field.

---

3) It may amuse the reader to calculate the time rate of change of the velocity operator for a Dirac particle by commuting the Dirac Hamiltonian $\vec{\alpha} \cdot \vec{p} + \beta m$ with the position vector $\vec{x}$. The calculation is almost trivial and shows that the velocity operator for a Dirac particle has eigenvalues $\pm c$ – as if the particle always goes at the speed of light but it mass changes the direction back and forth so that it appears to go slower. In the Standard Model, the Higgs field is like the thing that does the scattering back and forth, or couples the two helicities.

4) Incidentally, in noncommutative geometry [5] it is possible to think of the right and left handed fermions as living on two separate "sheets" of the world separated by a distance on the order of the weak scale. This is a sort of higher dimensional theory what we normally think of as a point is really two points – one for right handed particles and one for left handed ones. The Higgs field can then be given a geometrical interpretation as a connection between these two points. So far it's not clear if this model adds any deep insight into particle physics or not.

The same structure is assumed to hold for each generation:

$$\begin{pmatrix} \nu_e \\ e \end{pmatrix}_L \quad e_R \; ; \quad \begin{pmatrix} \nu_\mu \\ \mu \end{pmatrix}_L \quad \mu_R \; ; \quad \begin{pmatrix} \nu_\tau \\ \tau \end{pmatrix}_L \quad \tau_R$$

with similar expressions for the quarks.

How can one tell if such a strange theory is true? This is perhaps best answered by the following well-known quote from Michael Faraday, the great English physicist:

> "Nothing is too wonderful to be true, if it be consistent with the laws of nature, and in such things as these, experiment is the best test of such consistency."

LEP

LEP, the Large Electron Positron collider at CERN was designed specifically in order to shed light on this strange "electroweak" model [3]. In particular, it was designed to collide electrons and positrons first at centre-of-mass energies adequate to produce the neutral Z^0 boson (around 90 Gev), and later to produce pairs of W bosons (around 160 GeV). In fact, at the time of writing the machine has reached centre-of-mass energies in excess of 200 GeV.

A sketch of the 27 km circumference LEP ring and its accelerators is shown in figure 6.

A description of how accelerators such as LEP work is beyond the scope of these notes, but it might be instructive to think about what 200 GeV means in terms of something more familiar like an electrical spark. In dry air it takes between 10 and 20 thousand volts to make a 1 cm spark. Precision is difficult since this depends on details of electrode shape and other factors, but to get a feel for things, let's say conservatively that it takes about 20,000 volts to make a 1 cm spark. That means it takes 2 million volts to make a 1 m spark, and 2 billion volts to make a 1 km long spark. Now when LEP runs at 200 GeV, that means the energy in the centre-of-mass is like the energy that an electron would get crossing a 100 km long lightning bolt, which I think is pretty impressive!

LEP DETECTORS

Particle detectors [4] are designed to measure the particles emerging from the interaction point where the electron and positron beams collide. While we usually speak of "a LEP detector", each of the four detectors is in reality composed of numerous subdetectors.

While space limitations preclude a detailed description of how detectors work, the following is a brief summary of the most important ideas.

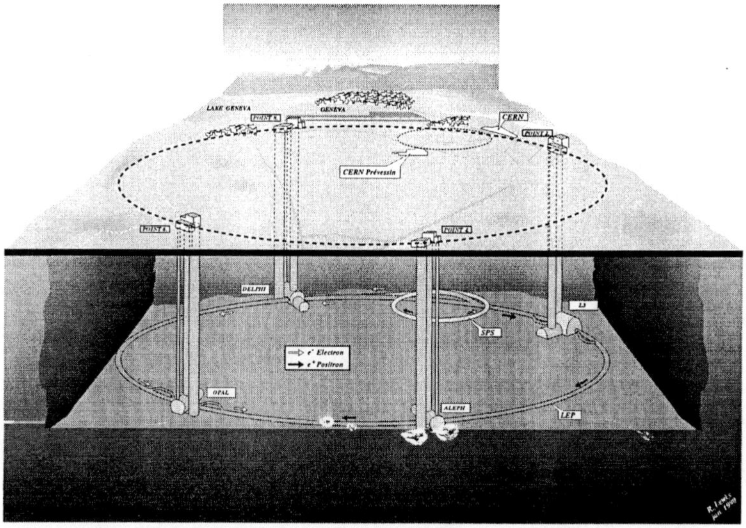

FIGURE 6. LEP and the LEP detectors.

Particles detectors, generally speaking, fall into two broad classes: 1) trackers, which try to measure the trajectory of a particle without disturbing it much, and 2) calorimeters, which try to stop a particle completely and measure its energy together with one point on its trajectory.

A generic LEP detector, schematically shown in figure 7, starts off with one or more central trackers which are usually silicon (close the interaction point) or gas (farther away). In either type of detector, charged particles cause ionization and the resulting ionization is turned into an electrical signal. Note that trackers, being based on ionization, will not register the passage of neutral particles.

After the trackers, an electromagnetic calorimeter is used to stop high energy electrons, positrons, and photons. An electromagnetic shower is initiated in which the energy of the incident particle is converted into a large number of secondary particles by a iterative sequence of branching processes:

- pair creation in the field of an atomic nucleus: $\gamma \to e^+e^-$
- Bremsstrahlung from the field of an atomic nucleus: $e^\pm \to e^\pm\gamma$

Some fraction of this energy is then converted into a readily detectable form (usually visible light and then an electrical signal).

Neutral particles which are not photons, and charged particles which are heavier than electrons have much lower probabilities of interacting in the electromagnetic calorimeter and continue on to a hadronic calorimeter. A hadronic caolorimeter does much the same thing that the electromagnetic calorimeter does, but now in place of repeated pair creations and Bremsstrahlung events, a strongly interacting particle ($\pi^{\pm}$,proton, *etc.*) triggers nuclear decay with fragments which may then cause more nuclear decays. Of course some of the particles produced will be photons, electrons, and positrons and thus an electromagnetic shower process inevitably accompanies the purely hadronic shower. Again, some fraction of the energy deposited is turned into an electrical signal.

Finally, particles which pass through the hadronic calorimeter reach the muon chambers, which are additional tracking chambers. Muons, which do not interact strongly and are too heavy to cause appreciable Bremsstrahlung, are the only particles likely to reach the muon chambers.

The whole detector is placed in a magnetic field directed parallel to the beam axis in order that the signs and magnitudes of the curvature of charged tracks can be used to determine the signs of the charges and magnitudes of the momenta of the particles that produce them.

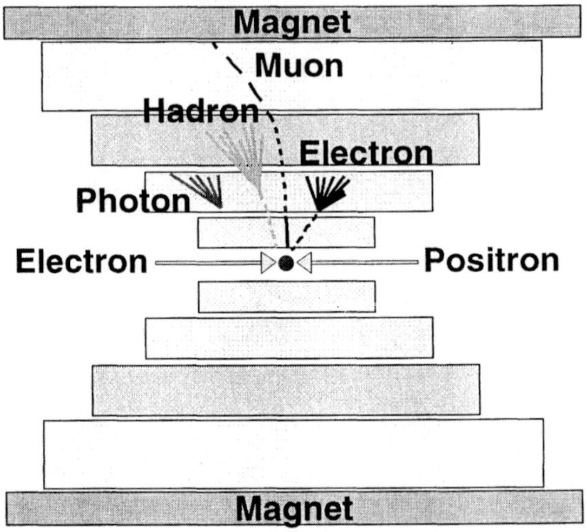

FIGURE 7. A schematic representation of a typical detector.

This combination of components allows almost all the particles which are produced in an e^+e^- interaction to be identified and measured. Even neutrinos can to some extent be "detected" (perhaps "inferred" would be a better term) using conservation of 4-momentum since one knows the intial energy-momentum 4-vector of the electron-positron pair and most detectors cover almost all of the solid angle around the interaction point.

There are other detectors which can be built - DELPHI, for example, also has a ring imaging Čerenkov detector which gives powerful additional particle identification. Each of the 4 LEP detectors, ALEPH, DELPHI, L3 and OPAL is optimized for one or another type of physics – doing better at one thing almost invariably means doing worse at something else. Clearly it is important to have multiple detectors in order to be able to independently cross-check and verify results claimed by any one experiment.

Of course no detector is perfect and there are always chances that particles are misidentified, energies and momenta are measured with errors, and so on – minimizing these and taking them into account in a quantitative way during data analysis are some of the main activities of an experimental particle physicist.

A concrete example of a LEP detector is L3, which is shown in figure 8. Note the enormous scale of the detector. The L3 magnet alone has more iron that the Eiffel tower in Paris!

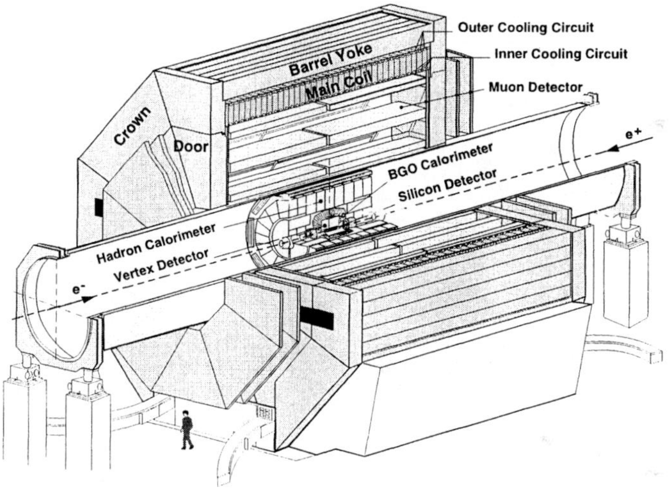

FIGURE 8. A perspective view of the L3 detector.

EVENTS

To give an indication of what events look like, a short picture gallery is presented in figures 9, 10, 11, 12, and 13. showing what various events look like in views which represent a slice through the detector perpendicular to the beam axis. These are all actually typical, and a lot of event classification could be done by eye were there not so much data! In each case the final state particles are produced predominantly from the decay of a Z^0 boson as the accelerator was running right at that resonance.

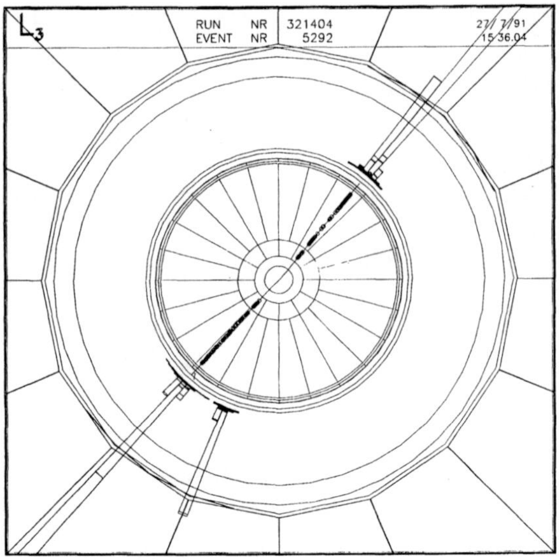

FIGURE 9. A typical e^+e^- event in L3. Note the charged tracks with hits shown, and the deposits of energy in the electromagnetic calorimeter.

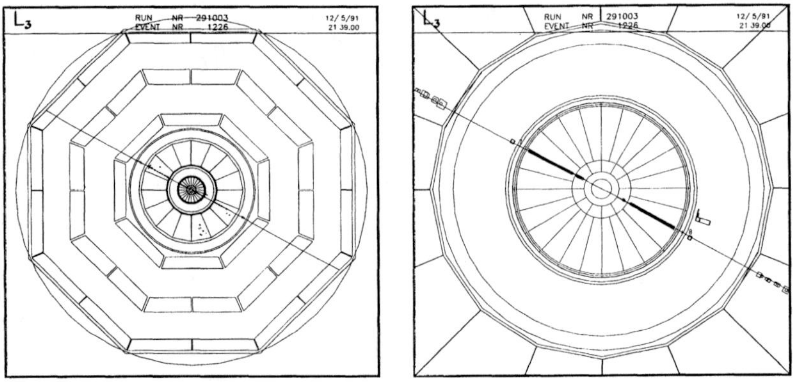

FIGURE 10. A typical $\mu^+\mu^-$ event in L3. Note the charged tracks with hits shown, minimal deposits of energy in the calorimeters, and tracks in the muon chambers. Two zoomed views are shown.

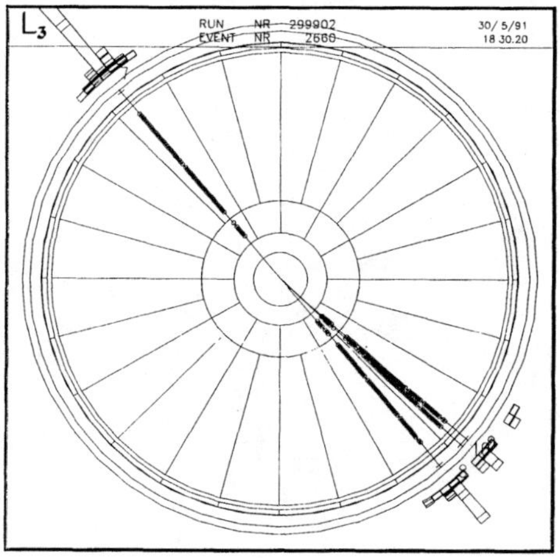

FIGURE 11. A typical $\tau^+\tau^-$ event in L3. One τ has decayed into three charged particles (likely charged pions) and a neutrino which is not detected, and the other into an electron and two neutrinos. Taus don't live long enough to reach the detector itself and must be detected from their decay products.

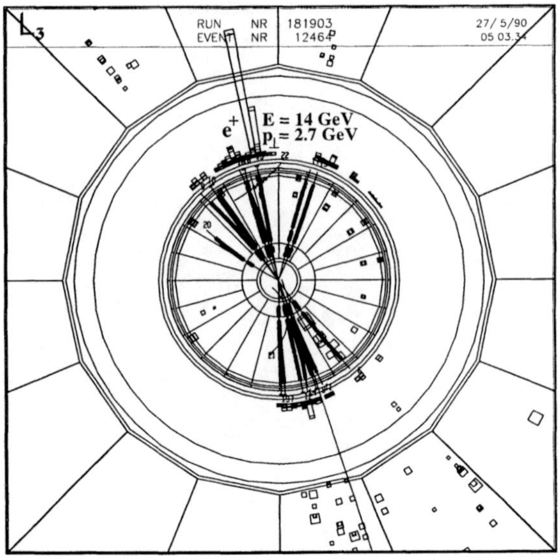

FIGURE 12. A typical $q\bar{q}$ event in L3. Quarks are not directly observable and hadronize, pulling quarks and gluons out of the vacuum to make "jets" of hadrons. In many cases just which quarks were produced is difficult to determine, but heavy quarks can be distinguished by the higher momenta of particles in the jets transverse to the jet axis, and from their longer lifetimes which can lead to the particles in the jets pointing back to a point displaced from the interaction point.

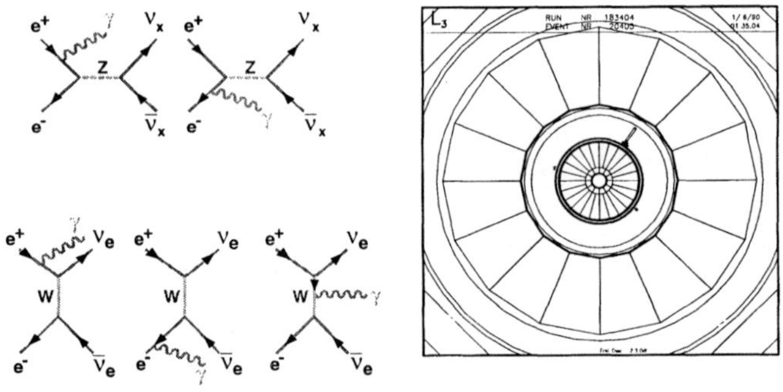

FIGURE 13. A typical $\nu\bar{\nu}$ event in L3 detected by a single photon emitted from the initial state or intermediate W boson. Unfortunately there is no way to know which sort of neutrino was produced.

Z PHYSICS

By studying the relative production probabilities of various particles and their distributions in space and momentum, detailed comparisons can be made with the electroweak theory that is meant to describe the Z^0 boson.

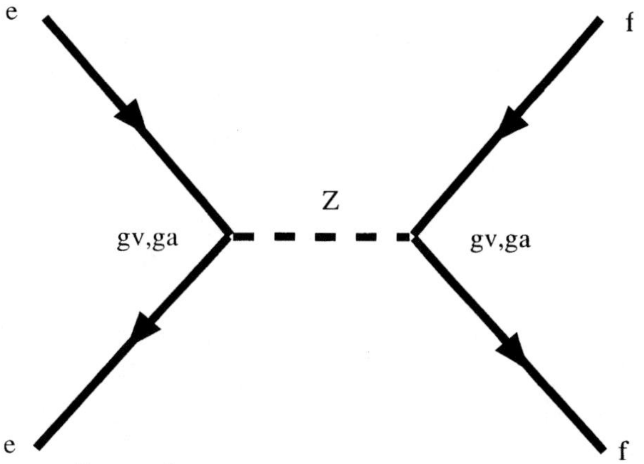

FIGURE 14. Feynman diagram for production of fermion-antifermion pairs from e^+e^- collisions via the Z^0.

Fermion pair production via the Z^0 can be thought of as a superposition of probabilities to produce various combinations of right and left handed fermions and antifermions. The admixtures of right and left are not in general equal, with the discrepancy being parameterized in terms of couplings constants to vector and axial vector pieces of the Z^0, g_v and g_a respectively. The difference between the two corresponds to the left-handed piece of the interaction, while the sum corresponds to the right-handed piece.

When we discuss the couplings of the Z^0 to different fermions, it is interesting to relax the standard assumption that these couplings are the same to all fermions, and instead of g_v and g_a, write v_f and a_f for the respective couplings to fermion f.

Various combinations of these can be measured (some much more easily than others) but all provide information on Z^0 couplings.

The sum total is simply the total cross section, and this can be measured directly. It is proportional to $(a_e^2 + v_e^2)(a_f^2 + v_f^2)$. An exact expression is difficult to write down in closed form, as radiative corrections both from the initial and final state particles are important.

A measurement of cross section versus energy then gives graphs like the one shown in figure 15 which shows the total cross section for the production of hadrons ($q\bar{q}$ pairs) as a function of energy. A few qualitative words about the figure are in order. First of all, note that the mass and width of the Z^0 can be determined from nothing more than cross section and beam energy measurements. That the width can be measured in this way has important implications, because it gives the total decay width (*i.e.* the finite lifetime, which via the uncertainty principle determines the uncertainty in energy). In other words, the width measured in this way knows about *all* possible decay modes including decay modes such as $Z^0 \to \nu\bar{\nu}$ which yield undetectable final states. By comparing this total width to the individual measured cross sections for producing the visible particles, the contribution to the width of invisible decay modes of the Z^0 can be determined! This has yielded the famous result that there are only three generations of light neutrinos.

Precise measurements of absolute cross sections are complicated by the need to know the luminosity of the accelerator. This must be determined at each experiment from measurements of $e^+e^- \to e^+e^-$ (Bhabha) scattering at very small angles where the cross section is dominated by known physics (QED) which does not depend much on Z^0's and W's.

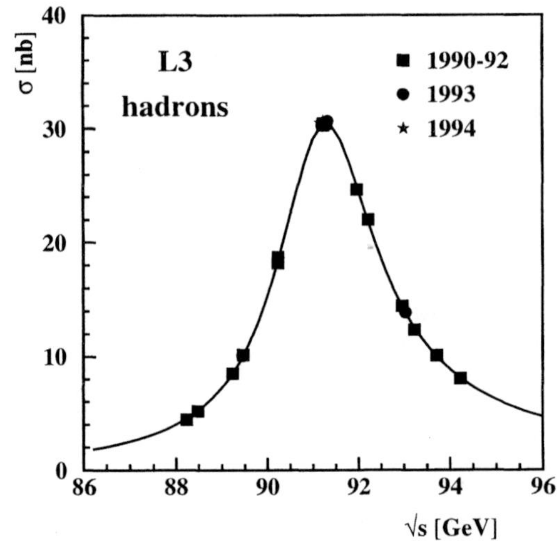

FIGURE 15. Lineshape results using LEP data together with predicted curves assuming various numbers of neutrinos.

A second important feature to note is that the distribution is not symmetrical about the maximum. This is due to initial state radiation, an important radiative correction which must be taken into account in order to make precise measurements (or even measurements at the 10 percent level!). If an electron or positron in the beam radiates a photon before interaction, the available centre-of-mass energy will be reduced. If the machine is running near or below the Z^0 mass, then this will result in a sharp drop in the probability that a Z^0 will be produced. If the machine is running above the Z^0 mass, the intial state photon may well reduce the centre-of-mass energy so that the cross section is higher. This explains why the cross section is higher at energies a little higher than the Z^0 mass than it is at energies which are a little lower.

Other radiative corrections involving virtual particles (such as photons which fail to escape and are recaptured by particles in the interaction) are also important. Their effects differ in different regions of phase space, and in practice the only reliable way to represent the predictions of the Standard Model is via Monte Carlo or semi-analytic expressions. The results for the Z^0 mass and width from detailed measurements from all the LEP collaborations are shown in figures 16 and 17.

Another important observable at the Z^0 is the forward-backward asymmetry. The idea here is that we're colliding particles and antiparticles (electrons and positrons) and it's interesting to measure whether or not the final state fermions tend to follow the direction of the electrons, or the positrons – sort of as if the "charge" had a sort of momentum associated with it. More formally,

$$A_{FB} = \frac{\sigma_F - \sigma_B}{\sigma_F + \sigma_B} \tag{6}$$

where the F and B subscripts represent "forward" and "backward" respectively. This quantity turns out to be proportional to $\frac{a_e v_e a_f v_f}{(a_e^2+v_e^2)(a_f^2+v_f^2)}$. Note that it suffices to merely determine the charges of the outgoing fermions and into which hemispheres of the detector (defined by a slice through the interaction point and perpendicular to it) the particles went. This measurement is very nice experimentally for a number of reasons, including the fact that measurements of luminosity and absolute detection efficiency cancel in the ratio defining A_{FB}.

The results of detailed measurements from all the LEP collaborations are shown in figure 18 including the energy-dependence of A_{FB}. The superscript 0 means that the asymmetries have been corrected so that the data is all as if it were taken right on the Z^0 peak since A_{FB} depends on the centre-of-mass energy.

The tau lepton is unique among the fermions produced at LEP as its decays can be used to measure its average polarization P_τ, giving new and independent information on the Z^0 couplings apart from the total cross section and forward-backward asymmetry. P_τ is proportional to $\frac{2a_f v_f}{a_f^2+v_f^2}$. It is often convenient to defined $A_\tau = -P_\tau$ for reasons which will become clear later.

The basic idea is simple: the tau decays via the weak interaction which is purely left-handed (V-A). Let's consider the case of the decay $\tau^- \to \pi^- \nu_\tau$ for concreteness. When the τ decays, the π^-, being spinless, carries away no helicity. That means that the helicity of the τ is carried entirely by the neutrino. But the neutrino is always left-handed (as far as we know), so the direction of the neutrino flight gives the direction of the neutrino spin and thus of the tau spin. Now we can't measure the neutrino, but we do know that the pion had to come off exactly opposite it in the tau rest frame in order to conserve linear momentum. That means that the pion direction tells you the neutrino direction which tells you the neutrino spin which tells you the tau spin – the tau reveals its spin direction from the direction in which is spits out the pion – it is its own polarimeter!

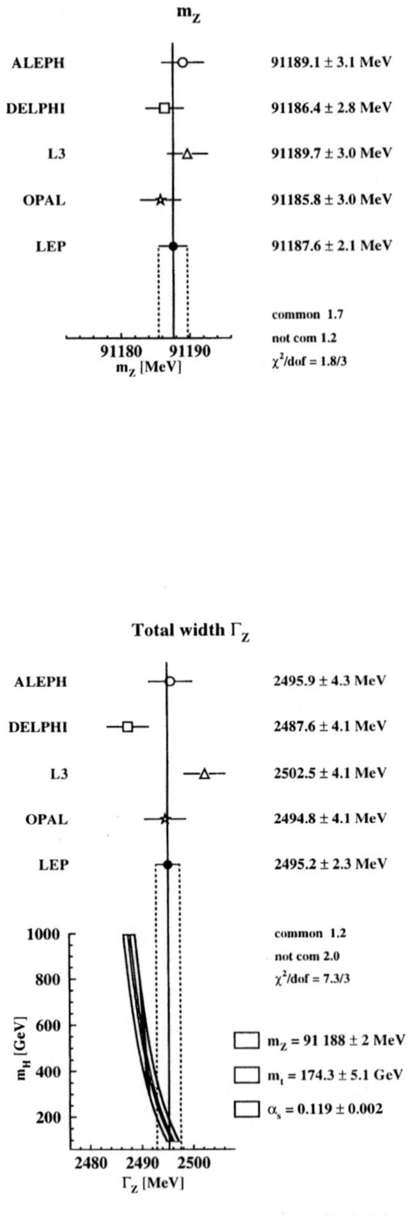

FIGURE 16. LEP results on the mass and width of the Z^0.

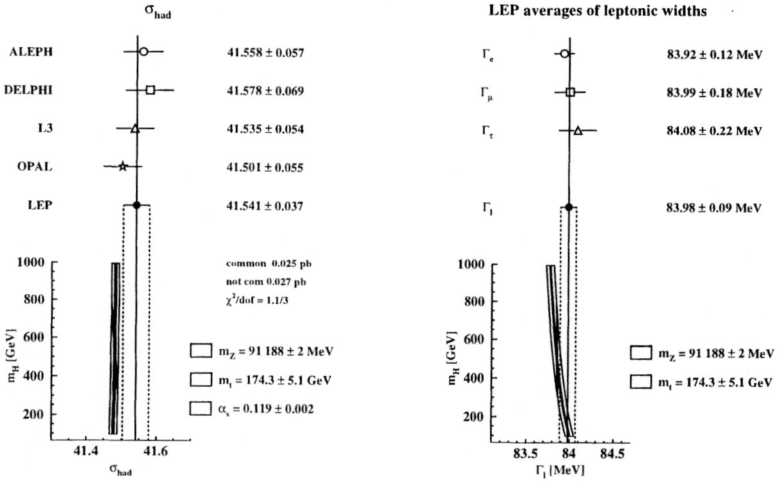

FIGURE 17. LEP results on the partial widths of the Z^0 into hadrons and into leptons.

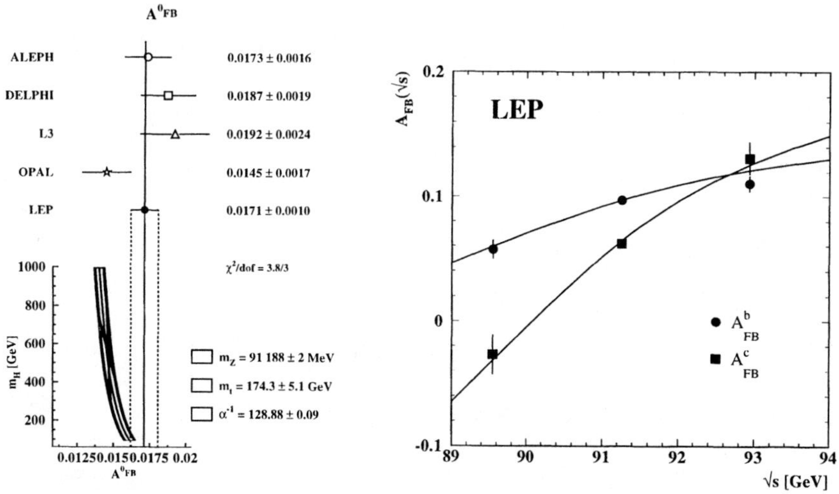

FIGURE 18. LEP results on A_{FB}

At LEP the taus are produced with considerable boosts, so what one measures in practice is the energy distribution of the pion, which tells you if the pion flew off with or against the direction of motion of the tau which we can take as a direction to define helicity. Arguments similar to the one in the previous paragraph show that the other decay modes of the τ also contain information about its polarization, though always a bit more diluted. In the cases of tau decays to muons or electrons one loses two neutrinos which decreases the sensitivity significantly, and in the case of the ρ, the fact that it is a vector particle means that it has two possible ways to come off (helicity 1 or helicity 0) and still conserve angular momentum. Sensitivity can be regained by using the distributions of the two pions from the ρ decay to help analyze the spin of the ρ itself. When all factors are taken into account, the ρ decay mode turns out to have the largest statistical weight.

The wary reader will have noticed that we use the assumed V-A structure of the charged weak interactions as a polarimeter, so this may look a bit like a swindle – we're using part of a theory to test the theory itself! There is actually a way to check the V-A assumption in tau decays and this is using the decay $\tau \to a_1 \nu_\tau$ where the a_1 decays into three pions. This decay is interesting since it is dominated by $\rho\pi$ but there are in fact two ways to make a ρ in the final state and the amplitudes for these two possibilities must be added quantum mechanically. The resulting interference is actually sensitive to the absolute sign of the neutrino helicity (*i.e.* can distinguish between V-A and V+A) and the assumption of V-A turns out to hold very well indeed! Possible admixtures of scalar, pseudoscalar and tensor interactions in the decay amplitude are strongly constrained by measurements of the "Michel parameters", a topic large enough to base several lectures on all by itself!

One final thing we can do with taus is to ask whether the polarization itself (we've looked at the *average* so far) has an asymmetry – in other words, does one tend to produce more right handed taus that go in the direction the electron went, or opposite to it? To this end we defined A_{pol}^{FB} which is approximately $-\frac{3}{2}\frac{a_e v_e}{a_e^2+v_e^2}$. This can be written as $\frac{3}{2}A_e$ where A_e is just like the A_τ we defined above, but with tau couplings replaced by electron couplings. Note that this quantity depends not on couplings of the Z^0 to the tau, but to the electron!

The results of detailed measurements from all the LEP collaborations are shown in figure 19.

What about measuring the polarizations of other particles? So far this has not been possible. Electrons lose all polarization information when they initiate electromagnetic showers, and muons would have to somehow be stopped without destroying their polarization in order to watch their spins precess in a magnetic field. Quarks generally turn into sprays of hadrons from which it is all but hopeless to expect any useful information on spin. The one exception could be the b quark, which in hadronizing to a Λ_b by picking up a (scalar) diquark, might retain some of its polarization. This topic has received some theoretical attention, but so far has not been very amenable to experiment, despite the fact b quarks from the Z^0

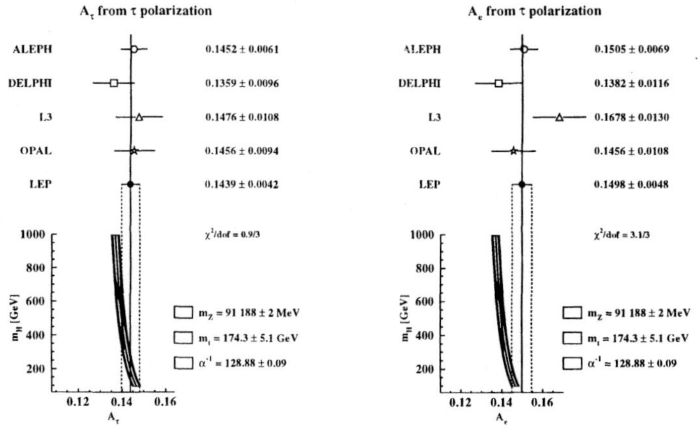

FIGURE 19. LEP results on tau polarization and polarization asymmetry.

are over 90% polarized before they hadronize.

The total LEP data sample is shown in tables 1 and 2.

LEP1, 1990-1995, Z-peak					
	ALEPH	DELPHI	L3	OPAL	LEP
$\int \mathcal{L}(pb^{-1})$	160	157	144	161	622
Num. $q\bar{q} \times 10^3$	4071	3705	3625	4066	15467
Num. $\ell^+\ell^- \times 10^3$	500	384	343	497	1724

Table 1: Data taken on the Z-peak at LEP.

LEP2, 1996-1999, WW production; Int. Lumin. in (pb^{-1})										
$\sqrt{s}$	L3		ALEPH		DELPHI		OPAL		LEP	
(GeV)	$N_{ev.}$	$\int \mathcal{L}$	$N_{ev.}$	$\int \mathcal{L}$	$N_{ev.}$	$\int \mathcal{L}$	$N_{ev.}$	$\int \mathcal{L}$	$N_{ev.}$	$\int \mathcal{L}$
161	21	10.9	-	-	29	10	28	9.9	78	30.8
172	113	10.3	119	10.6	99	10	120	10.4	451	41.3
183	844	55.5	851	56.8	738	50.3	877	57	3310	219.6
189	2663	176.4	2235	174.2	2209	155	3076	183	10183	688.6
192-202	3463	233.3	3538	237	3756	224	3764	218	14521	912.3

Table 2: Data for $W^\pm$ production taken at LEP.

A fit to the total LEP data sample gives values of the vector and axial vector couplings constant g_v and g_a shown in figure 20. Note that the largest errors are for the muons, since the tau analyses provide additional information on both τ and electron couplings. Arrows and a distorted rectangle indicate how the values would change with changes in the Higgs and top mass from the fitted values. The ellipse referred to as "combined" indicates the results if lepton universality is assumed (*i.e.* that the electroweak couplings are the same to each lepton). The band shows the results from SLD on the same quantities using the left-right asymmetry, an additional asymmetry which is only accessible to that experiment since SLC provides polarized *beams*.

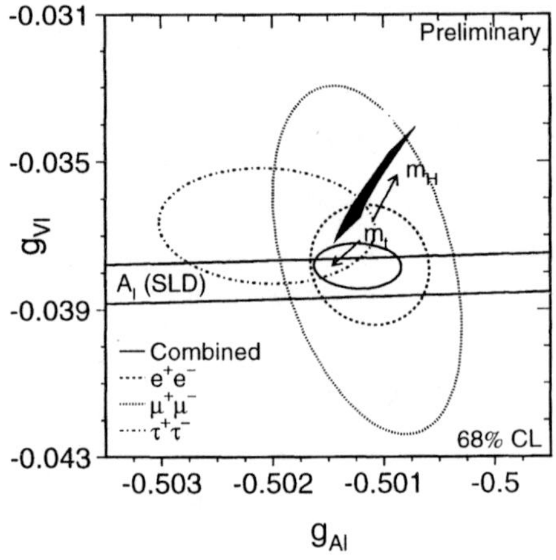

FIGURE 20. LEP results on g_a and g_v.

This can also be expressed in terms of the weak mixing angle as shown in figure 21. Also included is a charge-averaged forward-backward asymmetry $\langle Q_{FB} \rangle$.

W PHYSICS

At higher energies it is possible to produce not just Z^0 bosons, but also pairs of $W^\pm$ bosons. The corresponding (so-called CC03) Feynman diagrams are shown in figure 22.

The situation here is not quite as simple as it was in Z^0 production and decay as there are other processes which will give rise to the same final states and which interfere with the CC03 processes. These are shown in figure 23.

As in the case of the Z^0, we can only detect the heavy $W^\pm$ gauge bosons through their decay products. As can be seen from table 3, a large fraction of the time we wind up with final states with nontrivial backgrounds.

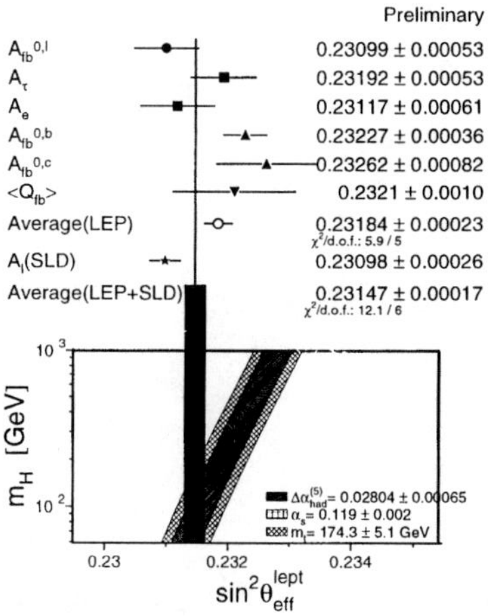

FIGURE 21. LEP results on the weak mixing angle.

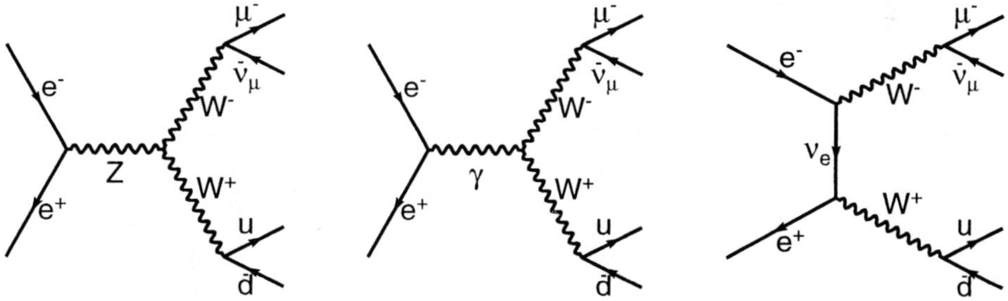

FIGURE 22. Diagrams for $W^\pm$ boson pair production.

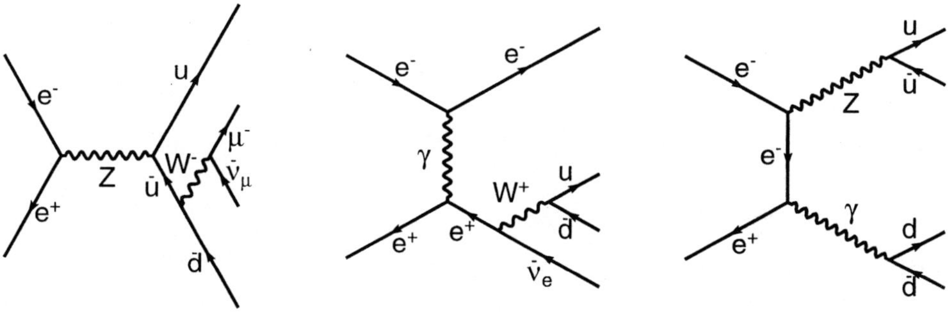

FIGURE 23. Diagrams which contribute to background to CC03 processes.

Final State	Class	Fraction
$q\bar{q}'q''\bar{q}''' \to 4$ Jets	hadronic	45.6%
$q\bar{q}'\ell\nu \to 2$ Jets $\ell\nu$	semileptonic	43.8%
$\ell\nu\ell\nu$	leptonic	10.6

Table 3: Table of fractions of final states for $W^{\pm}$ pair decays.

Figure 24 shows an additional problem which is the very much reduced cross section for the production of $W^{\pm}$ pairs compared to that for Z^0 bosons. There is no "W resonance" here!

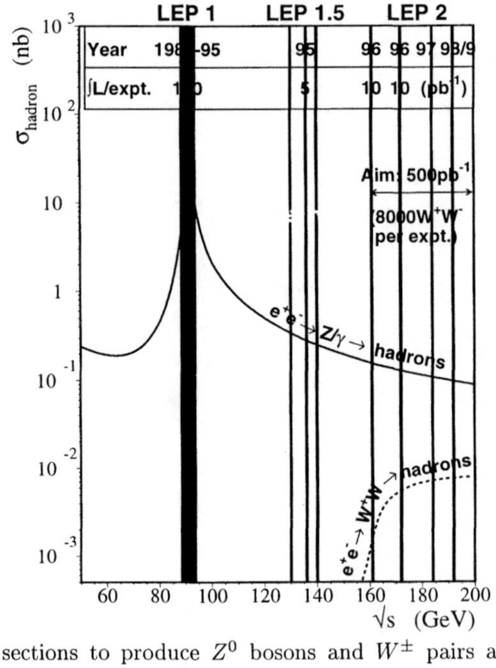

FIGURE 24. Cross sections to produce Z^0 bosons and $W^{\pm}$ pairs as a function of energy. Integrated luminosities are approximate predictions made some time ago.

The results of cross section measurements are shown in figures 25 and 26. Note that the agreement with various theoretical calculations is excellent, and attempts to reproduce the observed data omitting various subdiagrams are unsuccessful.

Masses and widths are shown in figure 27.

Branching ratios turn out to be just as expected from the Standard Model as shown in figure 28.

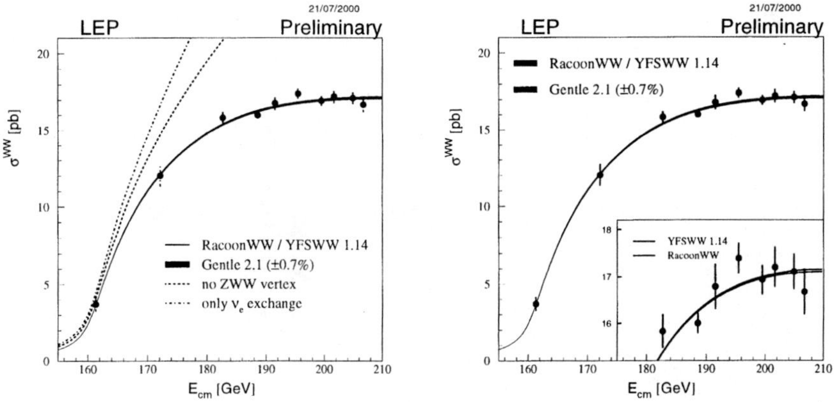

FIGURE 25. LEP results on $W^{\pm}$ production.

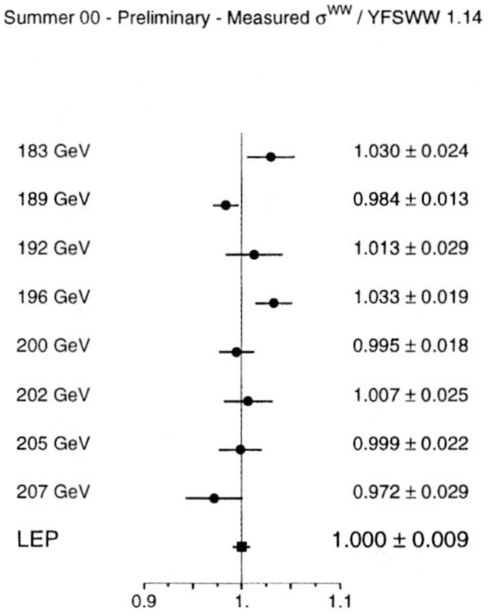

FIGURE 26. Cross section for W pair production as a function of energy compared to theory.

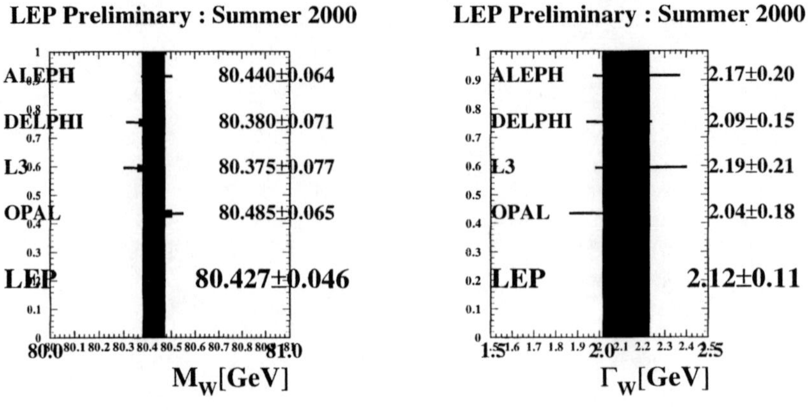

FIGURE 27. LEP results on W mass and width.

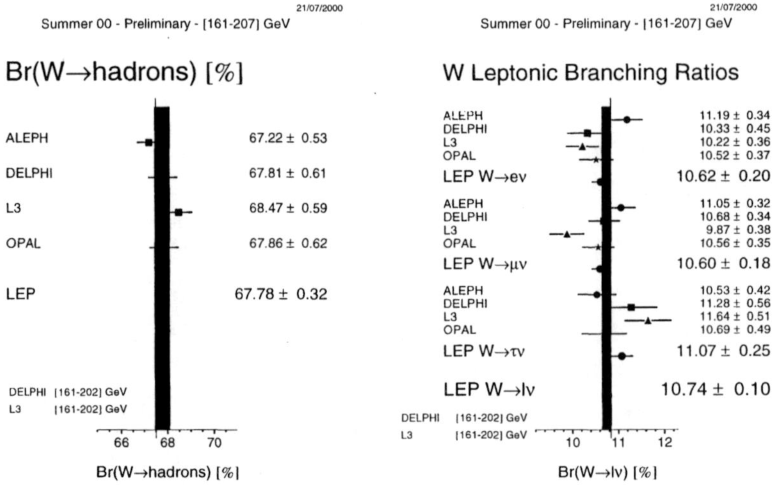

FIGURE 28. LEP results on W branching ratios.

THE HIGGS BOSON

The Higgs boson sector in the Standard Model remains largely unknown. The fact that the Higgs boson is meant to couple to mass means that the chance of producing it in e^+e^- collisions in the s-channel like the Z^0 is hopeless. The Higgs boson can be looked for at LEP in two different ways:

- Direct production, where it is radiated from a (heavy!) Z^0 boson and decays into (heavy!) b quarks, or can be reconstructed from the invariant mass of the Z decay products.

- From its contributions to radiative corrections where it appears in loop diagrams. While the use of radiative corrections to see virtual particles and probe beyond the actual energy available at LEP has been very successful in the case of the top quark, for which accurate determinations of its mass have been made at LEP without being able to make a single one on-shell, the Higgs is much more problematic. Its mass generally appears only logarithmically and thus radiative corrections are only weakly sensitive to it.

The status of searches for a Higgs boson are summarized in figures 29 and 30. Shaded regions show exclusions from direct searches while ovals or curves show the results of fits to electroweak data.

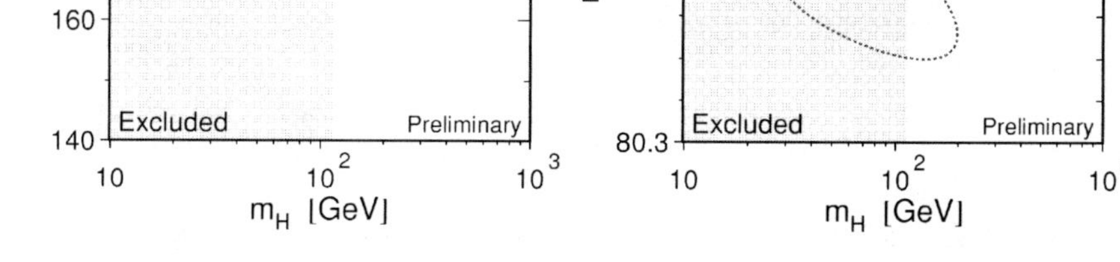

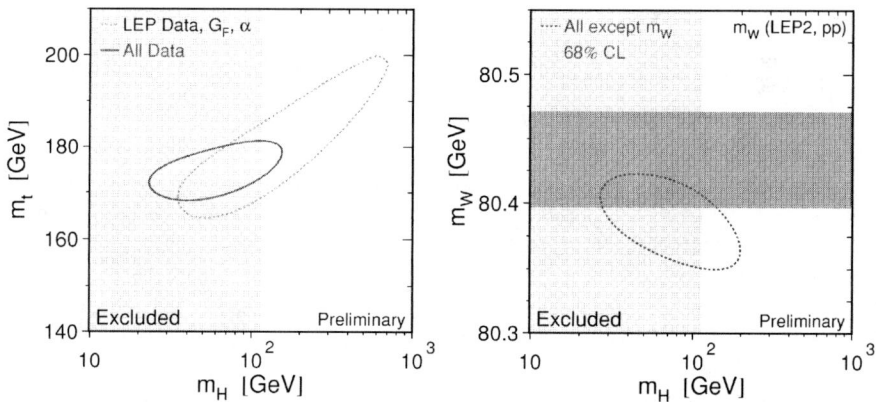

FIGURE 29. LEP results on Higgs mass versus top and W masses.

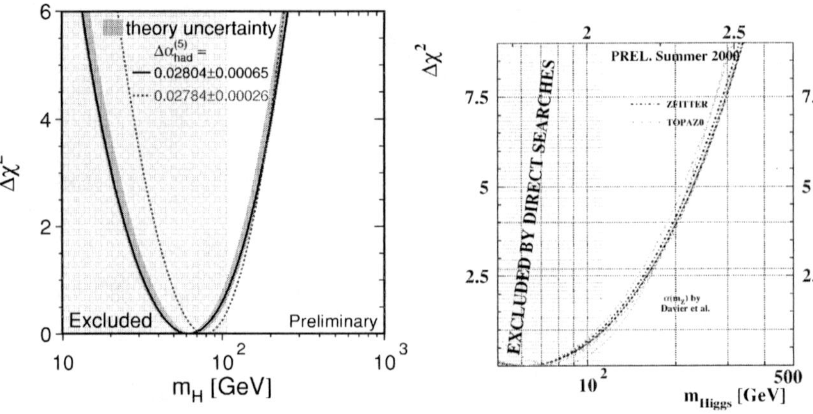

FIGURE 30. LEP results on Higgs mass. Note that the direct searches already rule out the value of Higgs mass most favoured from radiative corrections, although the errors are still very large.

MASS AND GRAVITY?

While in the Standard Model mass is just a coupling to some sort of omnipresent scalar field, if one asks beginning physics students what mass is, one invariably winds up getting some suggestion that it might have something to do with gravity.

In the Standard Model, gravity is left out completely, the idea being that it is usually negligible and can be added in at some point later on if need be, perhaps as some sort of (weak) perturbation. There are, in fact, various reasons to think that the gravity may play an important role and that the Higgs mechanism and in fact the whole renormalization program is really just a stop-gap or mathematical trick that allows us to do calculations. In this section I'd just like to make a few comments along these lines ... in a school it's perhaps more interesting to be left with questions than given answers!

Perhaps the oldest suggestion that gravity might have something to do with mass is that of Mach [6]. Suppose that the universe were completely empty – how would one know if a lone object were accelerated or not? With respect to *what* would one say it were accelerated? Usually one thinks of inertial frames as being unaccelerated with respect to the "distant stars" – but what if they weren't there?

In Mach's own words, "But what would become of the law of inertia if the whole of the heavens began to move and the stars swarmed in confusion? How would we apply it then? How would it have to be expressed then?"

If the inertial properties of objects were due to gravitational interactions with distant matter in the universe that would help us understand the fact that inertial and gravitational mass are the same – things are hard to accelerate because the rest of the mass in the universe tries to hold them in place gravitationally! Of course the Higgs mechanism has nothing to do with Mach's principle.

Let's have a look at another place where gravity may have something to do with mass. An interesting toy model to play with is a spherical shell of radius r with a charge e uniformly distributed on it. As the radius is taken to zero, the electromagnetic self-energy diverges as $1/r$ and this is usually taken to indicate some sort of pathology in the theory at short distances. Now let's try to do the calculation including gravity. I learned about this from [7].

The total mass (take $c = 1$) at finite radius r is that due to the bare mass m_0 of the shell itself, plus the electromagnetic self energy e^2/r.

$$m(r) = m_0 + \frac{e^2}{r} \tag{7}$$

Now let's add the gravitational self-energy.

$$m(r) = m_0 + \frac{e^2}{r} - G\frac{m_0^2}{r} \tag{8}$$

Note that while the charge tends to expand the sphere, gravity tends to shrink it. This is, however, not really any better since depending on the overall sign of the quantity multiplying $\frac{1}{r}$, this still blows up either negatively or positively as the radius tends to zero.

But shouldn't *all* energy gravitate? Taking a hint from general relativity, let's write instead

$$m(r) = m_0 + \frac{e^2}{r} - G\frac{m(r)^2}{r} \tag{9}$$

This may seem like a small change, but what has happened now is that we've got a quadratic equation for $m(r)$ which can be easily solved for any r. The amazing thing now is that $m(r)$ has a finite limiting value of $\frac{e}{\sqrt{G}}$ as r goes to zero! Not only that, but this is *independent* of the bare mass m_0.

Before you dismiss this as a naive toy, let me point out that a careful treatment in the context of general relativity also gets you a finite result – in other words, gravity doesn't make "ultraviolet divergences" worse; it makes them better!

Note that the result is also non-perturbative in G – there is no convergent power series for $\frac{1}{\sqrt{G}}$ about $G = 0$. It may amuse you to have a look at what goes wrong if you try to approximate the quadratic equation solution above with a series in G.

Suggestions that similar things are true in quantum corrections to the self-energy of a charged scalar particle have also been known for decades, though this line of investigation has not been followed much in recent years.

The warning, it seems, is that gravity may be deeply related to mass and it knows how to get finite answers out of otherwise hopeless-looking cases. However, it looks like it must be included from the outset and adding it in at the end of the day as a small correction is unlikely to do much good.

Incidentally, the whole (rather successful, it must be noted!) renormalization program calculates infinite corrections to particle masses and then *forces them by hand* to agree with the observed masses. Such an approach, it seems to me, has little hope of *predicting* or *explaining* any masses.

One final calculation which may amuse you is one of the width for a Higgs decay into gravitons [8]. Before I sketch this, let's be perfectly up-front about how dangerous it is to mix Higgs physics with general relativity. The scalar field corresponds to an enormous vacuum energy density which ought to curl the universe up into a small sphere which is wildly in disagreement with observation – you have to assume that the Higgs field doesn't do anything gravitationally which sort of flies in the face of the equivalence principle. Nevertheless, let's see what one might try to calculate.

Einstein tells us that (Ricci and scalar) curvature is given by the stress-energy-momentum tensor.

$$R_{\mu\nu} - \frac{1}{2}g_{\mu\nu}R = \frac{8\pi G}{c^4}T_{\mu\nu} \tag{10}$$

The Higgs field is supposed to be a scalar field that couples to mass, or, to put it a bit more relativistically, the trace of T. Now T is proportional to R, so we should get a coupling of the scalar field to R. If you put this all together, you find that G drops out (the equivalence principle at work) and you find $\Gamma(H \to gg) = \frac{\sqrt{2}}{16\pi}\frac{G_F M_H^2}{\hbar c}\frac{M_H c^2}{\hbar}$ which is about a million times bigger than the width for $\Gamma(H \to \gamma\gamma)$[5]. Can one really forget about gravity when one is thinking about mass?

THINGS THAT WEREN'T MENTIONED

Space is always at a premium when it comes to writing up notes from lectures at a school. Since there are so many excellent books available now on electroweak physics, I've tried to give a bit of a personal perspective, and that certainly biases the presentation a bit. A more insidious bias is what I left out, for "space reasons", as we tend to say. In the interest of being fair, here are a few things that could well have made it into these lectures, but didn't.

At the time of writing there is a hint from LEP of a possible Higgs boson at around 115 GeV. Whether this survives the test of time or not is an open question. A low mass Higgs is favoured by supersymmetry [9], an even more "ridiculous" symmetry than broken $SU(2)_L \times U(1)_Y$, which tries to see particles of different spins as different aspects of the same thing. Space limitations have also prevented me from discussing the CKM matrix (which is connected to mass and, of course, not predicted at all by the Higgs mechanism), electroweak tests with off-shell W's via

5) The two-photon decay mode occurs at one loop, while the gravitational one is already present at tree level. Forgetting this fact seems to have caused some confusion to some people!

weak decays, production of Z^0 pairs, searches for anomalous gauge boson couplings, rare Z decays, *etc.*

All over the world, and not only at LEP, physicists are desperately trying to find some weak point in the Standard Model, this strange theory of a "broken" symmetry. It won't have escaped your notice that I think there are many things that are deeply wrong with the structure we call the "Standard Model" that go beyond a simple aesthetic wish to have fewer parameters. However, despite some ugly features, mostly connected with the question of mass, the theory continues to survive every test. Where will it crack and what will that crack suggest? I wish I knew – perhaps one of you at this school will find out!

ACKNOWLEDGEMENTS

I would like to thank the organizers and all the students for a most stimulating time in Mexico and for all the excellent tequila! I would also like to thank my colleagues at Northeastern University and on LEP and CMS, and the National Science Foundation for its continuous support. LEP results quoted are the most recent ones from the LEP Electroweak Working Group [10] at the time of writing and available at `http://lepewwg.web.cern.ch/LEPEWWG`.

REFERENCES

1. L. H. Ryder, "Quantum Field Theory", Cambridge University Press, 1985.
2. E. P. Wigner, Annals of Mathematics, **40** (1939) 149.
3. Good general references for LEP are the following CERN "Yellow Reports": "Z Physics at LEP1", CERN 89-08, 3 volumes, eds. G. Altarelli, R. Kleiss, and C. Verzegnassi; and "Physics at LEP2", CERN 96-01, 2 volumes, eds. G. Altarelli, T. Sjöstrand, and F. Zwirner.
4. There are many good books on experimental particle physics. I particularly like C. Grupen, "Particle Detectors", Cambridge University Press, 1996.
5. A. Connes, " Noncommutative Geometry", Academic Press, 1994.
6. A good reference is J. Barbor and H. Pfister (eds.), "Mach's Principle: From Newton's Bucket to Quantum Gravity (Einstein Studies, Vol 6)", Springer Verlag, 1995.
7. A. Ashtekar, "Lectures on Non-Perturbative Canonical Gravity", World Scientific, 1991.
8. Y. Srivastava and A. Widom, `http://xxx.lanl.gov`, hep-ph/0003311
9. M. Carena and C. Wagner, these proceedings.
10. `http://lepewwg.web.cern.ch/LEPEWWG`

SHORT SEMINARS

Radiative Quark Mass Matrix Generation in a Model with a $U(1)_X$ Symmetry

E. García[1], A. Hernández-Galeana[3], D. Jaramillo[2], W. A. Ponce[2] and A. Zepeda[1]

[1] *Departamento de Física, Cinvestav*
Apartado Postal 14-740, 07000. México, D.F., México.

[2] *Departamento de Física, Universidad de Antioquia*
A.A. 1226, Medellín, Colombia.

[3] *Departamento de Física, Escuela Superior de Física y Matemáticas*
Instituto Politécnico Nacional, U. P. Adolfo López Mateos
México D.F., 07738, México

Abstract. In a model with a gauge group $G_{SM} \otimes U(1)_X$, where G_{SM} is the standard model gauge group and $U(1)_X$ is a horizontal local gauge symmetry, we propose a radiative generation of the spectrum of quark masses and mixing angles. The assignment of horizontal charges is such that at tree level only the third family is massive. The numerical fit results are shown.

I INTRODUCTION

A possible answer to the quark mass hierarchy problem is that the light quark masses arise by radiative corrections, while the top an possibly bottom quark masses are generated at tree level due to the breaking of an inter-family symmetry (horizontal symmetry) which can be either discrete or continuous ([1] and [2]).

Considering a continuous local horizontal symmetry, with symmetry group $U(1)_X$, which is broken spontaneously, we obtain the quark masses assuming only three families (the SM families) and assuming that only the top and bottom quarks are massive at tree level. The fermions are classified as in the SM in five sectors f = q,u,d,l and e, where q and l are the $SU(2)_L$ quark and lepton doublets, respectively, and u,d and e are the singlets, in an obvious notation. With $i = 1, 2, 3$ as a family index

$$X(f_i) = 0, \pm\delta_f \qquad \delta_q^2 - 2\delta_u^2 + \delta_d^2 = \delta_l^2 - \delta_e^2. \tag{1}$$

$$\delta_l = \delta_q = \pm\Delta \neq \delta_u = \delta_d = \delta_e = \pm\delta. \tag{2}$$

CP562, *Particles and Fields: Ninth Mexican School,* edited by G. Herrera Corral and L. Nellen

Eqs. 1 make the model free of anomalies and Eq. 2 guarantees that only the t and b quarks get tree level masses [3]. The assignment of horizontal charges to the fermions is then as given in Table 1(a). The $SU(3)_C \otimes SU(2)_L \otimes U(1)_Y$ quantum numbers of the fermions are the same as in the Standard Model. The

Sector \ Family	1	2	3
q	Δ	$-\Delta$	0
u	δ	$-\delta$	0
d	δ	$-\delta$	0
l	Δ	$-\Delta$	0
e	δ	$-\delta$	0

(a) Horizontal charges of fermions

	Class I		*Class II*					
	ϕ_1	ϕ_2	ϕ_3	ϕ_4	ϕ_5	ϕ_6	ϕ_7	ϕ_8
X	0	$-\delta$	0	Δ	0	δ	0	δ
Y	1	0	$-\frac{2}{3}$	$-\frac{2}{3}$	$\frac{4}{3}$	$\frac{4}{3}$	$-\frac{8}{3}$	$-\frac{8}{3}$
T	$\frac{1}{2}$	0	1	1	0	0	0	0
C	1	1	$\bar{6}$	$\bar{6}$	$\bar{6}$	$\bar{6}$	$\bar{6}$	$\bar{6}$

(b) Quantum numbers of scalar fields, C is the dimension under $SU(3)_C$

TABLE 1. Quantum numbers

new irreducible (*irreps*) representations of scalar needed to generate the light quark masses hierarchically are classified in classes I and II. Class I (II) contains scalar fields which get (do not get) vacuum expectation value (VEV). We introduce two irreps of scalars of class I and six irreps of scalars of class II, with the quantum numbers specified in Table 1(b).

The Yukawa couplings can be written as $L_Y = L_{Y_D} + L_{Y_M}$, where the D Yukawa couplings are

$$L_{Y_D} = Y^u \bar{q}_{L_3} \tilde{\phi}_1 u_{R_3} + Y^d \bar{q}_{L_3} \phi_1 d_{R_3} + h.c., \tag{3}$$

with $\tilde{\phi} \equiv i\sigma_2 \phi^*$, while the M couplings are

$$L_{Y_M} = Y_I [q_{1L}^{\alpha T} C \phi_{3\{\alpha\beta\}} q_{2L}^{\beta} + q_{3L}^{\alpha T} C \phi_{3\{\alpha\beta\}} q_{3L}^{\beta} + q_{2L}^{\alpha T} C \phi_{4\{\alpha\beta\}} q_{3L}^{\beta} + d_{2R}^T C \phi_5 d_{1R} + \\ d_{3R}^T C \phi_5 d_{3R} + d_{3R}^T C \phi_6 d_{2R} + u_{2R}^T C \phi_7 u_{1R} + u_{3R}^T C \phi_7 u_{3R} + u_{3R}^T C \phi_8 u_{2R}] + h.c. \tag{4}$$

where C represents the charge conjugation matrix and α and β are weak isospin indices. Color indices have not been written explicitly. Notice that $\phi_{3\{\alpha\beta\}}$ (ϕ_4) is represented as

$$\phi_3 = \begin{pmatrix} \phi^{-4/3} & \phi^{-1/3} \\ \phi^{-1/3} & \phi^{2/3} \end{pmatrix} \tag{5}$$

where the superscript denotes the electric charge of the field.

The most general scalar potential of dimension ≤ 4 that can be written is

$$
\begin{aligned}
-V(\phi_i) &= \sum_i \mu_i^2 \mid \phi_i \mid^2 + \sum_{i,j} \lambda_{ij} \mid \phi_i \mid^2 \mid \phi_j \mid^2 + \eta_{31} \phi_1^\dagger \phi_3^\dagger \phi_3 \phi_1 + \tilde{\eta_{31}} \tilde{\phi_1}^\dagger \phi_3^\dagger \phi_3 \tilde{\phi_1} \\
&+ \eta_{41} \phi_1^\dagger \phi_4^\dagger \phi_4 \phi_1 + \tilde{\eta_{41}} \tilde{\phi_1}^\dagger \phi_4^\dagger \phi_4 \tilde{\phi_1} \sum_{\substack{i \neq j \\ i,j \neq 1,2}} \eta_{ij} \mid \phi_i^\dagger \phi_j \mid^2 + (\rho_1 \phi_5^\dagger \phi_6 \phi_2 + \\
&\rho_2 \phi_7^\dagger \phi_8 \phi_2 + \lambda_1 \phi_5^\dagger \phi_1^\alpha \phi_{3\{\alpha\beta\}} \phi_1^\beta + \lambda_2 \phi_7^\dagger \tilde{\phi_1}^\alpha \phi_{3\{\alpha\beta\}} \tilde{\phi_1}^\beta + \\
&\lambda_3 Tr(\phi_3^\dagger \phi_4) \phi_2^2 + \lambda_4 \phi_5 \phi_6 \phi_7 \phi_2 + \lambda_5 \phi_5 \phi_6^\dagger \phi_7^\dagger \phi_8 + \lambda_6 \phi_2 \phi_8 \phi_5^2 + h.c.), \qquad (6)
\end{aligned}
$$

where Tr means trace and in $\mid \phi_i \mid^2 \equiv \phi_i^\dagger \phi_i$ an appropriate contraction of the $SU(2)_L$ and $SU(3)_C$ indices is understood.

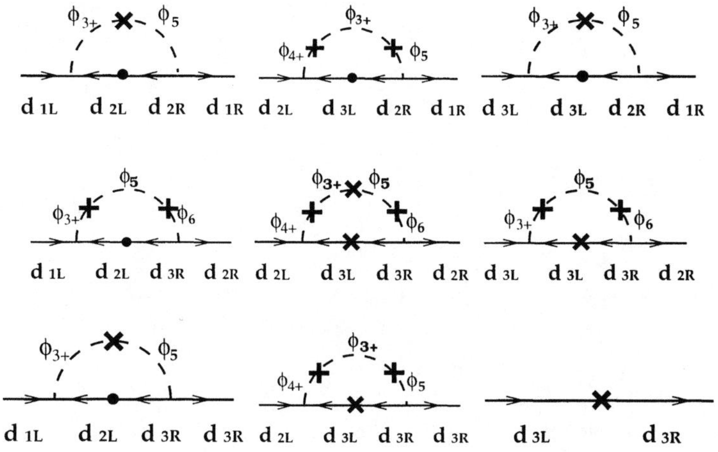

FIGURE 1. Mass matrix elements for d quarks

The contributions to the charge 1/3 quark sector are those diagrams of Fig. I, the cross means tree level mixing and the black circle means one loop mixing.

The VEVs of the class I scalar fields are $\langle \phi_1 \rangle = \frac{1}{\sqrt{2}} \begin{pmatrix} 0, & v_1 \end{pmatrix}^T, \langle \phi_2 \rangle = v_2, \qquad (7)$

and they achieve the breaking $G_{SM} \otimes U(1)_X \xrightarrow{\langle \phi_2 \rangle} G_{SM} \xrightarrow{\langle \phi_1 \rangle} SU(3)_C \otimes U(1)_Q. \qquad (8)$

After SSB scalar field mixings arise, producing the following mass matrices

$$
M_{2/3}^2 = \begin{pmatrix} s_4^2 & \lambda_3^* v_2^2 & 0 & 0 \\ \lambda_3 v_2^2 & s_3^2 & \frac{\lambda_1^* v_1^2}{\sqrt{2}} & 0 \\ 0 & \frac{\lambda_1 v_1^2}{\sqrt{2}} & u_5^2 & \frac{\rho_1 v_2}{\sqrt{2}} \\ 0 & 0 & \frac{\rho_1^* v_2}{\sqrt{2}} & u_6^2 \end{pmatrix} \text{and} M_{4/3}^2 = \begin{pmatrix} t_4^2 & \lambda_3^* v_2^2 & 0 & 0 \\ \lambda_3 v_2^2 & t_3^2 & \frac{\lambda_2 v_1^2}{\sqrt{2}} & 0 \\ 0 & \frac{\lambda_2^* v_1^2}{\sqrt{2}} & t_7^2 & \frac{\rho_2 v_2}{\sqrt{2}} \\ 0 & 0 & \frac{\rho_2^* v_2}{\sqrt{2}} & t_8^2 \end{pmatrix}, \qquad (9)
$$

where from Eq. (6) $t_i^2 = u_i^2 = \mu_i^2 + \lambda_{i1}v_1^2 + \lambda_{i2}v_2^2$ and $s_i^2 = t_i^2 + \eta_{i1}v_1^2$.

The non vanishing one loop contributions from the diagrams in Fig. I to the mass terms $\bar{d}_{iR}d_{jL}\Sigma_{ij}^{(1)} + h.c.$ have the form

$$\Sigma_{22}^{(1)} = 3m_b^{(0)}\frac{Y_I^2}{16\pi^2}\sum_k U_{1k}U_{4k}f(M_k, m_b^{(0)}), \tag{10}$$

where $m_b^{(0)}$ is the tree level contribution to the b quark mass, the 3 is a color factor, U is the orthogonal matrix which diagonalizes the mass matrix of the charge 2/3 scalars, and $f(a,b) \equiv \frac{1}{a^2-b^2}[a^2 ln\frac{a^2}{b^2}]$. The resulting second (third) rank mass matrix at this (next) level is thus

$$M_d^{(1)} = \begin{pmatrix} 0 & 0 & 0 \\ 0 & \Sigma_{22}^{(1)} & \Sigma_{23}^{(1)} \\ 0 & \Sigma_{32}^{(1)} & m_b^{(0)} \end{pmatrix}, \qquad M_d^{(2)} = \begin{pmatrix} \Sigma_{11}^{(2)} & \Sigma_{12}^{(2)} & \Sigma_{13}^{(2)} \\ \Sigma_{21}^{(2)} & m_2^{(1)} & 0 \\ \Sigma_{31}^{(2)} & 0 & m_3^{(1)} \end{pmatrix}. \tag{11}$$

At effective two loops we obtain expressions of the following form:

$$\Sigma_{11}^{(2)} = 3\frac{Y_I^2}{16\pi^2}\sum_{k,i} m_i^{(1)}(V_{dL}^{(1)})_{2i}(V_{dR}^{(1)})_{2i}U_{2k}U_{3k}f(M_k, m_i^{(1)}), \tag{12}$$

where the k (i) index goes from 1 to 4 (from 2 to 3), $V_{dL}^{(1)}$ and $V_{dR}^{(1)}$ are the unitary matrices which diagonalize $M_d^{(1)}$ of equation (11) and $m_i^{(1)}$ are the eigenvalues.

For the up sector the procedure to obtain the masses is completely analogous.

The CKM matrix takes the form $V_{CKM} = (V_{uL}^{(2)}V_{uL}^{(1)})^\dagger V_{dL}^{(2)}V_{dL}^{(1)}$, where the unitary matrices $V_{uL}^{(1)}$ and $V_{uR}^{(1)}$ diagonalize $M_u^{(1)}$, and $V_{uL}^{(2)}$ and $V_{uR}^{(2)}$ diagonalize $M_u^{(2)}$, with an analogous notation for the down sector.

II NUMERICAL EVALUATION

In order to test the model using the least possible number of free parameters, the scalar mass matrices are written as:

$$M_{2/3}^2 = \begin{pmatrix} a_+ & b & 0 & 0 \\ b & a_+ & c_+ & 0 \\ 0 & c_+ & a_+ & d_+ \\ 0 & 0 & d_+ & a_+ \end{pmatrix} \quad \text{and} \quad M_{4/3}^2 = \begin{pmatrix} a_- & b & 0 & 0 \\ b & a_- & c_- & 0 \\ 0 & c_- & a_- & d_- \\ 0 & 0 & d_- & a_- \end{pmatrix}. \tag{13}$$

Using the central value of the CKM elements in the PDG book [4] and the central values of the six quark masses at the top mass scale [5], we build the χ^2 function in the parameter space defined by $(a_+, a_-, b, c_+, c_-, d_+, d_-, Y_I, m_b^{(0)}, m_t^{(0)})$, where $m_b^{(0)}$ and $m_t^{(0)}$ are the tree level quark masses of the bottom and top quarks respectively. As can be seen from Table 2, even under the assumption that the CKM matrix elements are real, the numerical values are in good agreement with the allowed experimental results.

	RANGE	*INPUT*	*BEST FIT*
$m_d(m_t)$	3.85-5.07 MeV	4.46 MeV	3.63 MeV
$m_s(m_t)$	76.9-100 MeV	88.4 MeV	44.9 MeV
$m_b(m_t)$	2.74-2.96 GeV	2.85 GeV	2.91 GeV
$m_u(m_t)$	1.8-2.63 MeV.	2.21 MeV	2.22 MeV
$m_c(m_t)$	587-700 MeV	643 MeV	841 MeV
$m_t(m_t)$	159-183 GeV	171 GeV	166.5 GeV
CKM_{11}	0.9745-0.9760	0.9752	0.9761
CKM_{12}	0.217-0.224	0.2200	0.2179
CKM_{13}	0.0018-0.0045	0.0034	0.0032
CKM_{21}	0.217-0.224	0.2200	0.2214
CKM_{22}	0.9737-0.9753	0.9755	0.9742
CKM_{23}	0.036-0.042	0.0390	0.0382
CKM_{31}	0.004-0.013	0.0085	0.0117
CKM_{32}	0.035-0.042	0.0385	0.0365
CKM_{33}	0.9991-0.9994	0.9992	0.9992

TABLE 2. Experimentally allowed values for $m_q(m_t)$ and CKM matrix elements. We show the input and calculated values in the context of our model.

III CONCLUSIONS

By introducing a $U(1)_X$ gauge flavor symmetry and enlarging the scalar sector, we have presented a mechanism and an explicit model able to generate radiatively the hierarchical spectrum of quarks masses and CKM mixing angles.

Our results are encouraging; even under the assumption that the CKM matrix is real, and without knowing exactly the $U(1)_X$ mass scale, the numerical predictions are in the ballpark, implying also a value of order 10 for the Yukawa coupling Y_I, and masses for the exotic scalars being of order 10^3 TeV. Our model presents thus a clear mechanism able to explain the mass hierarchy and mixings of the quarks.

IV BIBLIOGRAPHY

REFERENCES

1. K. S. Babu and X.G. He, Phys. Rev. **D36**, 3484 (1987); E. Ma et al., Phys. Rev **D41**, 992 (1990); G.G. Wong et al., Phys. Rev. **D50**, 2962 (1994).
2. G. Degrassi et al., Phys. Rev. **D40**, 3066 (1989); P. Binetruy et al., Phys. Lett. **B350**, 49 (1995); K. Babu et al., Phys. Rev. **D43**, 2278 (1991); M. T. Yamawaki et al., Phys. Rev. **D43**, 2431 (1991); D. S. Shaw et al., Phys. Rev. **D47**, 241 (1993)
3. W. A. Ponce, A. Zepeda and J. M. Mira, Z. Phys. **C69**,683 (1996); W. A. Ponce, L. A. Wills and , A. Zepeda, Z. Phys. **C73**,711 (1997).
4. Particle Data Book, C.Caso *et all*, Eur. Phys. Journal **C3**,1 (1998).
5. H.Fusaoka and Y.Koide, Phys. Rev. **D57**, 3986 (1998).

New Physics with mirror fermions

Ricardo Gaitán
Centro de Investigaciones Teóricas,
Facultad de Estudios Superiores-Cuautitlan.
A. P. 142, Cuautitlán-Izcalli, Estado de México, México.
Enrique García
Departamento de Física, Cinvestav-IPN
Av. IPN 2508, A. P. 14-740, México, D.F.

Abstract. We study the conditions under which flavour violation arises in a left-right mirror model that includes additional fermions with mirror properties for vector and scalar interactions.

I INTRODUCTION

The nonconservation of parity (P) is incorporated in the Standard Model of electroweak interactions (EWSM) in an unpleasant way. A solution is [1] to restore P by including mirror fermions with a $V - A$ coupling between a charged lepton and a neutrino.

Mirror particles also appear in grand unified theories, string theories, extended supersymmetry, Kaluza-Klein theories [2]. On the other hand, the introduction of mirror fermions with masses between the weak scale and 1 TeV could offer a dynamical origin of EWSM symmetry breaking [4].

We analyze signals of flavor changing neutral currents (FCNC) and lepton flavor violation (LFV) in a mirror model with gauge group $SU(2)_L \otimes SU(2)_R \otimes U(1)$. In this model we can identify the FC transitions from: exchange of family changing neutral gauge bosons, mixing between mirror and standard fermions, and the Higgs sector.

II MODEL DESCRIPTION

In the left-right model with mirror fermions (LRMM) [3], the right-handed (left-handed) components of mirror fermions transform as doublets (singlets) under $\mathrm{SU(2)_R}$. Mirror and SM fermions will share hypercharge and color interactions.

CP562, *Particles and Fields: Ninth Mexican School,* edited by G. Herrera Corral and L. Nellen

Thus, the first family of leptons and quarks will be written as follows:

$$l_{eL}^{o}=\left(\begin{array}{c}\nu_e^o\\ e^o\end{array}\right)_L,\qquad e_{\mathrm{R}}^{o},\qquad \hat{l}_{eR}^{o}=\left(\begin{array}{c}\hat{\nu}_e^o\\ \hat{e}^o\end{array}\right)_R,\qquad \hat{e}_L^o \tag{1}$$

$$q_L^o=\left(\begin{array}{c}u^o\\ d^o\end{array}\right)_L,\qquad u_R^o,\qquad d_R^o,\qquad \hat{q}_R^o=\left(\begin{array}{c}\hat{u}^o\\ \hat{d}^o\end{array}\right)_R,\qquad \hat{u}_L^o,\qquad \hat{d}_R^o \tag{2}$$

The superscript (o) denote weak eigenstates, and the caret is associated with mirror particles. Because the model does not contain left-handed mirror neutrinos, they have to be massless.

The symmetry breaking is realized using two Higgs doublets, the SM one (ϕ) and its mirror partner $(\hat{\phi})$. The vacuum expectation values (VEV's) of the Higgs fields and the potential are

$$\langle\phi\rangle=\tfrac{1}{\sqrt{2}}\left(\begin{array}{c}0\\ v\end{array}\right),\qquad \langle\hat{\phi}\rangle=\tfrac{1}{\sqrt{2}}\left(\begin{array}{c}0\\ \hat{v}\end{array}\right). \tag{3}$$

$$V=-\left(\lambda_1\phi^\dagger\phi+\hat{\lambda}_1\hat{\phi}^\dagger\hat{\phi}\right)+\tfrac{\lambda_2}{2}\left[\left(\phi^\dagger\phi\right)^2+\left(\hat{\phi}^\dagger\hat{\phi}\right)^2\right]+\lambda_3\left(\phi^\dagger\phi\right)\left(\hat{\phi}^\dagger\hat{\phi}\right). \tag{4}$$

There is only one value λ_2 because we put by hand parity breaking through dim-2 mass terms in Higgs potential. The neutral Higgs boson square mass matrix that can be obtained from this potential is:

$$\mathsf{M}^2_{\mathsf{H}^0}=\left(\begin{array}{cc}2\lambda_1 v^2 & 2\lambda_2 v\hat{v}\\ 2\lambda_2 v\hat{v} & 2\lambda_1\hat{v}^2\end{array}\right) \tag{5}$$

Out of the eight scalar degrees of freedom, six become the Goldstone bosons required to give mass to $W^\pm$, $\hat{W}^\pm$, Z and$\hat{Z}$; thus only two neutral Higgs bosons remain; then neutral physical states are:

$$\begin{aligned} H&=\sqrt{2}\left[\left(\Re e\phi^o-v\right)\cos\alpha+\left(\Re e\hat{\phi}^o-\hat{v}\right)\sin\alpha\right],\\ \hat{H}&=\sqrt{2}\left[\left(v-\Re e\phi^o\right)\sin\alpha-\left(\hat{v}-\Re e\hat{\phi}^o\right)\cos\alpha\right]. \end{aligned} \tag{6}$$

where α is the mixing angle.

The mass matrix for the gauge bosons is obtained from the scalar Lagrangian

$$\mathcal{L}_{\mathrm{s}}=\left(\mathcal{D}^\mu\phi\right)^\dagger\left(\mathcal{D}_\mu\phi\right)+\left(\hat{\mathcal{D}}^\mu\hat{\phi}\right)^\dagger\left(\hat{\mathcal{D}}_\mu\hat{\phi}\right) \tag{7}$$

where D_μ $(\hat{D}_\mu)$ denotes the covariant derivative associated with the SM (mirror part). The mass matrix for the charged gauge bosons is diagonal ($M_W=\frac{1}{2}vg$ and $M_{\hat{W}}=\frac{1}{2}\hat{v}\hat{g}$), where g and $\hat{g}$ are the coupling constants associated with the $SU(2)_L$

and $SU(2)_R$ gauge group, respectively). The mass matrix for the neutral gauge bosons is not diagonal, and it has one massless and two massive eigenvalues

$$M_{Z,Z'} = \frac{1}{2}\left[\begin{array}{c} v^2\left(g^2+g'^2\right)+ \\ \hat{v}^2\left(\hat{g}^2+g'^2\right)\end{array}\right] \mp \frac{1}{2}\left(\begin{array}{c}\left[v^2\left(g^2+g'\right)+\hat{v}^2\left(\hat{g}^2+g'^2\right)\right]^2 - \\ 4v^2\hat{v}^2\left(g^2g'^2+g^2\hat{g}^2+\hat{g}^2g'^2\right)\end{array}\right)^{1/2}, \tag{8}$$

where g' is the coupling constant of the $U(1)$ gauge group, the matrix can be diagonalized by an orthogonal transformation R, relating the weak and mass eigenstates, given by [3]

$$R = \left(\begin{array}{ccc} c_{\theta_w}c_\Theta & c_{\theta_w}s_\Theta & s_{\theta_w} \\ -\frac{1}{c_{\theta_w}}\left(s_\Theta r_{\theta_w} + \frac{g}{\hat{g}}c_\Theta s^2_{\theta_w}\right) & \frac{1}{c_{\theta_w}}\left(c_\Theta r_{\theta_w} - \frac{g}{\hat{g}}s_\Theta s^2_{\theta_w}\right) & \frac{g}{\hat{g}}s_{\theta_w} \\ t_{\theta_w}\left(\frac{g}{\hat{g}}s_\Theta - r_{\theta_w}c_\Theta\right) & -t_{\theta_w}\left(\frac{g}{\hat{g}}c_\Theta + r_{\theta_w}s_\Theta\right) & r_{\theta_w}\end{array}\right), \tag{9}$$

with θ_w and Θ the rotations angles between the Z -A and Z - Z' gauge bosons, respectively. In the eq. (9), $r_{\theta_w} \equiv \sqrt{c^2_{\theta_w} - \frac{g^2}{\hat{g}^2}s^2_{\theta_w}}$.

To find the couplings of the Higgs fields from eq. (7), we use the expressions for D_μ and $\hat{D}_\mu$, substituting the physical states. For the Z-H-$\hat{H}$ Higgs interactions we get

$$\begin{aligned}\mathcal{L}_{ZZ\hat{\mathcal{H}}} = &\sqrt{2}\left(gM_W X(\Theta,\theta_w)\cos\alpha + \hat{g}M_{\hat{W}}Y(\Theta,\theta_w)\sin\alpha\right)HZ_\mu Z^\mu \\ &+\sqrt{2}\left(-gM_W X(\Theta,\theta_w)\sin\alpha + \hat{g}M_{\hat{W}}Y(\Theta,\theta_w)\cos\alpha\right)\hat{H}Z_\mu Z^\mu\end{aligned} \tag{10}$$

where

$$X(\Theta,\theta_w) = \left[c_{\theta_w}c_\Theta - \frac{g'}{g}t_{\theta_w}\left(\frac{g}{\hat{g}}s_\Theta - r_{\theta_w}c_\Theta\right)\right]^2, \tag{11}$$

$$Y(\Theta,\theta_w) = \left[-\frac{1}{c_{\theta_w}}\left(s_\Theta r_{\theta_w} + \frac{g}{\hat{g}}c_\Theta s^2_{\theta_w}\right) - \frac{g'}{g}t_{\theta_w}\left(\frac{g}{\hat{g}}s_\Theta - r_{\theta_w}c_\Theta\right)\right]^2 \tag{12}$$

and for the W-W-$\hat{H}$ interactions the result is

$$\mathcal{L}_{\mathrm{WH\hat{H}}} = \sqrt{2}\mathrm{g}\mathcal{M}_{\mathcal{W}}\mathrm{g}^{\mu\nu}\left(\mathcal{H}\cos\alpha - \hat{\mathcal{H}}\sin\alpha\right)\mathcal{W}^-_\mu\mathcal{W}^+_\nu \tag{13}$$

The gauge interaction of quarks and leptons can be obtained from the generic Lagrangian

$$\mathcal{L}^{\mathrm{int}} = \overline{\psi}\mathrm{i}\gamma^\mu\mathcal{D}^\mu\psi + \overline{\hat{\psi}}\mathrm{i}\gamma^\mu\hat{\mathcal{D}}_\mu\hat{\psi} \tag{14}$$

where ψ ($\hat{\psi}$) denote the standard (mirror) fermions.

To consider the mixing of fermions, including the mirror ones, we follow Ref. [5], grouping all fermions of a given electric charge q and a helicity $a = L, R$ into $n_a + m_a$ vector column of n_a standard (s) and m_a mirror (m) gauge eigenstates, i.e. $\psi_a^o = (\psi_s^o, \psi_m^o)_a^{\mathsf{T}}$.

The relation between the gauge eigenstates and the corresponding light (l) and heavy (h) mass eigenstates $\psi_a = (\psi_l, \psi_h)_a^{\mathsf{T}}$ is given by a unitary transformation [1]

$$\psi_a^o = \mathsf{U}_a \psi_a \qquad \text{where} \qquad \mathsf{U}_a = \begin{pmatrix} \mathsf{A}_a & \mathsf{E}_a \\ \mathsf{F}_a & \mathsf{G}_a \end{pmatrix}. \tag{15}$$

it is easy to see that the submatrix $\mathsf{A_a}$ is not unitary. The term $(\mathsf{F}^\dagger \mathsf{F})_a$, which is second order in the small mirror-standard fermion mixing, induce FC transitions in the light-light sector.

The renormalizable and gauge invariant interactions of the scalar doublets ϕ and $\hat{\phi}$ with the leptons are described by the Yukawa Lagrangian, and take the form

$$\mathcal{L}_{\mathcal{Y}}^{\mathrm{l}} = \sum_{\mathrm{i,j}} \lambda_{\mathrm{ij}} \bar{\mathrm{l}}_{\mathrm{iL}}^{\mathrm{o}} \phi \mathrm{e}_{\mathrm{jR}}^{\mathrm{o}} + \sum_{\mathrm{i,j}} \hat{\lambda}_{\mathrm{ij}} \overline{\hat{\mathrm{l}}_{\mathrm{iL}}^{\mathrm{o}}} \hat{\phi} \hat{\mathrm{e}}_{\mathrm{jR}}^{\mathrm{o}} + \sum_{\mathrm{i,j}} \mu_{\mathrm{ij}} \overline{\hat{\mathrm{e}}_{\mathrm{iL}}^{\mathrm{o}}} \mathrm{e}_{\mathrm{jR}}^{\mathrm{o}} + \mathrm{h.c.} \tag{16}$$

where $i, j = 1, 2, 3$ and λ_{ij}, $\hat{\lambda}_{ij}$, and μ_{ij} are unknown constants.

For the quarks fields, the corresponding Yukawa terms are written as

$$\begin{aligned} \mathcal{L}_{\mathcal{Y}}^{\mathrm{Q}} = & \textstyle\sum_{i,j} \lambda_{ij}^{d} \overline{Q}_{iL}^{\mathrm{O}} \phi D_{jR}^{o} + \sum_{i,j} \lambda_{ij}^{u} \overline{Q}_{iL}^{o} \tilde{\phi} U_{jR}^{o} + \sum_{i,j} \hat{\lambda}_{ij}^{d} \overline{\hat{Q}_{iR}^{o}} \hat{\phi} \hat{D}_{jL}^{o} + \\ & \textstyle\sum_{ij} \hat{\lambda}_{ij}^{u} \overline{\hat{Q}_{iR}^{o}} \tilde{\hat{\phi}} \hat{U}_{jL}^{o} + \sum_{i,j} \mu_{ij}^{d} \overline{\hat{D}_{iL}^{o}} D_{jR}^{o} + \sum_{i,j} \mu_{ij}^{u} \overline{\hat{U}_{iL}^{o}} U_{jR}^{o} + h.c. \end{aligned} \tag{17}$$

where the conjugate fields $\tilde{\phi}$ ($\tilde{\hat{\phi}}$) are obtained as $\tilde{\phi} = i\tau_2\phi^*$.

The VEV's of the neutral scalars produces the mass terms, which in the

gauge eigenstate basis reads $\qquad \mathcal{L}_{\mathrm{mass}} = \overline{\psi_{\mathcal{L}}^{\mathrm{O}}} \mathcal{M} \psi_{\mathrm{R}}^{\mathrm{O}} + \mathrm{h.c.} \qquad (18)$

The nondiagonal mass matrix M takes the form$\mathrm{M} = \begin{pmatrix} \mathsf{D} & 0 \\ \mu & \hat{\mathsf{D}} \end{pmatrix}$, $\qquad (19)$

where $\mathsf{D} = \frac{1}{2}\lambda \mathsf{v}$ and $\hat{\mathsf{D}} = \frac{1}{2}\hat{\lambda}\hat{v}$ are 3 x 3 matrices generated from the symmetry breaking VEV's , while μ corresponds to a gauge invariant 3 x 3 mixing mass matrix between ordinary and mirror fermions singlets.

The diagonal mass matrix $\mathrm{M_D}$ can be obtained through a biunitary rotation

acting on the L and R sectors, namely: $\qquad \mathrm{M_D} = \mathsf{U}_{\mathrm{L}}^{\dagger} \mathrm{M} \mathsf{U}_{\mathrm{R}} \qquad (20)$

---

[1] We shall name the light fermions as the SM ones, although this is not strictly true, then the heavy ones (ψ_h) will be consider the sector beyond the SM.

The neutral current term for the multiplet ψ of a given electric charge, including the contribution of the neutral gauge boson mixing, can be written as follows

$$-\mathcal{L}^{nc} = \sum_{a=L,R} \overline{\psi_a^O}\gamma^\mu \left(gT_{3a}, \hat{g}\hat{T}_{3a}, g'\frac{Y_a}{2}\right) \psi_a^O \begin{pmatrix} W^3 \\ \hat{W}^3 \\ B \end{pmatrix}_\mu \tag{21}$$

In terms of the mass eigenstates, using eqn. (15), one arrives to:

$$-\mathcal{L}^{nc} = \sum_{a=L,R} \overline{\psi_a}\gamma^\mu U_a^\dagger \left(gT_{3a}, \hat{g}\hat{T}_{3a}, g'\frac{Y_a}{2}\right) U_a \psi_a R \begin{pmatrix} Z \\ Z' \\ A \end{pmatrix}_\mu \tag{22}$$

where T_{3a}, $\hat{T}_{3a}$, and Y are the generators of the $SU(2)_L$, $SU(2)_R$, and $U(1)$, respectively.

On the other hand, eq. (20), we can express parameters λ, $\hat{\lambda}$, and μ, in terms of the mass-eigenvalues and the blocks of matrix U, as follows

$$\lambda = \frac{g}{\sqrt{2}}\left(A_L\frac{m_f}{M_W}A_R^\dagger + E_L\frac{m_{\hat{f}}}{M_W}E_R^\dagger\right), \qquad \hat{\lambda} = \frac{\hat{g}}{\sqrt{2}}\left(F_L\frac{m_f}{M_{\hat{W}}}F_R^\dagger + G_L\frac{m_{\hat{f}}}{M_{\hat{W}}}G_R^\dagger\right), \tag{23}$$

$$\mu = F_L m_f A_R^\dagger + G_L m_{\hat{f}} E_R^\dagger, \qquad 0 = A_L m_f F_R^\dagger + E_L m_{\hat{f}} G_R^\dagger \tag{24}$$

With these relations, and working within the Higgs mass-eigenstates basis, the tree-level interactions of the neutral Higgs bosons H and $\hat{H}$ with the light fermions are given by

$$\begin{aligned}\mathcal{L}_f = &\frac{g}{2\sqrt{2}}\overline{f_L}A_L^\dagger A_L\frac{m_l}{\mathcal{M}_W}\left(H\cos\alpha - \hat{H}\sin\alpha\right) + \\ &\frac{\hat{g}}{2\sqrt{2}}\overline{f_L}\frac{m_l}{M_{\hat{W}}}F_R^\dagger F_R f_R\left(H\sin\alpha + \hat{H}\cos\alpha\right) + \text{h.c.}\end{aligned} \tag{25}$$

One can see that the couplings are not diagonal in general, thus new phenomena associated with FCNC will be predicted in this model. The resulting phenomenological constraints and predictions will be discussed in future work [6].

For instance, once we obtain bounds on the coefficients of the couplings $\phi l_i l_j$, we can use these results to predict the Branching Ratio for the LFV decay modes of the Higgs themselves, namely $H \to \mu e$, $\tau\mu$, τe, which could be detected at the future colliders (Tevatron and LHC).

III ACKNOWLEDGEMENTS

This work was partially supported by CONACyT in Mexico and COLCIENCIAS in Colombia.

REFERENCES

1. T. D. Lee and C. N. Yang, Phys. Rev. **104**, 254 (1956).
2. J. Maalampi and M. Ross, Phys. Rev. **186**, 53 (1990).
3. V. Cerón, U. Cotti, J. L. Díaz-Cruz, and M. Maya, Phys. Rev. **D57**, (1998).
4. G. Triantaphyllou, J. Phys. G: Nucl. Part. Phys. **26**, 99 (2000).
5. P. Langacker and D. London, Phys. Rev. **D38**, 886 (1988).
6. J. L. Díaz, R. Gaitán, A. Hernández, and U. Cotti. (in preparation).

Multipole moments in the Δ^{++} resonance from bremsstrahlung and pion photoproduction

A. Mariano

Departamento Física, Centro de Investigación y de Estudios Avanzados del IPN, Apdo. Postal 14-740, 07000 México, D.F., México

Abstract. The Δ^{++} magnetic dipole moment is determined within a full dynamical and gauge invariant model. Using the same model the coupling constants that determine the M1 and E2 contributions to the pion photoproduction transition are also analyzed. First we study the elastic π^+p and radiative $\pi^+p\gamma$ scattering within a complex mass scheme which incorporates the finite width of the Δ^{++} resonance. The amplitudes for both processes are invariant under contact transformations that left invariant Lagrangians with the spin 3/2 fields and gauge-invariance is fulfilled for the radiative case. The pole parameters of the Δ^{++} obtained by fitting the experimental elastic cross section are $m_\Delta = (1211.9 \pm 0.5)$ MeV and $\Gamma_\Delta = (84.9 \pm 0.4)$ MeV. The anomalous magnetic dipole moment of the Δ^{++} is left as the only adjustable parameter in radiative $\pi^+p\gamma$ scattering. It was fixed by comparation with the experimental bremsstrahlung differential cross section in the case of most sensible photon geometries. In addition the coupling constants G1 and G2 of the $\gamma N \to \Delta$ vertex are fitted to reproduced pion photoproduction cross section.

INTRODUCTION

Within the hadronics resonances the isobar-$\Delta(1232)$ has been one of the most studied. It has been devoted a great effort in the determination of the Δ^{++} magnetic dipole moment(MDM) [1–3], being some of these previous determinations summarized in the Particle Data Group [4]. Nevertheless due to the large spread of central values, it is preferred to quote a rough estimate for this multipole which lie in the range $\mu_\Delta \sim 3.7$ to 7.5 in units of nuclear magnetons $e/2m_p$.

These various estimations of the Δ^{++} MDM are based on fits to the radiative π^+p scattering data of Ref. [5], SIN [7] and UCLA [6] experiments. Some of the works incorporate a dynamical model [1] but the use of form factors in the $\pi^+p \to \Delta^{++}$ vertex and of an energy-dependent width for the Δ^{++} resonance, lead to add terms out of the model to get electromagnetic gauge invariance. Other models [2] introduce an amplitude that depends on an arbitrary parameter associated to the

CP562, *Particles and Fields: Ninth Mexican School,* edited by G. Herrera Corral and L. Nellen

contact transformations used to eliminate irrelevant degrees of freedom in the spin 3/2 fields [8]; hence, the value of the MDM is quoted for an specific value of this arbitrary parameter. In Ref. [3] the MDM determination rely on the soft photon theorem [9], and on a specific parametrization of the off-shell elastic amplitude to fix the terms of order ω_γ^0 (ω_γ is the photon energy in the radiative process) by requiring gauge-invariance. In this work the emission of a photon from the Δ^{++} is explicitly included, but diagrams with vertices involving four particles and the effects of the Δ^{++} finite width are ignored.

On the other hand there is also another problem closely related with Δ resonance. That is the d-state admixture in the baryon ground state wave function, which results from the action of the quark-quark tensor hyperfine interaction [10]. This interaction induces a small violation of the Bechhi-Mopurgo selection rule, that the $\gamma N \to \Delta$ excitation is a pure M1 (magnetic dipole) transition, by introducing a non-vanishing E2 (electric quadrupole) amplitude. To observe a static deformation (d-state admixture) a target with spin at least 3/2(e.g.Δ) would be required. The experimental quantity of interest to compare with the different nucleon models is the ratio $R_{EM} = E2/M1$ in the region of the Δ resonance. Quite recently the precise experimental value of $R_{EM} = -(2.5 \pm 0.4)$ has been established [11]. However, there is a more serious problem on the theoretical side. For example in the effective Lagrangian approach the main problem is the unitarization of the amplitude that is put in by hand and leads to different separations between the resonant part and the background contributions [12]. On the other hand, models [13] which treat pion photoproduction dynamically (i.e., solving the Lippman-Schwinger equation to obtain final state interactions) are unitary but they need the use of form factors to regularize the driving terms of the interaction.

The aim of the present paper is to determine the MDM of the Δ^{++} resonance by using a *full* dynamical model which consistently describes the data on elastic and radiative $\pi^+ p$ scattering. In addition to analyze how does work this model in photoproduction reactions we will estimate the couplings G1 and G2 (M1 and M2 depend on them) of the $\gamma N \to \Delta$ vertex , by fitting the $\gamma p \to \pi^0$ cross section.

DYNAMICAL MODEL FOR ELASTIC AND RADIATIVE AMPLITUDE

The $\pi^+ N$ scattering amplitude for the range $T_{lab} = 100 - 300 MeV$ is dominated by a resonant behavior plus a background. The resonant part corresponds to the Δ^{++} formation and the background accounts for another non-resonant contributions. The dynamical model we use includes contributions to the scattering amplitude of intermediate states with nucleons and $\Delta^{++,\,0}$ resonances, and ρ, σ, ω mesons exchange. We will assume isospin symmetry for the masses, widths and strong couplings of the Δ's and nucleons. The effective lagrangian density relevant for our calculations are given in Refs. [14,15].

From the Feynman rules obtained from these Lagrangians the Δ^{++} contribution to the $\pi^+p \to \pi^+p$ amplitude reads:

$$\mathcal{M}_{\Delta^{++}}[\pi^+(q)p(p) \to \pi^+(q')p(p')] = -i\left(\frac{f_{\Delta\pi N}}{m_\pi}\right)^2 \overline{u}(p')q'_\mu G^{\mu,\nu}(p+q)q_\nu u(p), \quad (1)$$

being $u(p) \equiv u(p, ms)$ the proton spinor and $G^{\mu\nu}(P)$ is the Δ propagator [15]. We are using a complex mass scheme to include the unstable character of the resonance, that is $\tilde{m}^2_\Delta \equiv m^2_\Delta - im_\Delta\Gamma_\Delta$, being m_Δ and Γ_Δ the mass and width of the Δ^{++} respectively. Finally and as was mentioned previously the amplitude in Eq.(1), is free of ambiguities under contact transformations [16].

Since the Δ^{++} largely dominates the elastic scattering amplitude in the resonance region, we expect the contributions from σ and ρ exchange, and of crossed channels with intermediate nucleons and the Δ^0 to play the role of background contributions to the Δ^{++} resonance. These contributions can be evaluated from the hadronic Lagrangians in Ref. [15] Some of the couplings entering in the calculus of the amplitude can be fixed from low energy phenomenology: $(g^2_\rho, g^2_{\pi NN})/4\pi = (2.9, 14)$ and the magnetic ρNN coupling $\kappa_\rho = 3.7$. The masses of ρ and the nucleon were taken from [4], and the mass of the hypothetical σ meson was set to 650 MeV [22]. The couplings $g_{\sigma\pi\pi,\sigma NN}$ and $f_{\Delta N\pi}$ are left as free parameters to be determined from the elastic π^+p scattering total cross section [23].

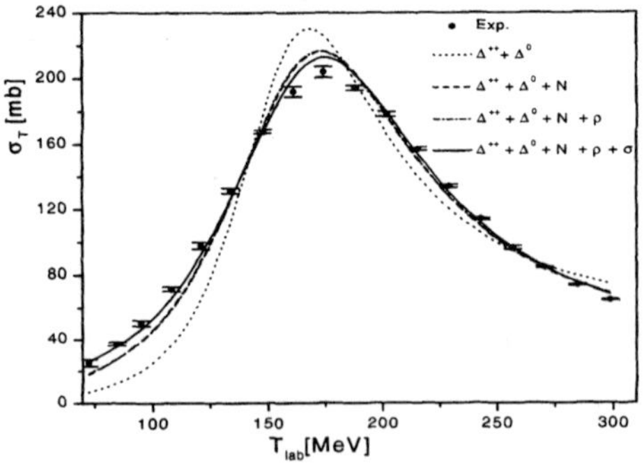

FIGURE 1. Elastic π^+p cross section as a function of incident π^+ kinetic energy.

In order to compare the size of the different contributions, we have chosen to fit the data by adding a new contribution in each fit. The results are shown in Figure 1. The values obtained for the mass and width of the Δ's, $m_\Delta = (1211.9 \pm 0.5)$ MeV and $\Gamma_\Delta = (84.9 \pm 0.4)$ MeV, are similar to those obtained from a model-independent analysis of the same data on the total cross section, namely [20]: $M = (1212.20 \pm 0.23)$ MeV and $\Gamma = (97.06 \pm 0.35)$.

Next we focus on the determination of the Δ^{++} MDM from the process $\pi^+ p \to \pi^+ p\gamma$. We are interested in the description of the differential cross section $d\sigma/d\omega_\gamma d\Omega_\pi d\Omega_\gamma$, as a function of the photon energy for fixed energies of incident pions and fixed photon and pion angles emission. Using the Lagrangians in Ref. [15] the following radiative amplitude for the Δ^{++} contribution is obtained:

$$\begin{aligned}\mathcal{M}[\pi^+ p \to \pi^+ p\gamma(k)] = -ie\left(\frac{f_{\Delta\pi N}}{m_\pi}\right)^2 q'_\mu q_\nu \overline{u}(p')\Bigg[G^{\mu\nu}(P')\left(\frac{q\cdot\epsilon}{q\cdot k}+\frac{p\cdot\epsilon - R(p)\cdot\epsilon}{p\cdot k}\right)\\ -\left(\frac{q'\cdot\epsilon}{q'\cdot k}+\frac{p'\cdot\epsilon - R(p')\cdot\epsilon}{p'\cdot k}\right)G^{\mu\nu}(P) + 2G^{\mu\alpha}(P')\Gamma_{\alpha\beta\rho}\epsilon^\rho G^{\beta\nu}(P)\\ + \frac{1}{q\cdot k}G^{\mu\rho}(P')(\epsilon_\rho k^\nu - \epsilon^\nu k_\rho) - \frac{1}{q'\cdot k}(\epsilon^\mu k_\rho - \epsilon_\rho k^\mu)G^{\rho\nu}(P)\Bigg]u(p)\end{aligned} \tag{2}$$

where

$$R_\mu(x) \equiv \frac{1}{4}[\not{k}, \gamma_\mu] + \frac{\kappa_p}{8m_N}\{[\not{k}, \gamma_\mu], \not{x}\},$$

$P = p+q$, $P' = p'+q'$, such that $P = P'+k$, and ϵ indicates the photon polarization. This amplitude is explicitly gauge-invariant. $\Gamma_{\alpha\beta\rho}$ in Eq.(2) is the electromagnetic vertex of the Δ^{++} [21], which depends on κ_Δ that is the Δ^{++} anomalous magnetic moment(related to the total moment as $\mu_\Delta = 2(1+\kappa_\Delta)(e/2m_\Delta)$). This is the only adjustable parameter in radiative $\pi^+ p$ scattering. We have chosen to fit a subset of data of Ref. [6] where photons are detected in angular configurations where the differential cross section is more sensitive to the effects of κ_Δ. The results are shown in Figure 2. The labels G7, G4 and G1 refer to photons detected at angles $(\theta_\gamma, \phi_\gamma) = (120^0, 0^0)$,$(140^0, 0^0)$, and $(160^0, 0^0)$, respectively. If we express the weighted average of the different geometries in units of nuclear magnetons we obtain:

$$\mu_\Delta = 2(1+\kappa_\Delta)\frac{m_p}{m_\Delta}\left(\frac{e}{2m_p}\right) = (6.22 \pm 0.53)\,\frac{e}{2m_p}\,. \tag{3}$$

This result is compatible with the prediction $\mu_\Delta = 5.58(e/2m_p)$ obtained in the SU(6) quark model [24], and it is somewhat larger than the prediction obtained from the bag-model corrections to the quark model, $\mu_\Delta = (4.41 \sim 4.89)(e/2m_p)$ [25]. This is to our knowledge, the first determination of the Δ^{++} MDM from a full dynamical model that consistently incorporates its finite width, that is free of ambiguities related to contact transformations, that does not use phenomenological form factors and that leads to a precise value.

PION PHOTOPRODUCTION

It is wellknown that a model must to work as well with hadronics probes as with electromagnetics probes, for the sake of consistence. Now we are going to use

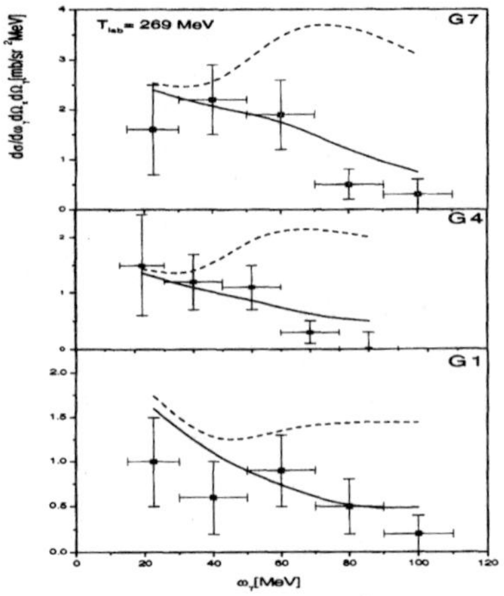

FIGURE 2. Differential cross section in radiative π^+p scattering for $T_{lab} = 269$ MeV. The G1, G4 and G7 geometries are defined in the text. The solid line corresponds to the best fit and the dashed line to $\kappa_\Delta = 1$.

the same dynamical model implemented to calculate the elastic pion-nucleon cross section in the calculus of photoproduction cross section. In this case the amplitude can be obtained from the Eq.(1) by replacing the initial $\pi N \to \Delta$ vertex by the $\gamma N \to \Delta$ vertex:

$$\Gamma[\gamma N \to \Delta]_\mu = G1/2m_N(\not{k}\epsilon_\mu - \not{\epsilon}k_\mu)\gamma_5 T_3 + G2/(2m_N)^2(p.\epsilon k_\mu - p.k\epsilon_\mu)\gamma_5 T_3, \quad (4)$$

where $\vec{T}$ is the isospin excitation operator. In terms of $G1$ and $G2$ the ratio R_{EM} can be obtained as

$$R_{EM} = -\frac{m_\Delta - m_N}{2m_N}\frac{G1 - G2(m_\Delta/2m_N)}{G1(3m_\Delta + m_N)/2m_N - G2\ m_\Delta(m_\Delta - m_N)/(2m_N)^2}. \quad (5)$$

In order to make an estimation of $G1$ and $G2$ one must to achieve a multipole analysis of the amplitude and calculate directly M1 and E2. In place of this we will get an estimation of these constants taking the experimental value of R_{EM} [11] and fitting G1 to reproduce the $\gamma p \to p\pi^0$ differential cross section. This procedure is not the usual one, but is more simple. In Figure 3 we show the results of this fitting, being the range of the obtained values $G1 = 2.14 - 2.6$, $G2 = 0.46G1$ for the different photon energies.

As result of this analysis we can conclude that the model also is consistent in the description of the photoproduction cross section. A more detailed analysis of the values of G1 and G2 should require the explicit calculation of M1 and E2.

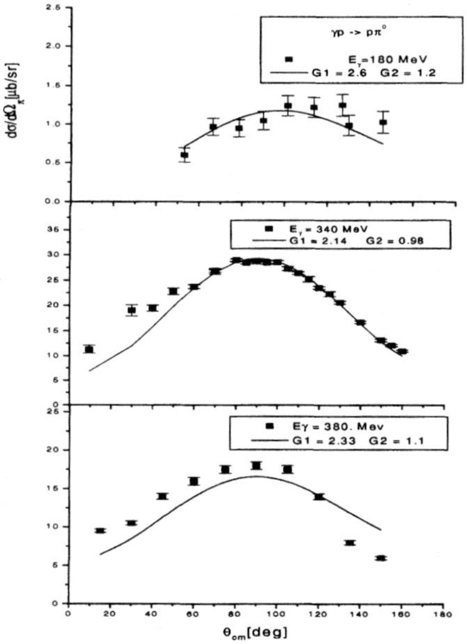

FIGURE 3. Differential π^0 photoproduction cross section as a function of incident photon energy

REFERENCES

1. M. M. Musakhanov, Sov. J. Nuc. Phys. **19**, 319 (1974).
2. R. Wittman, Phys. Rev. **C37**, 2075 (1988).
3. D. Lin , M.K. Liou and Z.M. Ding, Phys. Rev. **C44**, 1819 (1991).
4. C. Caso *et al*, Particle Data Group, Eur. Phys. J. **C3**, 1 (1998).
5. M. Arman te. al., Phys. Rev. Lett. **29**, 962 (1972).
6. B. M. K. Nefkens *et al*, Phys. Rev. **D18**, 3911 (1978).
7. A. Bosshard *et al*, Phys. Rev. **D44**, 1962 (1991).
8. L. M. Nath, B. Etemadi, and J. D. Kimel, Phys. Rev. **D3**, 2153 (1971); R. E. Behrends and C. Fronsdal, Phys. Rev. **106**, 277 (1958); J. Urías, Ph. D. Thesis, Université catholique de Louvain, Belgium (1976).
9. F. E. Low, Phys. Rev. **110**, 974 (1958).
10. N. Isgur, G. Karl, R. Koniuk, Phys. Rev. **D25**, 2394 (1982).
11. R. Beck,et.al., Phys. Rev. **C61**, 03524 (2000).
12. R.M. Davidson, N. C. Mukhopadhyay, and R.S. Wittman, Phys. Rev. **D43**, 71 (1991). R.M. Davidson, et.al., Phys. Rev. **C59**, 1059 (1999).
13. T. Sato and T.-S.H. Lee, Phys. Rev. **C54**, 2660 (1996).
14. J. Wess and B. Zumino, Phys. Rev. **163**, 1727(1967).
15. A. Mariano and G. López Castro, Phys. Rev. **62**, 014604(2000).

16. M. El-Amiri, G. López Castro and J. Pestieau, Nucl. Phys. **A543**, 673 (1992).
17. G. López Castro and G. Toledo Sánchez, Phys. Rev. **D61**, 033007 (2000).
18. G. López Castro, J. L. Lucio M., and J. Pestieau, Mod. Phys. Lett. **A**, (1991); Int. J. Mod. Phys. **A10**, (1996).
19. A. Pilaftsis and M. Nowakowski, Z. Phys. **C60**, 121 (1993).
20. A. Bernicha, G. López Castro and J. Pestieau, Nucl. Phys. **A597**, 623 (1996).
21. W. Rarita and J. Schwinger, Phys. Rev.**40**, 61(1941)
22. See for example: B.C. Pearce and B.K. Jennings, Nucl. Phys.,**A528**,655 (1991); C. Schütz, J.W. Durso, K. Holinde, and J. Speth, Phys. Rev. **C49**, 2671(1994).
23. E. Pedroni *et al*, Nucl. Phys. **A300**, 321(1978).
24. M. A. B. Beg, B. W. Lee and A. Pais, Phys. Rev. Lett. **13**, 514 (1964).
25. G. E. Brown, M. Rho and V. Vento, Phys. Lett. **B97**, 423 (1980).

Energy dependence of the quark masses and mixings

S.R. Juárez W.*, S.F. Herrera H.*, P. Kielanowski† and G. Mora H.†

*Departamento de Física, Escuela Superior de Física y Matemáticas, IPN, México
†Departamento de Física, Centro de Investigación y Estudios Avanzados, México
†Institute of Theoretical Physics, University of Białystok, Poland

Abstract. The one loop Renormalization Group Equations for the Yukawa couplings of quarks are solved. From the solution we find the explicit energy dependence on $t = \ln E/\mu$ of the evolution of the *down* quark masses $q = d, s, b$ from the grand unification scale down to the top quark mass m_t. These results together with the earlier published evolution of the *up* quark masses completes the pattern of the evolution of the quark masses. We also find the energy dependence of the absolute values of the Cabibbo-Kobayashi-Maskawa (CKM) matrix $|V_{ij}|$. The interesting property of the evolution of the CKM matrix and the ratios of the quark masses: $m_{u,c}/m_t$ and $m_{d,s}/m_b$ is that they all depend on t through only one function of energy $h(t)$.

PACS: 12.15.Ff, 12.15.Hh, 11.30.Hv

In a recent paper Ref. [1], a systematical investigation of the evolution of the CKM matrix and the quark Yukawa couplings $y_u(t)$ and $y_d(t)$ was performed. Exact solutions of the one loop Renormalization Group Equation (RGE) and some general properties of the RGE evolution for the quark masses m_q, $q = u, c, t, d, s, b$, and CKM matrix, compatible with the observed hierarchy Ref. [2], were obtained.

The objective of this talk is to present some complementary results to the previous ones. We show here the explicit solutions for the masses of the quarks of the down sector and how they are derived.

The quark masses are the eigenvalues of the Yukawa couplings obtained after its diagonalization by biunitary transformations with the help of the unitary matrices $(U_{u,d})_{L,R}$

$$\mathrm{Diag}(m_u, m_c, m_t) = (U_u)_L y_u (U_u)^\dagger_R, \ \mathrm{Diag}(m_d, m_s, m_b) = (U_d)_L y_d (U_d)^\dagger_R.$$

From the diagonalizing matrices we obtain the flavor mixing in the charged current described by the CKM matrix

$$V_{\mathrm{CKM}} = (U_u)_L (U_d)^\dagger_L.$$

CP562, *Particles and Fields: Ninth Mexican School,* edited by G. Herrera Corral and L. Nellen

The Yukawa couplings are scale dependent. Very frequently one makes various assumptions about their properties at the Grand Unification (GU) scale and then one has to compare the predictions with the measured values at low energies Ref. [3]. The RGE are an important tool for the search of the properties of the quark masses and the CKM matrix at different energy scales. The RGE have been worked out by various authors Ref. [4–7]. The structure of the one loop RGE for the gauge coupling constants g_k and the Yukawa couplings $y_{u,d,e,\nu}$ is the following

$$\frac{dg_k}{dt} = \frac{1}{(4\pi)^2} b_k g_k^3, \qquad \frac{dy_{u,d,e,\nu}}{dt} = \left[\frac{1}{(4\pi)^2}\beta^{(1)}_{u,d,e,\nu}\right] y_{u,d,e,\nu}.$$

Here $t \equiv \ln(E/\mu)$ is the energy scale parameter, the coefficients b_k are defined in Table 1 and the functions $\beta^{(1)}_{u,d,e,\nu}$ are defined for various models in the Appendix. The approximate form of the equations for the quark Yukawa couplings, neglecting all the terms of λ^4 and higher ($\lambda = 0.22$) have the following form

$$\begin{aligned}\frac{dy_u}{dt} &= \frac{1}{(4\pi)^2}[\alpha_1^u(t) + \alpha_2^u y_u y_u^\dagger + \alpha_3^u \mathrm{Tr}(y_u y_u^\dagger)]y_u,\\ \frac{dy_d}{dt} &= \frac{1}{(4\pi)^2}[\alpha_1^d(t) + \alpha_2^d y_u y_u^\dagger + \alpha_3^d \mathrm{Tr}(y_u y_u^\dagger)]y_d.\end{aligned} \tag{1}$$

The explicit solutions for $m_q(t)$ with $q = u, c, t$ and $y_d(t)$ previously obtained in Ref. [1] are:

$$\begin{aligned}m_{u,c}(t) = m_{u,c}(t_0)\sqrt{r_g(t)}(h(t))^{b/c}, \qquad m_t(t) = m_t(t_0)\sqrt{r_g(t)}\,(h(t))^{\frac{(b+2)}{c}},\\ y_d(t) = \sqrt{r_g'(t)}(h(t))^{2a/c}(U_u)_L^\dagger Z(t)(U_u)_L y_d(t_0),\end{aligned} \tag{2}$$

where the (a, b, c) are equal to $(0, 2, 2/3)$, $(0, 1, 1/3)$, $(1, 1, -1)$ in the MSSM, DHS and SM, respectively,

$$\begin{aligned}h(t) &= \exp\left(\frac{1}{(4\pi)^2}\frac{3c}{2}\int_{t_0}^t m_t^2(\tau)d\tau\right)\\ &= \left(\frac{1}{1 - \frac{3(b+2)}{(4\pi)^2} m_t^2(t_0)\int_{t_0}^t r_g(\tau)d\tau}\right)^{\frac{c}{2(b+2)}}.\end{aligned} \tag{3}$$

and

$$[Z(t)]_{ij} = \delta_{ij} + (h(t) - 1)\delta_{i3}\delta_{j3}.$$

These solutions Eqs. (2) do depend on the energy through the overall factors $r_g(t)$, $r_g'(t)$ (see Appendix) and the matrix $Z(t)$.

The procedure Ref. [8], to obtain the energy dependence of the masses for the down sector is the following:

Step 1.- Differentiate with respect to t the following equation

$$(U_u)_L y_d(t) y_d(t)^\dagger (U_u)_L^\dagger = r_g'(t)(h(t))^{(2\alpha_3^d/\alpha_2^d)} Z(t)(U_u)_L y_d(t_0) y_d(t_0)^\dagger (U_u)_L^\dagger Z(t),$$

which can be written in this way

$$(U_u)_L y_d(t) y_d(t)^\dagger (U_u)_L^\dagger = V_{\rm CKM}(t) M_d^2(t) V_{\rm CKM}^\dagger(t)$$

where $M_d^2(t)$ is the diagonal matrix with the squares of the physical down quarks Yukawa couplings on the diagonal, that become the squares of the down quark masses after the spontaneous symmetry breaking. Next we obtain the following matrix differential equation

$$V_{\rm CKM}^\dagger(t)\frac{dV_{\rm CKM}}{dt} = (M_d^2)^{-1} V_{\rm CKM}^\dagger(t)\frac{dV_{\rm CKM}}{dt} M_d^2$$
$$-(M_d^2)^{-1} V_{\rm CKM}^\dagger(t)\frac{d}{dt}((U_u)_L y_d(t) y_d(t)^\dagger (U_u)_L^\dagger) V_{\rm CKM}(t) + (M_d^2)^{-1}\frac{dM_d^2}{dt}. \qquad (4)$$

Eq. (4) becomes simpler after using the following relation (see Eq. (2))

$$(M_d^2)^{-1} V_{\rm CKM}^\dagger(t)\frac{d}{dt}((U_u)_L y_d(t) y_d(t)^\dagger (U_u)_L^\dagger) V_{\rm CKM}(t)$$
$$= \frac{h'}{h}(\mathbf{R}^\dagger\mathbf{R} + (M_d^2)^{-1}\mathbf{R}^\dagger\mathbf{R} M_d^2) + \frac{d\ln(r_g'(t)(h(t))^{(2\alpha_3^d/\alpha_2^d)})}{dt} I$$

where the vector $\mathbf{R} = (V_{td}, V_{ts}, V_{tb})$.

Step 2.- Extract the differential equations for the diagonal matrix elements of Eq. (4):

$$\frac{d}{dt}\ln m_d(t) = \frac{1}{2}\frac{d\ln\left[r_g'(t)(h(t))^{4a/c}\right]}{dt} + \left[\frac{d}{dt}\ln h(t)\right]|V_{td}(t)|^2,$$
$$\frac{d}{dt}\ln m_s(t) = \frac{1}{2}\frac{d\ln\left[r_g'(t)(h(t))^{4a/c}\right]}{dt} + \left[\frac{d}{dt}\ln h(t)\right]|V_{ts}(t)|^2,$$
$$\frac{d}{dt}\ln m_b(t) = \frac{1}{2}\frac{d\ln\left[r_g'(t)(h(t))^{4a/c}\right]}{dt} + \left[\frac{d}{dt}\ln h(t)\right]|V_{tb}(t)|^2. \qquad (5)$$

Step 3.- From the off diagonal matrix elements of Eq. (4) we deduce equations for the squares of the CKM matrix elements $|V_{ij}|^2$. We solve equations for the $|V_{td}|^2$, $|V_{ts}|^2$ and $|V_{tb}|^2$ matrix elements that are needed in Eq. (5):

$$\frac{d}{dt}|V_{td}(t)|^2 = -2\left[\frac{d}{dt}\ln h(t)\right]|V_{td}(t)|^2(1-|V_{td}(t)|^2),$$
$$\frac{d}{dt}|V_{tb}(t)|^2 = 2\left[\frac{d}{dt}\ln h(t)\right]|V_{tb}(t)|^2(1-|V_{tb}(t)|^2),$$
$$\frac{d}{dt}|V_{ts}(t)|^2 = -2\left[\frac{d}{dt}\ln h(t)\right]|V_{ts}(t)|^2\left[|V_{tb}(t)|^2-|V_{td}(t)|^2\right]. \qquad (6)$$

Obtaining

$$|V_{td}(t)|^2 = \frac{|V_{td}^0|^2}{h^2(t) + (1 - h^2(t))|V_{td}^0|^2}, \qquad |V_{tb}(t)|^2 = \frac{|V_{tb}^0|^2 h^2(t)}{1 + |V_{tb}^0|^2(h^2(t) - 1)},$$
$$|V_{ts}(t)|^2 = \frac{|V_{ts}^0|^2}{|V_{tb}^0|^2\left[1 - |V_{td}^0|^2\right]} \frac{h^2(t)}{\left[h^2(t) + \frac{1-|V_{tb}^0|^2}{|V_{tb}^0|^2}\right]\left[h^2(t) + \frac{|V_{td}^0|^2}{[1-|V_{td}^0|^2]}\right]}, \tag{7}$$

where

$$V_{ij}^0 = V_{ij}(t_0).$$

Step 4.- Finally, Eqs. (7) are used to solve in an analytical way the differential equations (5).

The formulas which give the explicit energy dependence of the quark masses for the down sector are

$$\frac{m_d(t)}{m_d(t_0)} = \sqrt{r_g'(t)} \frac{(h(t))^{(c+2a)/c}}{\sqrt{h^2(t) + |V_{td}^0|^2(1 - h^2(t))}},$$
$$\frac{m_s(t)}{m_s(t_0)} = \sqrt{r_g'(t)} \frac{\sqrt{h^2(t) + |V_{td}^0|^2(1 - h^2(t))}}{\sqrt{1 + |V_{tb}^0|^2(h^2(t) - 1)}} (h(t))^{2a/c},$$
$$\frac{m_b(t)}{m_b(t_0)} = \sqrt{r_g'(t)} \sqrt{1 + |V_{tb}^0|^2(h^2(t) - 1)}\, (h(t))^{2a/c}. \tag{8}$$

The results presented in this paper together with those of Ref. [1] form the complete set of the renormalization group evolution predictions for the observables derived from the quark Yukawa couplings. The consistent approximation scheme based on the hierarchy of the quark masses and the CKM matrix is strictly observed in all the derivations and it is shown that the final results for the ratios of the quark masses and the CKM matrix depend only on one function of energy in agreement with the theorem presented in Ref. [1].

As an illustration we show in Fig. 1 the evolution of the quark masses for the energy range from $E = m_t$ to $E = 10^{14}$ GeV for the Standard Model.

The explicit form of the evolution given here can be very useful for the phenomenological analysis of the models that specify the properties of the Yukawa interactions at the GU scale.

We acknowledge the financial support from CONACYT (México) projects 3512P-E9608 and 26247E. S.R.J.W. also thanks to "Comisión de Operación y Fomento de Actividades Académicas" (COFAA) from Instituto Politécnico Nacional.

APPENDIX

$$\beta_l^{(1)} = \alpha_1^l(t) + \alpha_2^l H_u^{(1)} + \alpha_3^l \mathrm{Tr}(H_u^{(1)}) + \alpha_4^l H_d^{(1)} + \alpha_5^l \mathrm{Tr}(H_d^{(1)})$$
$$+ \alpha_6^l H_e^{(1)} + \alpha_7^l \mathrm{Tr}(H_e^{(1)}) + \alpha_8^l H_\nu^{(1)} + \alpha_9^l \mathrm{Tr}(H_\nu^{(1)}), \quad H_l^{(1)} = y_l y_l^\dagger. \tag{A1}$$

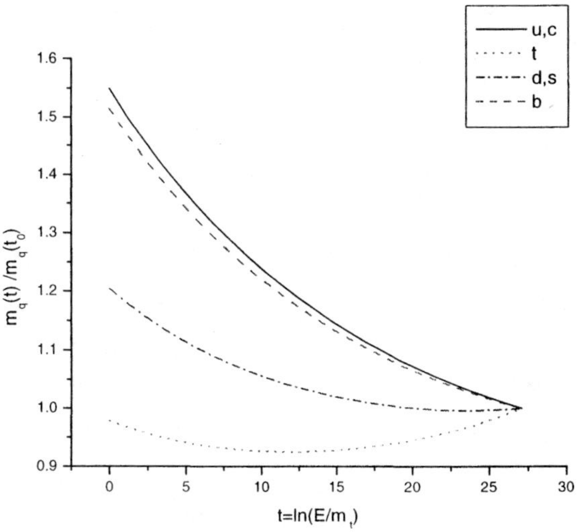

FIGURE 1. Evolution of the quark masses in the Standard Model

$l = u, d, e, \nu$. The $\alpha_1^l(t)$ and the α_i^l, $i = 2, ..., 9$ are given in Tables 2 and 3

$$r_g(t) = \exp(\frac{2}{(4\pi)^2}\int_{t_0}^{t} \alpha_1^u(\tau)d\tau) = \Pi_{k=1}^{k=3}\left[\frac{g_k^2(t_i)}{g_k^2(t)}\right]^{\frac{c_k}{b_k}},$$

$$r_g'(t) = \exp(\frac{2}{(4\pi)^2}\int_{t_0}^{t} \alpha_1^d(\tau)d\tau) = \Pi_{k=1}^{k=3}\left[\frac{g_k^2(t_i)}{g_k^2(t)}\right]^{\frac{c_k'}{b_k}}, \tag{A2}$$

where

$$g_k(t) = \frac{g_k(t_0)}{\sqrt{1 - \frac{2b_k g_k^2(t_0)(t-t_0)}{(4\pi)^2}}}.$$

The coefficients b_k, c_k and c_k' are defined in Table 1.

TABLE 1. The parameters for the various models.

Model	b_1	b_2	b_3	c_1	c_2	c_3	c_1'	c_2'	c_3'
MSSM	$\frac{33}{5}$	1	-3	$\frac{13}{15}$	3	$\frac{16}{3}$	$\frac{7}{15}$	3	$\frac{16}{3}$
DHM	$\frac{21}{5}$	-3	-7	$\frac{17}{20}$	$\frac{9}{4}$	8	$\frac{1}{4}$	$\frac{9}{4}$	8
SM	$\frac{41}{10}$	$\frac{-19}{6}$	-7	$\frac{17}{20}$	$\frac{9}{4}$	8	$\frac{1}{4}$	$\frac{9}{4}$	8

TABLE 2. The coefficients α_1^l for various models.

	SM and DHM	MSSM
$\alpha_1^u(t) =$	$-(\frac{17}{20}g_1^2 + \frac{9}{4}g_2^2 + 8g_3^2)$	$-(\frac{13}{15}g_1^2 + 3g_2^2 + \frac{16}{3}g_3^2)$
$\alpha_1^d(t) =$	$-(\frac{1}{4}g_1^2 + \frac{9}{4}g_2^2 + 8g_3^2)$	$-(\frac{7}{15}g_1^2 + 3g_2^2 + \frac{16}{3}g_3^2)$
$\alpha_1^e(t) =$	$-(\frac{9}{20}g_1^2 + \frac{9}{4}g_2^2)$	$-(\frac{9}{5}g_1^2 + 3g_2^2)$
$\alpha_1^\nu(t) =$	$-(\frac{9}{20}g_1^2 + \frac{9}{4}g_2^2)$	$-(\frac{3}{5}g_1^2 + 3g_2^2)$

TABLE 3. The coefficients α_k^l for various models; the constants (a, b, c) are equal to $(0, 2, 2/3)$, $(0, 1, 1/3)$, $(1, 1, -1)$ in the MSSM, DHS and SM.

l	α_2^l	α_3^l	α_4^l	α_5^l	α_6^l	α_7^l	α_8^l	α_9^l
u	$\frac{3}{2}b$	3	$\frac{3}{2}c$	$3a$	0	a	0	1
d	$\frac{3}{2}c$	$3a$	$\frac{3}{2}b$	3	0	1	0	a
e	0	$3a$	0	3	$\frac{3}{2}b$	1	$\frac{3}{2}c$	a
ν	0	3	0	$3a$	$\frac{3}{2}c$	a	$\frac{3}{2}b$	1

REFERENCES

1. P. Kielanowski, S.R. Juárez W. and G. Mora, *Phys. Lett.* **B479**, 181 (2000); S.R. Juárez W., P. Kielanowski and G. Mora, *New Properties of the Renormalization Group Equations of the Yukawa Couplings and CKM Matrix* in the Proceedings of the Eighth Mexican School *Particles and Fields*, Oaxaca, México, AIP Conference Proceedings **490**, 351–354 (1998), ed. J.C. D'Olivo, G. López C. and M. Mondragón, L.C. Catalog Card No. 99-067150, ISBN 1-56396-895-9, ISSN 0094-243X., DOE CONF-981188.
2. L. Wolfenstein, *Phys. Rev. Lett.* **51**, 1945 (1983).
3. R.D. Peccei and K. Wang, *Phys. Rev.* **D53**, 2712 (1996); H. González *et al.*, *Phys. Lett.* **B440** 94 (1998); H. González *et al.*, *A New symmetry of Quark Yukawa Couplings* in the Proceedings of the *International Europhysics Conference on High Energy Physics*, Jerusalem 1997, Eds. Daniel Lellouch, Giora Mikenberg, Eliezer Rabinovici, Springer-Verlag 1999, p. 755.
4. K.S. Babu, *Z. Phys.* **C 35**, 69 (1987); P. Binetruy and P. Ramond, *Phys. Lett.* **B350**, 49 (1995); K. Wang, *Phys. Rev.* **D54**, 5750 (1996).
5. B. Grzadkowski and M. Lindner, *Phys. Lett.* **B193**, 71 (1987); B. Grzadkowski, M. Lindner and S. Theisen, *Phys. Lett.* **B198**, 64 (1987); M. Olechowski and S. Pokorski, *Phys. Lett.* **B257**, 388 (1991).
6. H. Arason *et al.*, *Phys. Rev.* **D46**, 3945 (1992).
7. H. Fusaoka and Y. Koide, *Phys. Rev.* **D57**, 3986 (1998).
8. K. Sasaki, *Z. Phys.* **C 32**, 146 (1986).

Limits on excited tau lepton from $W \to \tau\nu_\tau$ decay

R. Diaz Sanchez and R. Martinez
Depto de Fisica
Universidad Nacional
Bogota, Colombia
and
O. A. Sampayo
Universidad de Mar del Plata
Argentina

Abstract. We evaluate the compositeness effects of leptons on the vertex $W\tau\nu_\tau$ in the context of an effective Lagrangian approach and get the corrections to the non universal coupling g_τ where we consider that only the third family is composed. We find the allowed region from the experimental g_τ/g_e quotient for (Λ versus m^*) plane when the masses of the excited states of the third generation are considered equal, i.e., $m_\tau^* = m_{\vartheta_\tau}^* = m^*$.

Owing to the precision reached by experimental results nowadays, we are able to impose significant constraints to physics beyond the Standard Model (SM). Compositeness is a very important alternative of new physics which could solve some problems of SM, among them the family mass hierarchy. Compositeness theories lie on the idea that known fermions and perhaps bosons possess an underlying structure, characterized by the scale Λ. The fundamental constituents of fermions and bosons could generate the mass spectra if we knew the confinement mechanism, so the mass hierarchy problem would be solved.

Some attempts to generate a proper confinement mechanism have been done by Seiberg, Harari, Terazawa [1] and others [2], [3]. However there is not any satisfactory confinement mechanism able to generate the whole mass spectrum from preons hitherto, as a consequence we should resort to effective Lagrangian techniques in order to describe the behavior of excited states [5]. Such states should become manifest at a certain energy scale Λ, and the SM is seen as an effective theory of a more fundamental one.

Measurements of anomalous magnetic moment of muon and electron have not shown any track of substructure for the first and second lepton family, conse-

CP562, *Particles and Fields: Ninth Mexican School,* edited by G. Herrera Corral and L. Nellen

quently we will make the assumption that only the third family shows excited states. The other families are either elementary or exhibit a much higher scale factor i.e. $\Lambda_\tau << \Lambda_e, \Lambda_\mu$. From this fact, excited states of the third family have appealed the attention of many collaborations such as L3, DELPHI and OPAL [4] whose analyses are based on an effective $SU(2) \times U(1)$ invariant Lagrangian proposed by Hagiwara et.al. [5]. Moreover, theoretical constraints have been derived from the contribution of anomalous magnetic moment of leptons and the Z scale observables at the CERN e^+e^- collider LEP. Bounds from an effective Lagrangian approach have been extracted from leptonic branching ratios as an allowed region on the $(m^*, f/\Lambda)$ −plane, based on different experiments [6] as well as bounds coming from the anomalous weak magnetic moment of the tau lepton and precision measurement on the Z peak.

Additionally, another information about excited states could come from non universal coupling constants g_l from $Wl\nu_l$ vertices by evaluating the contribution of this new physics to them. Since the electrons are considered elementary, no corrections from compositeness are made to the coupling g_e from the SM, then we can evaluate the correction to τ couplings by the ratio g_τ/g_e and taking $g_e = g$.

We assume that the excited fermions acquire masses before the $SU(2) \times U(1)$ breaking, so that both left handed and right handed states belong to weak isodoublets (vector-like model). The effective dimension five Lagrangian that describes the coupling of excited leptons with ordinary ones can be written as

$$\pounds_{eff} = -\sum_{V=\gamma,Z,W} T_{VLl}\overline{L}\sigma^{\mu\nu}P_L l\partial_\mu V_\nu - i\sum_{V=\gamma,Z} Q_{VLl}\overline{L}\sigma^{\mu\nu}P_L lW_\mu V_\nu \qquad (1)$$

where

$$L = \begin{pmatrix} \nu_\tau^* \\ \tau^* \end{pmatrix} \quad l = \begin{pmatrix} \nu_\tau \\ \tau \end{pmatrix}_L \qquad (2)$$

represent the excited states and ordinary leptons of the third generation, respectively. The coupling constants T_{VLl} and the quartic interaction couplings Q_{VLl} are given by

$$\begin{aligned}
T_{\gamma\tau^*\tau} &= -\frac{e}{2\Lambda}(f+f') \quad , \\
T_{\gamma\vartheta^*\vartheta} &= \frac{e}{2\Lambda}(f'-f) \quad , \\
T_{Z\tau^*\tau} &= -\frac{e}{2\Lambda}(f'\cot\theta_W - f\tan\theta_W) \quad , \\
T_{Z\vartheta^*\vartheta} &= -\frac{e}{2\Lambda}(f'\cot\theta_W - f\tan\theta_W) \quad , \\
T_{W\tau^*\vartheta} = T_{W\vartheta^*\tau} &= \frac{e}{\sqrt{2}\sin\theta_W\Lambda}f' \quad , \\
Q_{\gamma\tau^*\vartheta} = -Q_{\gamma\vartheta^*\tau} &= -\frac{e^2\sqrt{2}}{2\sin\theta_W\Lambda}f' \quad ,
\end{aligned}$$

$$Q_{Z\gamma\tau^*\vartheta} = -Q_{Z\vartheta^*\tau} = -\frac{e^2\sqrt{2}\cos\theta_W}{2\sin^2\theta_W\Lambda}f' \tag{3}$$

The couplings of gauge bosons with excited states of leptons in a vector like model are given by the following $SU(2)\times U(1)$ invariant, renormalizable four dimensional Lagrangian

$$\pounds_{ren} = -\sum_{V=\gamma,Z,W} A_{VLL}\overline{L}\gamma^{\mu}V_{\mu}L \tag{4}$$

where the coupling constants are

$$A_{\gamma\tau^*\tau^*} = -e \;;\; A_{\gamma\vartheta^*\vartheta^*} = 0 \;;\; A_{Z\tau^*\tau^*} = \frac{\left(2\sin^2\theta_W - 1\right)e}{2\sin\theta_W\cos\theta_W}$$
$$A_{Z\vartheta^*\vartheta^*} = \frac{e}{2\sin\theta_W\cos\theta_W} \;;\; A_{W\tau^*\vartheta^*} = \frac{e}{\sqrt{2}\sin\theta_W} \tag{5}$$

We do not consider oblique contributions because they are cancelled out in the g_τ/g_e quotient. Additionally, we assume that $m_Z, m_W, m_\tau << m_\tau^*$. We evaluate the diagrams by using dimensional regularization with dimension $D = 4-2\epsilon$ where the pole $D=4$ was identified with $\ln\Lambda^2$; and we imposed the on-shell renormalization scheme. The most important contribution comes from the diagrams containing two excited leptons into the loop. The contributions from self-energy diagrams have been neglected in the on-shell scheme.

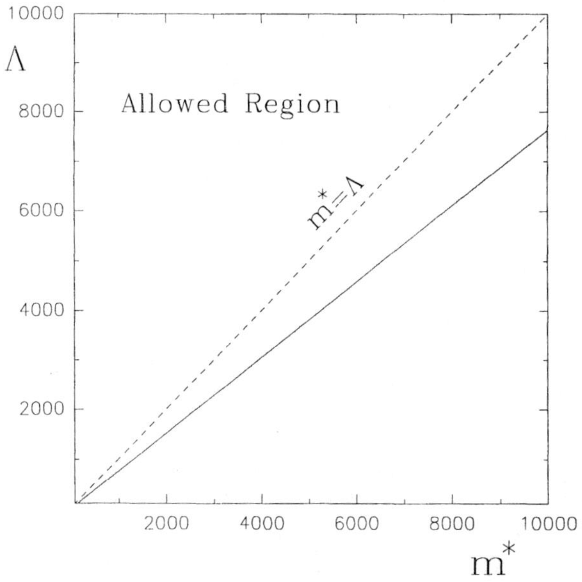

FIGURE 1. Allowed region from g_τ/g_e quotient.

Taking into account all diagrams, we get the following result for the amplitude:

$$\Gamma_\mu = -\frac{ig}{\sqrt{2}}\gamma_\mu P_L \left\{1 + \delta_{NP}\right\} \tag{6}$$

with

$$\delta_{NP} = \frac{g^2}{32\pi^2}\left(\frac{fm^*}{\Lambda}\right)^2 \left[\frac{118}{5\xi_W} - \frac{323}{75\xi_Z} - \frac{s_W}{3\xi_W} - \frac{17}{8c_W^2} - \frac{1}{8c_W^2\xi_Z} + \right.$$
$$\left. + \left(-\frac{8}{s_W} + \frac{629}{96\xi_Z} - \frac{3}{8c_W^2\xi_Z} + \frac{2s_W}{\xi_W} - \frac{3}{4c_W^2}\right)\ln\xi_\Lambda\right] \tag{7}$$

where $s_W = \sin\theta_W$, $c_W = \cos\theta_W$, $\xi_x = (m^*/m_x)^2$. Moreover, we are supposing that $f = f'$ and $m^*_\tau = m^*_{\tau_\nu}$. Then, the effective coupling for $W\tau\nu_\tau$ is given by:

$$g_\tau \equiv g\left\{1 + \delta_{NP}\right\}. \tag{8}$$

In the figure 1, we plot the allowed region on Λ versus m^* plane coming from the experimental value

$$\frac{g_\tau}{g_e} = 1.004 \pm 0.019 \pm 0.026 \tag{9}$$

and the expressions (7) and (8).

In conclusion, we evaluated the contributions to the non-universal coupling g_τ coming from compositeness effects by means of the $W \to \tau\nu_\tau$ decay in the framework of an effective Lagrangian approach. Assuming $g_e = g$ we get that $fm^*/\Lambda \simeq 1.03225 \pm 2.24974$ by taking only the main contribution and neglecting the term coresponding to $\ln\xi_\Lambda$.

We thank COLCIENCIAS (Colombia), CONACyT (Mexico) and CONICET (Argentina) for their financial support.

REFERENCES

1. H. Harari, *Phys. Lett.* **86B**,83 (1979); H. Harari, N. Seiberg, *Nucl. Phys.* **B204**, 141 (1982); H. Terazawa, Y. Chikashige, K. Akama, *Phys. Rev.* **D15**, 480 (1977).
2. L.F. Abbott, E. Farhi, *Phys. Lett.* **101B**,69 (1981); *Nucl. Phys.* **B189**,547 (1981); B. Schermpp, F. Schermpp, *Nucl. Phys.* **B231**,109 (1984); **B242**,203 (1984).
3. S. Fajfer, D. Tadic, *Phys. Rev.* **D38**, 962 (1988).
4. L3 Collaboration, M. Acciarri *et.al.*, *Phys. Lett.* **B401**,139 (1997); DELPHI Collaboration, P. Abreu *et.al.*, *ibid.* **B393**, 245 (1997). OPAL Collaboration, K. Acherstaff *et.al.*, *ibid.* **B391**, 197 (1997).
5. K. Hagiwara, S. Komamiya, and D. Zeppenfeld, *Z. Phys.* **C29**, 115 (1985)
6. Jorge Isidro Aranda, R. Martínez, O. A. Sampayo *Phys. Rev.* **D62**, 013010 (2000).

Pion dispersion relation at finite temperature in the linear sigma model

Sarira Sahu[1]

Instituto de Ciencias Nucleares, Universidad Nacional Autonoma de Mexico, Circuito Exterior C.U., A. Postal 70-543, 04510 Mexico D.F.

Abstract. We develop one-loop effective vertices and propagators in the linear sigma model at finite temperature satisfying the chiral Ward identities and use these in turn to compute the pion dispersion relation in a pion medium for small momentum and temperatures on the order of the pion mass at next to leading order in the parameter $m_\pi^2/4\pi^2 f_\pi^2$ and to zeroth order in the parameter $(m_\pi/m_\sigma)^2$. We show that this expansion reproduces the result obtained from chiral perturbation theory at leading order. The main effect is a perturbative, temperature-dependent increase of the pion mass.

The propagation properties of pions in a thermal hadronic environment are a key ingredient for the proper understanding of several physical processes taking place in dense and hot plasmas. The scenarios include dense astrophysical objects such as neutron stars –where the inclusion of mesonic degrees of freedom is essential to determine the equation of state of matter in the inner core– and the evolution of the highly interacting region formed in the aftermath of a high-energy heavy-ion collision.

In order to account for the hadronic degrees of freedom at temperatures and densities below the QCD phase transition to a QGP, one needs to resort to effective chiral theories whose basic ingredient is the fact that pions are Goldston bosons, originated in the spontaneous breakdown of chiral symmetry. Chiral perturbation theory (χPT) is one of such effective theories that has been successfully employed to show the well known result that at leading perturbative order and at low momentum, the modification of the pion dispersion curve in a pion medium is just a constant, temperature dependent, increase of the pion mass [1].

The simplest realization of chiral symmetry is nevertheless provided by the much studied linear sigma model which possesses the convenient feature of being a renormalizable field theory, both at zero [2] and at finite temperature [3]. The linear sigma model also reproduces –as we will show– the leading order modification to the pion mass in a thermal pion medium (for temperatures on the order of the pion

[1] This work is done in collaboration with A. Ayala, ICN-UNAM and Mauro Napsuciale, IF Universidad de Guanajuata, León

CP562, *Particles and Fields: Ninth Mexican School,* edited by G. Herrera Corral and L. Nellen

mass), when use is made of a systematic expansion in the ratio $(m_\pi/m_\sigma)^2$ at zeroth order, where m_σ, m_π are the vacuum sigma and pion masses, respectively. In this work, we use the linear sigma model to construct effective vertices and propagators to one loop. We use these in turn to compute the one loop modification to the pion propagator in a pion medium.

The Lagrangian for the linear sigma model, including only the mesonic degrees of freedom [2]

$$\mathcal{L} = \frac{1}{2}\left[(\partial\pi)^2 + (\partial\sigma)^2 - m_\pi^2\pi^2 - m_\sigma^2\sigma^2\right] - \lambda^2 f_\pi \sigma(\sigma^2+\pi^2) - \frac{\lambda^2}{4}(\sigma^2+\pi^2)^2\,, \quad (1)$$

where π and σ are the pion and sigma fields, respectively, and the coupling λ^2 is given by $\lambda^2 = \frac{m_\sigma^2 - m_\pi^2}{2f_\pi^2}$, .

From the Lagrangian in Eq. (1), one obtains the Green's functions and the Feynman rules to be used in perturbative calculations, in the usual manner. The bare pion and sigma propagators $\Delta_\pi(P)$, $\Delta_\sigma(Q)$ and the bare one-sigma two-pion and four-pion vertices Γ_{12}^{ij}, Γ_{04}^{ijkl} are given by (hereafter, capital letters are used to denote four momenta)

$$\begin{aligned} i\Delta_\pi(P)\delta^{ij} &= \frac{i}{P^2 - m_\pi^2}\delta^{ij}, \quad i\Delta_\sigma(Q) = \frac{i}{Q^2 - m_\sigma^2} \\ i\Gamma_{12}^{ij} &= -2i\lambda^2 f_\pi \delta^{ij}, \quad i\Gamma_{04}^{ijkl} = -2i\lambda^2(\delta^{ij}\delta^{kl} + \delta^{ik}\delta^{jl} + \delta^{il}\delta^{jk})\,. \end{aligned} \quad (2)$$

These Green's functions are sufficient to construct the modification to the pion propagator, both at zero and finite temperature, at any given perturbative order. An alternative approach consists on exploiting the relations that chiral symmetry imposes among different n-point Green's functions, known as chiral Ward identities (χWIs), are a direct consequence of the fact that the divergence of the axial current may be used as an interpolating field for the pion. For example, two of the χWIs satisfied –order by order in perturbation theory– by the functions $\Delta_\pi(P)$, $\Delta_\sigma(Q)$, Γ_{12}^{ij} and Γ_{04}^{ijkl} are [2]

$$\begin{aligned} f_\pi\Gamma_{04}^{ijkl}(;0,P_1,P_2,P_3) &= \Gamma_{12}^{kl}(P_1;P_2,P_3)\delta^{ij} + \Gamma_{12}^{lj}(P_2;P_3,P_1)\delta^{ik} + \Gamma_{12}^{jk}(P_3;P_1,P_2)\delta^{il} \\ f_\pi\Gamma_{12}^{ij}(Q;0,P) &= \left[\Delta_\sigma^{-1}(Q) - \Delta_\pi^{-1}(P)\right]\delta^{ij}\,, \end{aligned} \quad (3)$$

where momentum conservation at the vertices is implied, that is $P_1 + P_2 + P_3 = 0$ and $Q + P = 0$. Therefore, any perturbative modification of one of these functions introduces modifications in other, when the former are related to the latter through χWIs.

At one loop and after renormalization, the sigma propagator is modified by finite terms. At finite temperature the modification results on real and imaginary parts. The real part modifies the sigma dispersion curve whereas the imaginary part represents a temperature dependent contribution to the sigma width.

In the imaginary-time formalism of Thermal Field Theory (TFT), the expression for sigma self-energy is given as $6\lambda^4 f_\pi^2 I(\omega, q)$, where the function I is defined by

$$I(\omega, q) = T \sum_n \int \frac{d^3k}{(2\pi)^3} \frac{1}{\omega_n^2 + k^2 + m_\pi^2} \frac{1}{(\omega_n - \omega)^2 + (\mathbf{k} - \mathbf{q})^2 + m_\pi^2} . \tag{4}$$

Here $\omega = 2m\pi T$ and $\omega_n = 2n\pi T$ (m, n integers) are discrete boson frequencies and $q = |\mathbf{q}|$. From Eq. (4) we obtain the time-ordered version I^t of the function I, after analytical continuation to Minkowski space. The imaginary part of I^t is given by

$$\begin{aligned} \mathrm{Im} I^t(q_0, q) &= \frac{\varepsilon(q_o)}{2i} \left[I(i\omega \to q_o + i\epsilon, q) - I(i\omega \to q_o - i\epsilon, q) \right] \\ &= -\frac{1}{16\pi} \left\{ a(Q^2) + \frac{2T}{q} \ln \left(\frac{1 - e^{-\omega_+(q_0,q)/T}}{1 - e^{-\omega_-(q_0,q)/T}} \right) \right\} \Theta(Q^2 - 4m_\pi^2) , \end{aligned} \tag{5}$$

where $Q^2 = q_0^2 - q^2$, ε and Θ are the sign and step functions, respectively, and the functions a and $\omega_\pm$ are given by $a(Q^2) = \sqrt{1 - \frac{4m_\pi^2}{Q^2}}$, $\omega_\pm(q_0, q) = \frac{|q_0| \pm a(Q^2) q}{2}$, whereas the real part of I^t at $Q = 0$ is given by

$$\mathrm{Re} I^t(0) = -\frac{1}{8\pi^2} \int_0^\infty \frac{dk}{E_k} \left[1 + 2f(E_k) \right] , \tag{6}$$

where $E_k = \sqrt{k^2 + m_\pi^2}$ and the function f is the Bose-Einstein distribution. Therefore, the one-loop effective sigma propagator becomes

$$i\Delta_\sigma^\star(Q) = \frac{i}{Q^2 - m_\sigma^2 + 6\lambda^4 f_\pi^2 I^t(Q)} , \tag{7}$$

In order to preserve the χWIs expressed in Eq. (3), the corresponding one-loop effective one-sigma two-pion and four-pion vertices become

$$\begin{aligned} i\Gamma_{12}^{\star\, ij}(Q; P_1, P_2) &= -2i\lambda^2 f_\pi \delta^{ij} \left[1 - 3\lambda^2 I^t(Q) \right] \\ i\Gamma_{04}^{\star\, ijkl}(; P_1, P_2, P_3, P_4) &= -2i\lambda^2 \left\{ \left[1 - 3\lambda^2 I^t(P_1 + P_2) \right] \delta^{ij}\delta^{kl} \right. \\ &+ \left. \left[1 - 3\lambda^2 I^t(P_1 + P_3) \right] \delta^{ik}\delta^{jl} + \left[1 - 3\lambda^2 I^t(P_1 + P_4) \right] \delta^{il}\delta^{jk} \right\} . \end{aligned} \tag{8}$$

The functions in Eq. (8) arise from considering all of the possible one-loop contributions to the one-sigma two-pion and four-pion vertices, when maintaining only the zeroth order terms in a systematic expansion in the parameter $(m_\pi/m_\sigma)^2$. We now use the above effective vertices and propagator to construct the one-loop modification to the pion self-energy. The leading contributions come from keeping only the leading order terms when considering m_π, T and P as small compared to $m_\sigma$xi. Then pion self-energy can be written as

$$\Pi(P) = \left(\frac{m_\pi^2}{2f_\pi^2}\right) T\sum_n \int \frac{d^3k}{(2\pi)^3}\frac{1}{K^2+m_\pi^2}\left\{5+2\left(\frac{P^2+K^2}{m_\pi^2}\right)\right.$$
$$\left. - \left(\frac{m_\pi^2}{2f_\pi^2}\right)\left[9I^t(0)+6I^t(P+K)\right]\right\}, \tag{9}$$

where we carry out the calculation in the imaginary-time formalism of TFT with $K=(\omega_n,\mathbf{k})$ and $P=(\omega,\mathbf{p})$. The pion dispersion relation is thus obtained from the solution to

$$P^2 + m_\pi^2 + \mathrm{Re}\Pi(P) = 0\,, \tag{10}$$

after the analytical continuation $i\omega \to p_o + i\epsilon$.

As it stands, Eq. (9) contains a temperature-dependent infinity coming from the product of the vacuum piece in $\mathrm{Re}I^t(0)$ and the temperature-dependent piece in the indicated integral, as well as vacuum infinities. However, it is well known that finite temperature does not introduce new divergences and that whatever regularization and renormalization is needed at zero temperature, it will also be necessary and sufficient at finite temperature. Let us recall that at one loop, renormalization is carried out by introducing appropriate counterterms into the original Lagrangian and that at this level, no temperature-dependent infinities arise.

Let us first look at the dispersion relation at leading order. After analytical continuation all the terms are real. The integrals involved are

$$T\sum_n\int\frac{d^3k}{(2\pi)^3}\frac{1}{K^2+m_\pi^2} \to \frac{1}{2\pi^2}\int_0^\infty \frac{dk\;k^2}{E_k}f(E_k) \equiv \frac{m_\pi^2}{2\pi^2}\,g(T/m_\pi)$$
$$T\sum_n\int\frac{d^3k}{(2\pi)^3}\frac{K^2}{K^2+m_\pi^2} \to -\,m_\pi^2\left(\frac{m_\pi^2}{2\pi^2}\right)g(T/m_\pi)\,, \tag{11}$$

where g is a dimensionless function of the ratio T/m_π and the arrows indicate only the temperature dependence of the expressions. Thus, the dispersion relation results from

$$[1+2\,\xi\;g(T/m_\pi)]\,(p_0^2-p^2) - [1+3\,\xi\;g(T/m_\pi)]\,m_\pi^2 = 0\,, \tag{12}$$

with $\xi = m_\pi^2/4\pi^2 f_\pi^2 \ll 1$. For $T\sim m_\pi$, $g(T/m_\pi)\sim 1$, therefore, at leading order and in the kinematical regime that we are considering, Eq. (12) can be written as

$$p_0^2 = p^2 + m_\pi^2\,[1+\xi\;g(T/m_\pi)]\,, \tag{13}$$

which coincides with the result obtained from χPT [1].

We now look at the next to leading order terms in Eq. (9). The first of these is purely real and represents a constant, second order shift to the pion mass squared

$$-\,9\left(\frac{m_\pi^2}{2f_\pi^2}\right)^2 T\sum_n\int\frac{d^3k}{(2\pi)^3}\frac{I^t(0)}{K^2+m_\pi^2} \to \frac{9}{2}\,\xi^2\;g(T/m_\pi)\;h(T/m_\pi)\;m_\pi^2\,, \tag{14}$$

where h is a dimensionless function of the ratio T/m_π defined by

$$h(T/m_\pi) \equiv \int_0^\infty \frac{dk}{E_k} f(E_k)\,. \tag{15}$$

The remaining term in Eq. (9) shows a non-trivial dependence on P. It involves the function S defined by

$$S(P) \equiv T \sum_n \int \frac{d^3k}{(2\pi)^3} \frac{I^t(P+K)}{K^2+m_\pi^2}\,. \tag{16}$$

The sum is performed by resorting to the spectral representation of I^t and $(K^2 + m_\pi^2)^{-1}$. Thus, the real part of the retarded version of S, after analytical continuation is

$$\begin{aligned}
\mathrm{Re}S^r(p_0,p) &\equiv \frac{1}{2}\left[S(i\omega \to p_0 + i\epsilon, p) + S(i\omega \to p_0 - i\epsilon, p)\right] \\
&= -\,\mathcal{P} \int \frac{d^3k}{(2\pi)^3} \int_{-\infty}^{\infty} \frac{dk_0}{2\pi} \int_{-\infty}^{\infty} \frac{dk_0'}{2\pi} \left[1 + f(k_0) + f(k_0')\right] \\
&\quad \frac{2\pi\ \varepsilon(k_0')\ \delta[{k_0'}^2 - (\mathbf{k}-\mathbf{p})^2 - m_\pi^2]\ 2\ \mathrm{Im}I^t(k_0,k)}{p_0 - k_0 - k_0'}\,,
\end{aligned} \tag{17}$$

where $\mathcal{P}$ represents the principal part of the integral. The integration in Eq. (17) can only be performed numerically. It shows that this term is non-vanishing at $p = 0$ and thus it also contributes to the perturbative increase of the pion mass but the increase is very small.

Including all the terms, the dispersion relation up to next to leading order, for $T \sim m_\pi$ and in the small momentum region is obtained as the solution to

$$p_0^2 = p^2 + \left\{1 + \xi g(T/m_\pi) + \frac{\xi^2}{2} g(T/m_\pi)\left[9h(T/m_\pi) - 4g(T/m_\pi)\right]\right\} m_\pi^2 + \xi^2 \tilde{S}(p_0,p)\,, \tag{18}$$

where $\tilde{S}(p_0,p) = -(24\pi^4)\mathrm{Re}S^r(p_0,p)$. Figure 1 shows the dispersion relation obtained from Eq. (18) for $T = m_\pi$ (solid curve) where we also display the solution without the term $\xi^2\tilde{S}(p_0,p)$ (dotted curve). As mentioned before, inclusion of this last term does not alter the shape of the dispersion relation in this kinematical regime.

In conclusion, we have shown that in the linear sigma model at finite temperature, the one-loop modification of the sigma propagator induces a modification in the one-sigma two-pion and four-pion vertices in such a way as to preserve the χWIs. Strictly speaking, chiral Ward identities fully constrain vertices only in the limit where all the external momenta are zero [2]. We have used these objects to compute

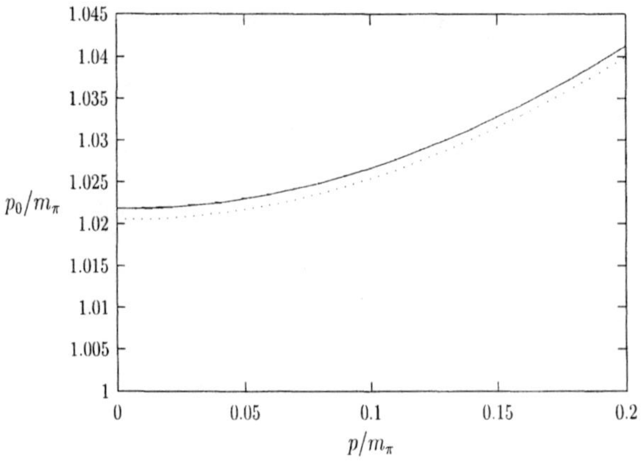

FIGURE 1. Pion dispersion relation obtained as the solution to Eq. (18) for $T = m_\pi$ (upper curve). Shown is also the dispersion relation obtained by ignoring the term $\tilde{S}(p_0, p)$ (lower curve).

the next to leading order correction to the pion propagator in a pion medium for small momentum and for $T \sim m_\pi$. We have shown that the linear sigma model yields the same result as χPT at leading order in the parameter $\xi = m_\pi^2/4\pi^2 f_\pi^2$ when use is made of a systematic expansion in the parameter $(m_\pi/m_\sigma)^2$ at zeroth order. The main modification to the pion dispersion curve in the considered kinematical regime is a perturbative increase of the in-medium pion mass. The shape of the curve is not significantly altered [4]. For a sigma meson with a mass on the order of 600 MeV and in the mentioned kinematical regime, corrections in powers of the parameter $(m_\pi/m_\sigma)^2$ are small. If a lighter sigma meson exists, corrections of order $(m_\pi/m_\sigma)^2$ could be important.

REFERENCES

1. J. Gasser and H. Leutwyler, *Phys. Lett.* **B** 184, 83 (1987).
2. B. W. Lee, *Chiral Dynamics* (Gordon and Breach, 1972).
3. L. R. R. Mohan, *Phys. Rev.* **D** 14, 2670 (1976).
4. A. Ayala, Sarira Sahu and M. Napsuciale, *Phys. Lett. B* **479**, 156 (2000).

CP violation with hypermagnetic fields and electroweak baryogenesis

Gabriel Pallares and Alejandro Ayala

Instituto de Ciencias Nucleares
Universidad Nacional Autónoma de México
Apartado Postal 70-543, México Distrito Federal 04510, México.

Abstract. We show that in the presence of primordial magnetic fields, it is possible to produce a large enough amount of CP violation to possibly explain the baryon asymmetry of the universe. This can happen during the reflection of fermions off the true vacuum bubbles nucleated during a first order phase transition. The chiral nature of the fermion coupling to the background field in the symmetric phase can be used to construct two-fermion interference processes in analogy to the Bohm-Aharanov effect. By describing the transport of the CP asymmetry into the symmetric phase, it is possible to generated a baryon to entropy ratio $\rho_B/s = (3-6) \times 10^{-11}$, which is close to the observed value based on nucleosynthesis considerations.

There has been much work which suggests that the baryon asymmetry of the universe might have been produced at the electroweak phase transition (EWPT) [1]. Any theory that aims to explain such asymmetry has to meet the three well know Sakharov conditions [2], namely: (1) Violation of baryon number; (2) C y CP violation; (3) departure from thermal equilibrium. The above conditions are met in the standard model (SM) of electroweak interactions. However the minimal SM as such, cannot explain the observed baryon number. The reason is that the EWPT turns out to be only too weakly first order. Moreover, the maximal amount of CP violation coming from the CKM matrix is insufficient to explain the baryon to entropy ratio.

Nevertheless, it has been recently pointed out that, provided enough CP violation exists, the above scenario could significantly change in the presence of large scale primordial magnetic fields [3–5], which can be responsible for a stronger first order EWPT. There are no compelling reasons why magnetic fields should not have been present in the early universe. Moreover it can be argued that the existence of these fields at high temperatures is a quite natural phenomenon. These fields could have had some influence on physical processes which occurred in the early universe, like big-bang nucleosynthesis and electroweak baryogenesis.

Recall that for temperatures above the EWPT, the $SU(2) \times U(1)_Y$ symmetry is

CP562, *Particles and Fields: Ninth Mexican School,* edited by G. Herrera Corral and L. Nellen

restored and the propagating, non-screened vector modes that represent a magnetic field correspond to the $U(1)_Y$ hypercharge group. Thus, in the unbroken phase, any primordial magnetic fields belong to the hypercharge group instead of to the $U(1)_{em}$ group and are properly called *hypermagnetic* fields.

In this work we show that the existence of such primordial magnetic fields also provides a mechanism to produce a large enough amount of CP violation during the EWPT to possibly explain the observed baryon to entropy ratio in the SM. This can happen during the reflection of fermions off the true vacuum bubbles nucleated during a first order phase transition through an interference process equivalent to the Bohm-Aharnov effect, given that in the symmetric phase, fermions couple chirally to hypermagnetic fields with the hypercharge.

In a first order phase transition, the conversion from one phase to another occurs through nucleation of the true phase in the false phase. This happens when the system is either supercooled or superheated. The bubbles of the true phase expand rapidly, absorbing the region of the false phase. For the EWPT, this true phase eventually fills the entire volume with no intermediate mixed phase. It is at the bubble wall that matter is strongly out of equilibrium and where the baryon asymmetry is generated.

To understand the generation of the baryon asymmetry, the effect of the bubble wall on the propagation of fermions should be understood. If fermions pass through the bubble, they acquire mass, given their coupling with the Higgs field, proportional to its finite temperature vacuum expectation value (VEV). This VEV is determined from the equations of motion of the finite temperature effective potential. In the approximation where the bubble wall is treated as a planar interface and for the case where the energy densities of the two phases is the same, the solution for the Higgs field looks like

$$\varphi(x) = 1 + \tanh(x), \tag{1}$$

where the dimensionless position coordinate x is $x = \delta T/\sqrt{2\lambda}\, r$. The parameter δ represents the width of the domain wall [6]. It is also the parameter responsible for the first order nature of the phase transition. It can also be checked that δ becomes smaller in the presence of hypermagnetic fields. In terms of Eq. (1), we can see that at $x = -\infty$, the system is in the symmetric phase, that is outside the bubble; at $x = +\infty$, the system is in the broken phase, that is, inside the bubble. In practice, the wall velocity is determined by a complicated analysis which involves computing the effects of transport process and is not well determined.

Given a background scalar field, the problem of fermion reflection and transmission through the domain wall can be cast in terms of solving the Dirac equation. In the presence of an external magnetic field, the equation becomes

$$\left(\not{p} - a\not{A} - \frac{\delta T}{\sqrt{2\lambda}} \xi \varphi(x) \right) \Psi(x) = 0, \tag{2}$$

where $\xi = 2m/m_H$, m and m_H being the fermion and Higgs masses at zero temperature, respectively and Ψ the fermion wave function. $A^\mu = (0, \mathbf{A})$ is the four-vector potential related to the external magnetic field by $\mathbf{B} = \nabla \times \mathbf{A}$. In the symmetric phase, $\mathbf{A} = \mathbf{A}_Y$ corresponds to the hypermagnetic field vector potential whereas in the broken phase, where only the Maxwell projection is unscreened [3,4], $\mathbf{A} = \mathbf{A}_{em}$ corresponds to the ordinary photon vector potential. $a = g'Y/2$ is the fermion coupling to the external field in the symmetric phase, with Y the fermion's hypercharge and g' the $U(1)_Y$ coupling constant; in the broken phase $a = e$, the electric charge. In the absence of a magnetic field, the solution to Eq. (2) has been found in Ref. [7]. Let such a solution, with a definite helicity ($h = L, R$), be $\tilde{\Psi}_h(x)$. It can easily be shown that in the presence of a magnetic field, the solution to Eq. (2) with definite helicity is given by

$$\Psi_h(x) = \tilde{\Psi}_h(x) e^{-i\frac{g'}{2}Y_h \int^{\mathbf{X}} \mathbf{A}_Y \cdot d\mathbf{x}'}, \tag{3}$$

where Y_h is the hypercharge of the corresponding helicity mode. We will be interested in fermion solutions in the symmetric phase, where particles are massless and thus helicity is a good quantum number.

Let us consider a situation in which the fermion amplitude at a space point a distance l away from the the bubble wall receives contributions from fermions coming from the symmetric phase which are reflected from two points on the bubble wall separated by a distance d. For simplicity, we take $d \ll l$. In this case, we can consider that the fermion propagation is approximately directed along the direction perpendicular to the bubble wall.

The coupling of the given helicity mode with the hypermagnetic field is chiral, then the square of the amplitude representing the interference of left-handed fermions $|M_L|^2$ will differ from the square of the amplitude for right-handed interfering antifermions $|\bar{M}_R|^2$. Consequently, an axial asymmetry $\mathcal{A}$ is built in front of the bubble wall. Within the above approximation, it is easy to show that the explicit expression for the axial, CP violating asymmetry is given by

$$\mathcal{A} \equiv |M_L|^2 - |\bar{M}_R|^2 = \mathcal{R} \sin(\Upsilon_+/2) \sin(\Upsilon_-/2), \tag{4}$$

where

$$\Upsilon\pm = \frac{g'}{2}(Y_R \pm Y_L)\Theta. \tag{5}$$

$\mathcal{R}$ is the fermion reflection coefficient [7], common to left and right-handed modes and depends on the parameter ξ and the fermion energy, being equal to 1 for energies below twice the zero temperature mass of fermion. Θ is the hypermagnetic flux through the area defined by the two trajectories of the interfering fermions. If the constant hypermagnetic field is perpendicular to this area then $\Theta = B_Y l d/2$, with $B_Y = |\mathbf{B_Y}|$. From Eq. (4) the asymmetry disappears for a vanishing hypermagnetic field.

We now construct the net axial charge flux $\mathcal{F}$ in front of the expanding bubble wall in the symmetric phase. This flux receives contributions both from the reflected fermions within the symmetric phase and from those fermions transmitted from the bubble's interior. From CPT theorem the complementary processes above are related. The explicit expression for $\mathcal{F}$ is given by [8]

$$\mathcal{F} = \frac{1}{2\pi^2\gamma}\int_0^\infty dp_l \int_0^\infty dp_t p_t [f^s(-p_l, p_t) - f^b(p_l, p_t)]\mathcal{A}, \tag{6}$$

where the wall velocity in the fluid frame is u and $\gamma = 1/\sqrt{1-u^2}$ is the Lorentz factor. p_l, p_t are the longitudinal and transverse components of the fermion's momentum and f^s, f^b are the fermion equilibrium fluxes in the symmetric and broken phases, respectively

$$f^{s,b} = \frac{p_l/E^{s,b}}{\exp[\gamma(E^{s,b} \mp up_l)/T] + 1}, \quad E^s = \sqrt{p_l^2 + p_t^2}, \quad E^b = \sqrt{p_l^2 + p_t^2 + m^2}. \tag{7}$$

Baryon violation in this model occurs through a quantum anomaly, the rate of baryon number density production is given in terms of the rate of baryon number violation per unit volume Γ_B and the partial derivate of the free energy with respect to baryon number $\partial F/\partial B$,

$$\dot{\bar{\rho}}_B = -\frac{\Gamma_B}{T}\frac{\partial F}{\partial B}. \tag{8}$$

The quantity $\partial F/\partial B$ is also the baryon number chemical potential μ_B and represents the force pushing the universe towards its equilibrium baryon number value.

From Ref. [9], one finds

$$\mu_B = \frac{\partial F}{\partial B} = -\frac{4\rho_Y}{(1+2n)T^2}, \tag{9}$$

where ρ_Y is the hypercharge density, n is the number of scalar doublets in the theory (for the SM, n=1). We now integrate Eq. (8) to get the net baryon number density in terms of the axial flux

$$\bar{\rho}_B = -\frac{\Gamma_B}{3T^3}\frac{\tau}{u}\mathcal{F}. \tag{10}$$

here τ is the time called the *transport time*: only those processes which happen fast enough with respect to the time that the reflected fermions in the symmetric phase (or those that passed from the broken to the symmetric phase) spend before being retaken by the expanding wall, will contribute to drive baryon number towards equilibrium. τ is of order $\sim 100/T$.

Finally the probability η of such process, to find a second fermion in the trajectory of the first one, assuming that there is not initial correlation between these two particles, is given by [10]

$$\eta = \frac{2}{3\pi\zeta(3)}\frac{\lambda^2\, l\, d/2}{[l^2 + (d/2)^2]^2}\frac{1}{1+e}. \tag{11}$$

From Eqs. (10) and (11), we can find, with $\Gamma_B = 3\kappa\alpha_w^4 T^4$ and with the expression for the entropy density at the EWPT epoch [11], the net baryon to entropy ratio produced [10]

$$\rho_B/s = (3-6) \times 10^{-11} \tag{12}$$

for $u = 0.1c$, $l = 2d \sim 9/T$ and $B_Y = (0.3 - 0.5)T^2$, which means that the estimates based on the present analysis are within the experimental value based on nucleosynthesis, at least for slowly expanding walls.

CONCLUSION

In conclusion we have shown that in the presence of strong, large scale primordial hypermagnetic fields, it is possible to generate a large amount of CP violation that combined with a stronger first order EWPT, also produced by the hypermagnetic field, could account for the observed baryon number to entropy ratio, even within the SM.

ACKNOWLEDGMENTS

Support for this work has been received in part by CONACyT-México under grant number 32279-E and by PROBETEL-UNAM.

REFERENCES

1. M. Trodden, Rev. Mod. Phys. 71, 1463 (1999).
2. A. D. Sakharov. JETP Lett. 6, 24, (1967).
3. M. Giovannini and M. E. Shaposhnikov, Phys. Rev. D 57, 2186 (1998).
4. P. Elmfors, K. Enqvist and K. Kainulainen, Phys. Lett. B 440, 269 (1998).
5. For recent reviews on the origin, evolution and some cosmological consequences of primordial magnetic fields see: K, Enqvist, Int. J. Mod. Phys. D 7, 331 (1998); R. Maartens, "Cosmological magnetic fields", International Conference on Gravitation and Cosmology, India, Jan. 2000, **astro-ph/0007352** and references therein.
6. B. H. Liu, L. McLerran and N. Turok, Phys. Rev. D 46, 2668 (1992).
7. A. Ayala, J. Jalilian-Marian, L. McLerran and A. P. Vischer, Phys. Rev. D 49, 5559 (1994).
8. E. Torrente-Lujan, Phys. Rev. D 60, 085003 (1999).
9. A. G. Cohen, D. B. Kaplan and A. E. Nelson, Phys. Lett. B 263, 86 (1991).
10. A. Ayala and G. Pallares **hep-ph/0008142**
11. E. Kolb and M. Turner, "The Early Universe" (Adison-Wesley, New York 1990).

Charged Current Cross Section ν-N to Low Energy and Their Match with Observations

Klara Goiz Hernández and J.J. Godina Nava

Departamento de Física, CINVESTAV-IPN, Ap. Postal 14-740, 07000 México, D.F., México

Abstract. The purpose of this talk is reanalyze the charged current (CC) cross section neutrino-nucleon with particular attention to the energy range $E_\nu \leq 10GeV$ introducing new experimental data concerning to form factors involved in the calculations, and discuss possible consequences for the interpretation of the measurements of the atmospheric fluxes obtained by deep underground detectors.

INTRODUCTION

It has been argued in recent papers [1] that it is possible to improve the description of the cross sections used in several recent analysis of neutrino-induced upward going muon [2], including a more careful treatment of the lowest multiplicity channels (quasielastic scattering (QE) and single pion production). This careful description have as a main effect the increase of the flux of low energy upward going muons. We are going to take this recipe from those papers and reanalyze the calculation of charged current (CC) cross section $\nu - N$ incorporating new experimental data concerning the form factors for the nucleon, and compare with those results. The flux of upward going muons of energy E_μ, E_μ^{min} and direction Ω can be calculated through

$$\Phi_{\mu^\mp}(E_\mu^{min}, \Omega) = \int_{E_\mu^{min}}^{\infty} dE_\nu \phi_{\nu_\mu}(\bar{\nu}_\mu)(E_\nu, \Omega) n_{\nu_\mu}(\bar{\nu}_\mu) \rightarrow \mu^\mp(E_\mu^{min}, E_\nu) \tag{1}$$

where $\phi_{\nu_\mu}(\bar{\nu}_\mu)(E_\nu, \Omega)$ is the differential flux of $\nu_\mu(\bar{\nu}_\mu)$ and $n_{\nu_\mu}(\bar{\nu}_\mu) \rightarrow \mu^\mp(E_\mu^{min}, E_\nu)$ is the average number of muons above a threshold energy E_μ^{min} produced by a neutrino of energy E_ν, which is evaluated with the expression

$$n_{\nu_\mu}(\bar{\nu}_\mu) \rightarrow \mu^\mp(E_\mu^{min}; E_\nu) = N_A \int_{E_\mu^{min}}^{E_\nu} \frac{d\sigma_{\nu_\mu}(\bar{\nu}_\mu)}{dE_0}(E_0, E_\nu)\Big[R(E_0) - R(E_\mu^{min})\Big] \tag{2}$$

here N_A represent the Avogadro's number and the cross section refers to the CC cross section $\nu - N$ whose evaluation we are going to dedicate this talk. The energy E_0 of the muon at the production point is weighted by a factor $d\sigma/dE_0$, the

CP562, *Particles and Fields: Ninth Mexican School,* edited by G. Herrera Corral and L. Nellen

relevant cross section, and a factor $R(E_0) - R(E_\mu^{min})$, [R(E) is some like the penetration intensity in rock measured in gr/cm^2 of a muon of energy E] that takes into account the larger effective target available with increasing muon energy. The calculated muon flux depends on the inclusive cross section for muon production. In the literature this cross section has been evaluated using the deep inelastic scattering formalism (DIS). However is observed that the DIS formulas are expected to be valid only for Q^2 sufficiently large, and that using them for calculating the cross section of low energy neutrino implies made an extrapolation into the region where nonperturbative effects are important neglecting further evolution [2]. More or less things happen like this: Taking a typical spectrum with mean energy $\langle E_\nu \rangle \simeq 13\ GeV$, 5% of charged current neutrino interactions are quasi-elastic, 13% are resonance production, and 82% are DIS scattering. This "medium" energy range is somewhat problematic because it lies at the transition region between our two intuitive models for neutrino interactions. At low energies, which are well understood phenomenologically, neutrino scattering are predominantly QE, in which the target is taken to be an entire nucleon. At high energies, neutrino interactions are mainly DIS, where the target is one of the constituent partons inside the nucleon. Again in this region the theoretical and experimental situations are well in hand. In the medium energy ($\approx 10 GeV$) range, the concept of a well defined "target" is more tenuous, as both QS and DIS interations can occur. This uncertainty can be viewed as a consequence of the fact that we are "pushing" the limits of perturbative QCD, and the assumptions underlying the derivation of DIS formulae are less valid. These theoretical difficulties are particularly evident in a lack of a uniform treatment of neutrino resonance production. Hadronic final states consisting of a nucleon and a single pion are primarily the result of the decay of a low mass resonance (such as the $\Delta(1232)$). The related problem is determining the total CC cross section in this medium energy range in terms of these 3 process. Fortunatelly, data exist in this energy range and can be used to check both the overall normalization, σ_T, and the relative contribution of low multiplicity channels. So, like is suggested in [1], we are going to consider separately the contributions of the exclusive channels of lowest multiplicity, i.e., QS scattering and single pion production, and describe the additional channels collectively using the DIS formulas. Thus we decompose the CC $\nu - N$ cross section as the sum of three contributions

$$\sigma_{\nu(\bar{\nu})}^{CC} = \sigma_{QS} + \sigma_{1\pi} + \sigma_{DIS} \qquad (3)$$

Next, we are going to speak about how is evaluated each one of this contributions.

Quasi-Elastic Scattering

At low neutrino energies, CC neutrino-hadron interactions are predominantly QS and single pion production, in which the neurino scatters off an entire nucleon rather than the constituent partons. These processes have been studied with in low energy (100 MeV-10GeV) bubble chambers experiments. The general hadronic current can be decomposed in terms of its Lorentz structure [2]

$$\langle P(p') \mid J_\mu^- \mid N(p)\rangle = \overline{P}(p')[F_V^1\gamma_\mu + \frac{i\sigma_{\mu\nu}q^\nu\xi F_V^2}{2M} + \frac{q_\mu F_S}{M} + F_A\gamma_5\gamma_\mu + \frac{F_P\gamma_5 q_\mu}{M} + \frac{\gamma_5(p+p')_\mu F_T}{M}]N(p) \tag{4}$$

where F_i(i=S,P,V,A,T) describes the scalar, pseudoscalar, vector, axial-vector, and tensor form factors of the nucleon, respectively; $q = (p-p')$ is the momentum transfer, where p is the 4-momentum of the target nucleon and p' is the 4-momentum outgoing nucleon.

The differential cross section for the QS process can be written in the laboratory frame of reference, where the neutrino has energy E_ν, the taget nucleon of mass M is at rest, and the Mandelstam invariants are $s, t = q^2$, and u [3]

$$\frac{d\sigma}{d\mid q^2\mid}\begin{pmatrix}\nu n \to \ell^- p\\ \bar{\nu}p \to \ell^+ n\end{pmatrix} = \frac{M^2G^2cos^2\theta_c}{8\pi E_\nu^2}[A(q^2) \mp B(q^2)\frac{s-u}{M^2} + C(q^2)\frac{(s-u)^2}{M^4}] \tag{5}$$

where $s - u = 4ME_\nu + q^2 - M_\ell^2$. Here, G is the Fermi coupling constant($G = 1.1663910^{-5}GeV^{-2}$) and Θ_c ($cos^2\theta_c \approx 0.941$) is the Cabibbo mixing angle. The functions A,B, and C are convenient combinations of the nucleon form factors. Contractions of the hadronic and leptonic currents yields:

$$\begin{aligned} A = \frac{(M_\ell^2 - q^2)}{4M^2}[&(4 - \frac{q^2}{M^2}) \mid F_A \mid^2 -(4 + \frac{q^2}{M^2}) \mid F_V^1 \mid^2 \\ &- \frac{q^2}{M^2} \mid \xi F_V^2 \mid^2 (1 + \frac{q^2}{4M^2}) - \frac{4q^2 \Re e F_V^{1*}\xi F_V^2}{M^2} + \frac{q^2}{M^2}(4 - \frac{q^2}{M^2}) \mid F_T \mid^2 \\ &- \frac{M_\ell^2}{M^2}(\mid F_V^1 + \xi F_V^2 \mid^2 + \mid F_A + 2F_P \mid^2 +(\frac{q^2}{M^2} - 4)(\mid F_S \mid^2 + \mid F_P \mid^2))] \end{aligned} \tag{6}$$

$$B = -\frac{q^2}{M^2}\Re e F_A^*(F_V^1 + \xi F_V^2) - \frac{M_\ell^2}{M^2}\Re e[(F_V^1 + \frac{q^2}{4M^2}\xi F_V^2)^* F_S - (F_A + \frac{q^2 F_P}{2M^2})^* F_T] \tag{7}$$

$$C = \frac{1}{4}(\mid F_A \mid^2 + \mid F_V^1 \mid^2 - \frac{q^2}{M^2} \mid \frac{\xi F_V^2}{2} \mid^2 - \frac{q^2}{M^2} \mid F_T \mid^2) \tag{8}$$

where M_ℓ is the final state lepton mass. Now ignoring second-class currents allows us to set the scalar and tensor form factors to zero. According to the CVC hypothesis, the vector part of the weak current and the isovector part of the electromagnetic current form a isotriplet of conserved current [4]. This hypothesis allows us to relate F_V^1 and F_V^2 to the electromagnetic form factors, which are better measured.

$$F_V^1(q^2) = (1 - \frac{q^2}{4M^2})^{-1}\left[G_E^V(q^2) - \frac{q^2}{4M^2}G_M^V(q^2)\right] \tag{9}$$

$$\xi F_V^2(q^2) = (1 - \frac{q^2}{4M^2})^{-1} \left[G_M^V(q^2) - G_E^V(q^2)\right] \tag{10}$$

The electromagnetic form factors are determined from electron scattering experiments:

$$G_E^V(q^2) = \frac{1}{(1 - \frac{q^2}{M_V^2})^2}, \quad G_M^V(q^2) = \frac{1 + \mu_P - \mu_n}{(1 - \frac{q^2}{M_V^2})^2} \tag{11}$$

where $\xi = \mu_P - \mu_N$, with μ_P, μ_N the anomalous magnetic moment for the proton and neutron, and $M_V^2 = 0.71\ GeV^2$. The situation is slightly more complicated for the hadronic axial current. $F_A(q^2 = 0) = -1.261 \pm 0.004$ which is well known from neutron beta decay. The q^2 dependence has to be inferred or measured. So, by analogy with the vector case is assumed the same dipole term:

$$F_A(q^2) = \frac{-1.23}{(1 - \frac{q^2}{M_A^2})^2} \tag{12}$$

The only remaining parameters needed to describe the QS cross section are thus M_V and M_A. M_V is determined with high accuracy through electron scattering experiment [5]. M_A needs to be extracted from neutrino scattering data. Analyzing the total cross sections and kinematical distributions for QS $\nu - N$ scattering, a global average value of the axial-vector mass is found to be $M_A = 1.032 \pm 0.036 GeV$ [5]. With the purpose of compare with result of ref[1] we are going to use the value $M_A = 1$. The predicted QS cross section normalized with energy as a function of energy is shown in Figure 1.

Resonance Production

Several models has been proposed in the literature to study this process. We are going to use like the most people a isobaric model [6], in which the resonance is considered like a "elementary" particle, which is describing the resonant part of the amplitude in those channels where resonance is dominant. We should be careful to employ the Feynman's rules and propagators consistent with a covariant treatment for spin 3/2 particles [7]. Then the expression for the amplitude becomes:

$$\mathcal{M} = \frac{G}{\sqrt{2}} cos\theta_c \mathcal{J}^\lambda \mathcal{L}_\lambda \tag{13}$$

$$\mathcal{J}^\lambda = \bar{\Psi}_\rho(k)\Gamma_{\rho\lambda}\Psi(p), \qquad \mathcal{L}_\lambda = \bar{\Psi}(\ell)\gamma_\lambda(1 - \gamma_5)\Psi(\nu) \tag{14}$$

Ψ_ρ, $\Psi(p)$: are Rarita-Schwinger and Dirac spinors for Δ and nucleon of momentum k and p. The square of the amplitude can be witten as:

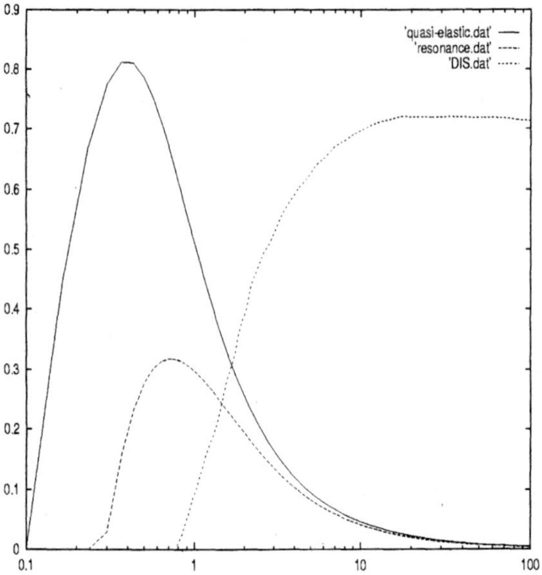

FIGURE 1. All exclusive channels contributions to CC cross section $\nu - N$ versus energy for an isoscalar target.

$$| \mathcal{M} |^2 \sim Tr\mathcal{S}_{\beta\rho}(k)\Gamma_{\rho\lambda}(\not{p}+m)\bar{\Gamma}_{\beta\alpha} \otimes Tr \ \not{\ell}\gamma^{\lambda}(1-\gamma_5) \ \not{\nu}\gamma^{\alpha}(1-\gamma_5) \tag{15}$$

$$\mathcal{S}_{\beta\rho}(k) = [-g_{\beta\rho} + \frac{1}{3}\gamma_\beta\gamma_\rho - \frac{1}{3M'}(\gamma_\beta k_\rho - \gamma_\rho k_\beta) + \frac{2}{3M'^2}k_\beta k_\rho](\not{k} + M_p) \tag{16}$$

$\mathcal{S}_{\beta\rho}(k)$ is the spin 3/2 propagator, M' is the mass of the resonance, M_P is the nucleon mass and the structure constant $\Gamma_{\rho\lambda}$ are used from two different sources in the literature:
a) Following to Fogli and Nardulli [8]:

$$\Gamma^{\rho\lambda} = \left[(\not{Q}g^{\rho\lambda} - Q^\rho\gamma^\lambda)\frac{C_3^V}{M_p} + (g^{\rho\lambda}k\cdot Q - Q^\rho k^\lambda)\frac{C_4^V}{M_P^2}\right]\gamma_5 - g^{\rho\lambda}C_5^A \tag{17}$$

b) Using reference [9]:

$$\Gamma^{\rho\lambda} = [(\not{Q}g^{\rho\lambda} - Q^\rho\gamma^\lambda)\frac{C_3^V}{M_p} + (g^{\rho\lambda}k\cdot Q - Q^\rho k^\lambda)\frac{C_4^V}{M_P^2} \tag{18}$$

$$+ (g^{\rho\lambda}k'\cdot Q - Q^\rho k'^\lambda)\frac{C_5^V}{M_P^2}]\gamma_5 + (\not{Q}g^{\rho\lambda} - Q^\rho\gamma^\lambda)\frac{C_3^A}{M_p} \tag{19}$$

$$+ (g^{\rho\lambda}k\cdot Q - Q^\rho k^\lambda)\frac{C_4^A}{M_P^2} + C_5^A g^{\rho\lambda} + \frac{C_6^A}{M_P^2}Q^\rho Q^\lambda \tag{20}$$

Here $Q = \ell - \nu = \underline{p_\ell - p_\nu} = -q$: the momentum transfer, and employing the fact that $\overline{\gamma}_5 = -\gamma_5$ and $\overline{\gamma^\lambda \gamma_5} = \gamma^\lambda \gamma_5$, is simple to see that the conjugate of the structure factor is given by

$$\Gamma^{\alpha\beta} = \left[(\not{Q} g^{\alpha\beta} - Q^\alpha \gamma^\beta) \frac{C_3^V}{M_p} - (g^{\alpha\beta} k \cdot Q - Q^\alpha k^\beta) \frac{C_4^V}{M_P^2} \right] \gamma_5 - g^{\alpha\beta} C_5^A \tag{21}$$

The weak $N - \Delta$ transition then will be described in terms of 3 or 8 form factors. Both calculations were performed in order to obtain some comparison with previous results [1]. $C_i^V (i = 3 - 6)$ are obtained using the conserved vector current (CVC) hypothesis, which requiere that $C_6^V = 0$. From the experimental data is obtained [10]:

$$C_5^V = 0, \quad C_4^V = \frac{M_p}{M'} C_3^V, \quad C_3^V = \frac{2.05}{(1 - \frac{q^2}{0.54 GeV^2})^2} \tag{22}$$

The weak axial form factor $C_i^A (i = 3 - 5)$ are determined by fitting the available data on the differential cross section $d\sigma / dq^2$ in neutrino scattering [11], mainly from deuteron target, in order to minimize the nuclear corrections. As a way of help, these values of the form factors are also compatible with the data on neutrino scattering from nuclear target [11]. C_6^A is proportional to the lepton mass, then it is neglected in the analysis. The values of the axial form factors most often used in the analysis of the neutrino experiment are [10–12]:

$$C_{i=3,4,5}^A (q^2) = C_i^A (0) \left[1 - \frac{a_i q^2}{b_i - q^2} \right] \left(1 - \frac{q^2}{M_A^2} \right)^{-2} \tag{23}$$

with $C_3^A(0) = 0$, $C_4^A(0) = -0.3$, $C_5^A(0) = 1.2$, $a_4 = a_5 = -1.21$, $b_4 = b_5 = 2\ GeV^2$ and M_A is treated as a free parameter. For our present purpose $M_A = 1.28\ GeV$ [10]. The results are to show up in Figure 1.

Deep Inelastic Scattering

In the formalism of DIS, the neutrino scattering takes place off of the partons inside the nucleon. This process has been studied with high precision in the 20-200 of GeV energy range by a number of experiments (for example CCFRR, CHARM, CDHS) [13]. So, for such processes one can takes the most general form for the hadronic vertex summed over spins:

$$\begin{aligned} \mathcal{W}_{\alpha\beta} = \Big[& W_1 M^2 \delta_{\alpha\beta} + W_2 p_\alpha p_\beta + W_3 \epsilon_{\alpha\beta\gamma\delta} p_\gamma \frac{Q_\delta}{2} \\ & + W_4 Q_\alpha Q_\beta + W_5 \frac{(p_\alpha Q_\beta + p_\beta Q_\alpha)}{2} \Big] \end{aligned} \tag{24}$$

Here we have terms like W_j, which are the hadron form factors, p and p' are the initial and final momenta of the hadrons and $Q = (p' - p)$.
The differential cross section can then be written as:

$$\frac{d^2\sigma}{dQ^2 d\nu} = \frac{G^2 E'}{2\pi E}\left(\frac{M_W^2}{M_W^2+Q^2}\right)^2 \left[\frac{W_2}{2}\left(1+\frac{p_\ell}{E'}cos\theta\right) + W_1\left(1-\frac{p_\ell}{E'}cos\theta\right)\right.$$

$$\mp \frac{W_3}{2M}\left(-\frac{M_\ell^2}{E'} + (E+E')\left(1-\frac{p_\ell}{E'}cos\theta\right)\right) \tag{25}$$

$$\left. + \frac{W_4 m_\ell^2}{2M^2}\left(1-\frac{p_\ell}{E'}cos\theta\right) - \frac{W_5 M_\ell^2}{2ME'}\right]$$

E,k, and E',k' are the incoming neutrino and scattered lepton energies and 4-momenta, and θ is the scattering angle between the incoming neutrino and the outgoing lepton in the laboratory frame. Also $\nu = E - E'$ and $-Q^2 = q^2 = (k-k')^2$. Altenatively, one can consider the process at the parton level. For do that, we can write the differential cross section for the neutrino-quark scattering, and then summing over all quarks participating, the connnection can be made between the form factors $W_1 - W_5$ above mentioned and the parton distributions: $F_1 = MW_1$; $F_2 = \nu W_2 = \nu x W_5$; $F_3 = -\nu W_3$; $W_4 = 0$; the variables x and y are related to q^2, ν by $x = \frac{q^2}{2M\nu}$, $y = \frac{\nu}{E}$; M is the nucleon mass. The differential cross section then becomes:

$$\frac{d\sigma}{dxdy} = \frac{G^2 ME}{\pi}\left(\frac{M_W^2}{M_W^2+Q^2}\right)^2 \left[F_2(1-y-\frac{Mxy}{2E}) + F_1 xy^2 \pm xF_3(y-\frac{y^2}{2})\right. \tag{26}$$

$$\left. + \frac{M_\ell^2}{ME}\left(-F_2\left(\frac{M}{4E}+\frac{1}{2x}\right) + \frac{F_1 y}{2} \mp \frac{F_3 y}{4}\right)\right]. \tag{27}$$

Using the Callan-Gross relation $2xF_1 = F_2$ we can simplifies the above expression further. It is then dependent on only two form factors, F_2 and F_3, which are given in terms of the parton distributions function by

$$F_2 = \Sigma_i x(q_i + \bar{q}_i) \tag{28}$$

$$F_3 = \Sigma_i x(q_i - \bar{q}_i) \tag{29}$$

where q_i and $\bar{q}_i$ are the quark distribution functions [15].

$$q(x,Q^2) = \frac{u_v(x,Q^2)+d_v(x,Q^2)}{2} + \frac{u_s(x,Q^2)+d_s(x,Q^2)}{2} \tag{30}$$

$$+ s_s(x,Q^2) + b_s(x,Q^2) \tag{31}$$

$$\bar{q}(x,Q^2) = \frac{u_s(x,Q^2)+d_s(x,Q^2)}{2} + c_s(x,Q^2) + t_s(x,Q^2) \tag{32}$$

the subscrits v and s label valence and sea contributions, and u, d, c, s, t, b denote the distributions for varios quark flavors in a proton. The parton distribution are calculated in the usual way, with the Q^2 behavior described by the Altarelli-Parisi equations. As I mentioned at the begining, the situation is somewhat more complicated in the lower energy regimen (which we need), where perturbative QCD is of questionable validity. Most parton distribution have a range of validity which extends down toward some cutoff Q^2 around 4 or 5 GeV^2. The Q^2 dependence of the partons distributions is then "frozen"at this cutoff Q^2, i.e. for $Q^2 < Q^2$ the parton distribution is evaluated with $Q^2 = Q_0^2$. For a Q^2 cutoff of Q_0^2 and incident neutrinos of energy E_ν, a fraction $f = (Q_0^2/2ME_\nu)(1 - ln(Q_0^2/2ME_\nu))$ of phase space $x, y \in [0,1]$ has $Q^2 \leq Q_0^2$. For $E_\nu = 13\ GeV$, $Q_0^2 = 4\ GeV$ then 44% of DIS events will have $Q^2 < Q_0^2$. So, one remedy si to use a set of parton distributions which are valid to lower Q^2. In this case we use the set of parton distribution function of Owens [14], the results are plotted in Figure 1, and compared with data on one pion production [8] and QS scattering [13]. The resultant CC cross section $\nu - N$ is plotted in Figure 2 .

The exclusive channels contribute in 89% (12%) of the cross section for $E_\nu = 1\ (10)\ GeV$. There does not seem to be a consensous as to the best way to combine these three channels at a fixed energy to form the total cross section. However with our choice, the phase space integration for the calculation of DIS is limited to region $W_0 < W$ to avoid double counting. The fluxes obtained for lower muon energies when the new experimental data in the evaluation of cross section is included gives a result of $\approx$ 20% if we use $E_\mu = 3\ GeV$ and $\approx$ 35% for $E\mu = 1\ GeV$. So, is obtained a improved values from those result in [1] for about 10% which is well accomodated in the reference [1] experimental expectatives.

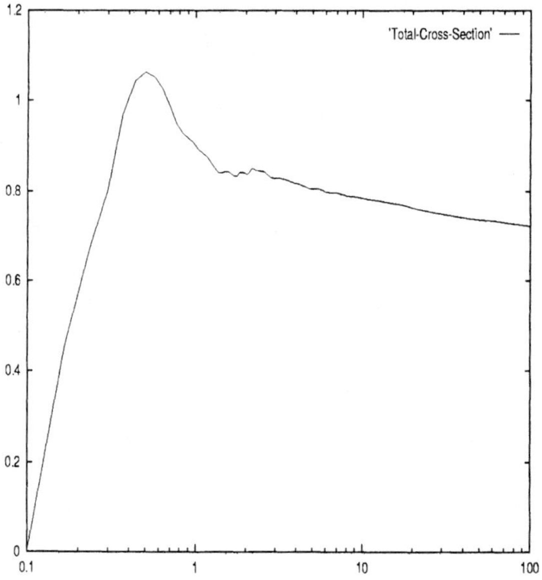

FIGURE 2. Total Charged Current cross section $\nu - N$.

ACKNOWLEDGEMENTS

The authors would like to thanks to CONACYT through the grant number J32220E.

REFERENCES

1. Paolo Lipari, Maurizio Lusignoli, and Francesca Sartogo, Phys. Rev. Lett. **74** (1995) 4384; M.C. Gonzalez-Garcia, H. Nunokawa, O.L.G. Perez, T. Stanev, J.W.F. Valle, Phys. Rev. **D58**:033004,1998.
2. R. Becker-Szendy, et. al. (IMB Collaboration), Phys. Rev. Lett. **69**, (1992) 1010; E. Akhmedov, P. Lipari, M. Lusignoli, Phys. Lett. **B300**, (1993) 128; W. Frati, T. K. Gaisser, A.K. Mann, and T. Stanev, Phys. Rev. **D48** (1993) 1140.
3. C.H. Llewellyn Smith, Phys. Rep. 36, (1971) 261.
4. R.P. Feynman and M. Gell-Mann, Phys. Rev. **109** (1958) 193.
5. N.J. Baker et. al. Phys. Rev. **D23**, (1981) 2499; F. Boehm and P. Vogel, Physics of Massive neutrinos. Cambridge University Press, Cambdrige, 46 (1992).
6. S M. Berman and M. Veltman, Nuovo Cimento **38**, (1965) 993;C.H. Albright and S. L. Liu, Phys. Rev. **140B**, (1965) 748; C. W. Kim, Nuovo Cimento **37** (1965) 142; P. Salin, Nuovo Cimento **48A**, (1967) 506.
7. M. El Amiri, G. López Castro, and J. Pestieu, Nucl. Phys. **A543**, (1993) 673.
8. G.L. Fogli and G. Nardulli, Nucl. Phys. **B160**, (1979) 116.
9. L. Alvarez-Ruso, S. K. Singh and M. J. Vicente Vacas, Nucl-th/9804007.
10. S. J. Barish et. al., Phys. Rev. **D19**, (1979) 2521; T. Kitagaki et. al., Phys. Rev. **D42**, (1990) 1331; G.M. Radecky et. al. Phys. Rev. **D25**, (1982) 1161; P.A. Schreiner and F. Von Hipped, Nucl. Phys. **B58**, (1973) 333.
11. P. Allen et. all., Nucl. Phys. **B21**, (1970) 158.
12. S.K. Singh, M.J. Vicente Vacas and E. Oset, Phys. Lett. **B416**, (1998) 23; S.L. Adler, Phys. Rev. **D12**, (1975) 2644; J. Bijtebier, Nucl. Phys. **B21**, (1970) 158.
13. R. Blair et. al. Phys. Rev. Lett. **51**, (1983) 343; M. Junker, et. al. Phys. Lett. **B99**, (1981) 265; Phys. Lett. **B100**, (1981) 520; E.P. Berge, et. al., Z. Physics, **C38**, (1988) 403.
14. J.F. Owens, Phys. Lett **B266**, (1991) 126; ibid, Phys. Rev. **D30**, (1984) 49.
15. Raj Gandhi, Chris Quigg, Ina Sarcevic and M. Hall Reno, Phys. Rev. **D58** 093009, 1998.

Signals of new physics and Higgs detection

J. Lorenzo Diaz-Cruz
Theory Group, Lawrence Berkeley Laboratory
Berkeley, CA, 92710
Instituto de Fisica, BUAP, 72570 Puebla, Pue, Mexico
October 24, 2000

Abstract. The production of Higgs boson at Tevatron by gluon fusion, can be enhanced by the presence of heavy colored particles that receive their mass from the SM Higgs. This can be used to constraint additional standard or mirror families, chiral colored sextets and octet quarks. The possibility to detect the lepton-flavour violating Higgs $h \to \tau\mu/\tau e$, at the first stages of RUN-II is also discussed.

1.- The discovery of the Higgs boson is one of the most cherised goals of present and future high-energy experiments. Within the minimal standard model (SM), the range of Higgs masses favored by present data is $113 < m_h < 220$ GeV [1], however new physics can allow heavier masses [2]. The characteristic Higgs couplings determine the strategies employed for its search at present and future colliders. For instance, the Higgs-fermion couplings can be studied either by open production of $t\bar{t}h, b\bar{b}h$ or by the loop-induced coupling with gluon and photon pairs $(hgg, h\gamma\gamma)$ [3]. Any additional heavy particle that receives its mass from the SM Higgs mechanism, will also contribute to the 1-loop vertices [4]. However, there is another important effect not considered previously, namely that the presence of these new particles could also induce non-decoupling corrections to the tree vertices hff, hWW and hZZ [5], which can affect the decay rate of detectable signatures [6]. Lepton flavour violation is another phenomena that could be tested in the Higgs sector [11], namely it is possible that the lepton-flavour violating Higgs decays $h \to \tau\mu/\tau e$, could reach detectable levels, and in fact these could be the only modes by which a non-standard Higgs boson could be discovered at the first stages of RUN-II. We comment on several models where these modes could be detectable.

2.- In order to probe heavy scales through their effect on the Higgs coupling, we shall consider extensions of the SM that include additional particles that receive their mass from the SM Higgs mechanism, which can only be doublets and singlets. Present global fits for electroweak data seem to exclude more than one additional

CP562, *Particles and Fields: Ninth Mexican School,* edited by G. Herrera Corral and L. Nellen

SM family [7], however this conclusion relies on the assumption that no other kind of physics occurs at the energy scale of the new fermion masses, which may not be the case in SUSY models. Thus, we shall consider first the situation when the effects of heavy colored particles are weakly bounded by electroweak precision meassurements, i.e. we assume that their contribution to the Peskin-Takeuchi parameters S, T is within experimental range [8]. Then, we shall discuss the case of a heavy fourth family, for which we find that the corrections to the vertices hgg and hWW/hZZ are correlated, and in fact can be written in terms of the parameter T. In fact, the effect of heavy majorana neutrinos can also help to make the parameter S to be within experimental ranges.

To describe the effects of heavy quanta on the higgs couplings to gauge bosons and fermions, we shall write them as: $g_{hXX} = g^{SM}_{hXX}(1+\epsilon_X)$, where $X = W, Z, g, \gamma, t, b$. Since we are interested in the limit $m_h << M_{heavy}$, one can use the low-energy theorems [9], to relate $\epsilon_{g,\gamma}$ to the beta-function coefficients for the corresponding coupling constant $(\beta_{I,X})$. Thus, the higgs-gluon coupling is given by $\epsilon_g = \beta_{3,X}/\beta_{3,t}$, where $\beta_{3,t(X)}$ denote the contribution of top and heavy quanta to strong beta function. The values of ϵ_g arising from a pair of heavy color triplets, sextets and octet quarks are : $\epsilon_g = 2$, 5 and 6, respectively, whereas a new SM family plus its mirror partner gives $\epsilon_g = 4$.

The contribution coming from the new particles will modify the cross section for gluon fusion, as follows: $\sigma(pp \to h + X) = \sigma_{SM}(1+\epsilon_g)^2$ where the SM cross-section σ_{SM} can be written in terms of the Higgs decay width $\Gamma(h_{SM} \to gg)$; details of the equations can be found in the literature [10], Since the new particles modify the decay width into $h \to gg$, it will enhance the production by gluon fusion reaction, but this enhancement could work against the final signal rate. Both effects can be taken properly into account in the analysis by writing the product of the cross-section times the branching ratio of the signal as $[\sigma \times B.R.(h \to VV)]_{new} = R_V \times [\sigma \times B.R.(h \to VV)]_{SM}$, where R_V is given by ,

$$R_V = \frac{(1+\epsilon_g)^2(1+\epsilon_V)^2}{[1 + \sum_Y(\epsilon_Y^2 + 2\epsilon_Y) * B.R.(h_{SM} \to YY)]} \qquad (1)$$

where the sum in the denominator runs over $Y = t, b, g, A, W, Z$.

We can obtain bounds on the parameters $\epsilon_{g,W}$ using gluon fusion and the decay $h \to WW^*$ at Tevatron, which was studied in detail in ref. [12], this work concluded that it is possible to detect a SM Higgs boson with an integrated luminosity of 30 fb^{-1}, provided that an optimized selection of cuts is implemented. For the case when the corrections to ϵ_W can be neglected (Scenario-I), those cuts already allow us to get bounds on the parameter ϵ_g, that will test the presence of heavy colored particles, provided that the Higgs mass lays in the intermediate mass range; this include colored triplets, sextets and octest [6].

On the other hand, when the new physics also modifies the parameters $\epsilon_{W,Z}$, the decay rates into $h \to WW^*$ will receive further modification due to loops involving the heavy fermions. This happens for the case of a fourth heavy family; here one finds an interesting relation between $\epsilon_{f,W,Z}$ and the corresponding expression for the parameter T [6]; the corrections to the vertices hVV are negative (as compared with the positive tree-level value) and grow with the mass of the heavy quanta. For instance for fermion masses of order 500 (700) GeV the deviations from the SM tree-level couplings are of order - 27 (-52) %, which seems to imply that the signal will no longer be detectable. With 10 fb^{-1} of integrated luminosity, it will be possible to exclude only a limited range of fermion masses, up to about 900 GeV at best, as it is shown in table 1.

3.- Lepton flavour violation (LFV) is another phenomena that could provide new signatures that may allow detection of the Higgs boson, namely it is possible that the LFV decays of the Higgs boson $H \to l_i^+ l_j^-$ $(l_i = e, \mu, \tau)$ can reach detectable levels. We have studied several extensions of the SM, including both the (linear) effective lagrangian lagrangian extension of the SM and several specific models [11]. Within the effective lagrangian case, we focused on a set of dimension-6 operators that induce LFV vertices for the Higgs and Z boson. For those operators whose coefficients can not be constrained by present data, we obtain a Branching Ratio (B.R.) of order $10^{-1} - 10^{-2}$ for the LFV Higgs decays, which could be detected at future hadron colliders. For the other operators that are bounded by current limits on LFV transition, there are strong bounds on $e - \mu$ transitions, which imply $B.R.(H \to e\mu) \simeq 10^{-9}$; however, the decay modes $H \to \tau\mu/\tau e$ are still allowed to have a B.R. of the order 10^{-1}. In the case of the general Two-Higgs doublet model, we can also obtain $B.R.(H \to \tau\mu/\tau e) \simeq 10^{-1}$, which can be detected at Tevatron Run-II. For the extension of the SM that incldes massive neutrinos and the Minimal SUSY-SM, the corresponding LFV Higgs decays are induced at 1-loop, and turn out to be strongly suppressed. Further studies of LFV Higgs decays have appeared very recently, both at future muon collider [13] and at hadron colliders [14].

Work supported by CONACYT-SNI (Mexico); the hospitality of LBL (Berkeley) is also acknwoledged.

REFERENCES

1. For a recent review of higgs mass bounds see: M. Kado, talk at Recontres de Moriond (March, 2000), hep-ex/0005022.
2. J.L Diaz-Cruz et al., hep-ph/9911222; C. Kolda and L. Hall, Phys. Lett. B459 (1999) 213 [hep-ph/9904236].
3. J. Gunion, H. Haber, G. Kane and S. Dawson, 'The Higgs Hunters Guide', Adison Wesley, Reading 1990.
4. M. Chanowitz, Phys. Rev. Lett. 69 (1992) 2037; I.F. Ginzburg, I.P. Ivanov and A. Shiller, Phys. Rev. D60 (1999) 095001.
5. M. Veltman, Nucl. Phys. B123 (1977) 89; M. Chanowitz, M. Furman and I. Hinchliffe, Nucl. Phys. B153 (1979) 402.
6. J.L Diaz-Cruz, hep-ph/0008001;

7. J. Erler and P. Langacker, in Review of Particle Properties, Eur. Phys. J C3 (1998) 1.
8. M. Peskin and T. Takeuchi, Phys. Rev. Lett. 65 (1990) 964.
9. A. Vainshtein et al. Sov. J. Nucl. Phys. 30 (1979) 711; B. Kniehl and M. Spira Z. Phys. C69 (1995) 77.
10. For a review see: M. Spira, Fortsch. Phys. 46 (1998) 203.
11. J.L. Diaz-Cruz and J.J. Toscano, hep-ph/9910233 (Phys. Rev. D, in press).
12. T. Han, A. Turcot and R.J. Zhang, hep-ph/9812275.
13. M.Sher, Phys. Lett. B487 (2000) 151 (hep-ph/0006159).
14. T. Han and D. Marfatia, hep-ph/0008141.

TABLE CAPTION

Table 1. Upper limits to 4th generation fermion masses that can be obtained from Higgs search at Tevatron RUN-II, with 10 fb^{-1}.

Table. 1

m_h [GeV]	m_{4th}^{max} [GeV]
120.	350.
130.	520.
140.	650.
150.	730.
160.	870.
170.	710.
180.	670.
190.	600.
200.	510.

"Eightfold Way" From Instanton Dynamics

M. Napsuciale[1] and M. Kirchbach [2]

[1] *Instituto de Fisica, Universidad de Guanajuato, AP E-143, 37150, Leon, Guanajuato, Mexico*
[2] *Escuela de Fisica, Univ. Aut. de Zacatecas AP C-580, Zacatecas, ZAC 98068 Mexico.*

Abstract. The instanton–induced determinantal 't Hooft interaction is built into a three-flavor linear sigma model which is considered in the OZI–rule–respecting basis $\{(\bar{s}s), \frac{1}{\sqrt{2}}(\bar{u}u + \bar{d}d)\}$. The mixing of the strange and non-strange quarkonia, which emerges in the presence of instantons, is shown to be ideal thus leading to the formation of an octet–flavor state.

INTRODUCTION

Flavor symmetry is presently understood on the basis of QCD and the structure of the quark mass matrix. In the zero quark mass limit QCD acquires a $U(3)\times U(3)$ symmetry which at the level of hadrons is assumed to be realized in the Goldstone phase and we identify the would be Goldstone bosons with the lightest pseudoscalar mesons. The first problem one encounters this way is the large mass of the η'. The way out this problem is to take into account quantum corrections which spoil the conservation of the singlet axial current. Particularly relevant to this problem is the existence of Euclidean solutions with non-trivial topological properties (instantons) which also break this symmetry. In fact, the most appealing explanation to the problem of the large η' mass is the instanton induced quark-quark interaction [1].

The situation in the pseudoscalar sector is in contrast with that in the vector sector. Isoscalar vector mesons closely follow a flavor basis structure: $\phi = \bar{\psi}\lambda_{\rm s}\psi = \bar{s}s$ and $\omega = \bar{\psi}\lambda_{\rm ns}\psi = (\bar{u}u + \bar{d}d)/\sqrt{2}$, instead of the singlet-octet pattern followed by isoscalar pseudoscalars.

In the absence of symmetry breaking terms we can freely use any basis for the generators of the group. The physically interesting basis are those whose generators still reflect a residual symmetry of the system in the presence of symmetry breaking terms. In this concern, we have three sources of breaking for $U(3)\times U(3)$ symmetry. The first one is the $U_A(1)$ symmetry breaking by quantum effects. Although the instanton induced interaction is small at high energies due to the factor

CP562, *Particles and Fields: Ninth Mexican School,* edited by G. Herrera Corral and L. Nellen

$exp(-8\pi^2/g^2)$, it is always present and becomes determinant at low energies. The second source is the spontaneous breaking of chiral symmetry occurring at the scale $\lambda_\chi \approx 4\pi f_\pi$. The last source is the presence of the quark mass matrix.

Isospin is a good symmetry for hadron strong interactions and in the isospin limit the quark mass matrix is $\mathcal{M}_q = Diag(m, m, m_s)$. This matrix can be written in terms of the $U(3)$ (not of $SU(3)$) generators. Using SU(3) would require to simultaneously consider two irreducible representations, the singlet and the octet.

In [2], based on the OZI rule [3], group theoretical arguments have been given for the $SU(3)$ symmetry as an artifact of the $U(1)_A$ anomaly. It is the goal of the present study to show that unitary spin is an accidental symmetry which is manufactured by the instanton dynamics. To this end we will consider the effect of the instanton induced quark-quark interaction in the pseudoscalar and scalar sectors, considering from the very beginning the flavor basis.

"EIGHTFOLD WAY" FROM INSTANTON DYNAMICS.

To support the thesis concluding the previous section by an explicit example, we introduce here a chirally symmetric $[U(3)_L \otimes U(3)_R]$ meson Lagrangian [4–6] which describes a scalar and a pseudoscalar nonet, in turn denoted by (σ_i) and (P_i),

$$\mathcal{L} = \mathcal{L}_{sym} + \mathcal{L}_{INST} + \mathcal{L}_{SB}. \tag{1}$$

Here,

$$\mathcal{L}_{sym} = \langle \frac{1}{2}(\partial_\mu M)(\partial^\mu M^\dagger)\rangle - \frac{\mu^2}{2}X(\sigma,P) - \frac{\lambda}{4}Y(\sigma,P) - \frac{\lambda'}{4}X^2(\sigma,P), \tag{2}$$

$M = \sigma + iP$, and X, Y stand in turn for the left-right symmetric traces

$$X(\sigma,P) = tr\left(MM^\dagger\right), \quad Y(\sigma,P) = tr\left((MM^\dagger)^2\right). \tag{3}$$

The pseudoscalar and scalar matrix fields P and σ are written in terms of a specific basis spanned by seven of the standard Gell-Mann matrices, namely λ_i $(i = 1, \ldots, 7)$, *and* by two non-standard matrices $\lambda_{\rm ns}$=diag(1,1,0), and $\lambda_{\rm s} = \sqrt{2}$ diag(0,0,1), respectively. The decomposition obtained in this way reads $P \equiv \frac{1}{\sqrt{2}}\lambda_i P_i$ with $i = ns, s, 1, \ldots, 7$ and similarly for the scalar field. The conventions are such that the isoscalar fields in the diagonal of the P and S matrix are $\chi_{ns}/\sqrt{2}$ and χ_s with $\chi \in \{P, S\}$, respectively. The instanton-induced interaction in (1) is

$$\mathcal{L}_{INST} = -\beta\left\{\det(M) + \det(M^\dagger)\right\} \tag{4}$$

It stands for 't Hooft's effective quark-quark interaction which is determinantal and flavor dependent [1]. It is induced by instantons and is reminiscent of the $\mathrm{U}(1)_A$ breaking at the quantum level of QCD. Finally, there is the standard term

$$\mathcal{L}_{SB} = tr\,(c\sigma) \tag{5}$$

which breaks the left-right symmetry explicitly. The c matrix is spanned by the same basis $c \equiv \frac{1}{\sqrt{2}}\lambda_i c_i$, where the nine expansion coefficients c_i are independent constants. The most general c-matrix that preserves isospin, respects PCAC and is consistent with the quark mass matrix, has $c_{\rm s}$ and $c_{\rm ns}$ as the only non-vanishing entries. While the $c_{\rm s}$ term explicitly causes $[U(3)_L \times U(3)_R]/[U(2)_L \times U(2)_R]$ breaking, the $c_{\rm ns}$ leads to $[U(2)_L \otimes U(2)_R]/U(2)_I$ breaking. Furthermore, the linear σ term in Eq. (5) induces σ-vacuum transitions which supply the scalar fields with non-zero vacuum expectation values (v.e.v) (hereafter denoted by $\langle\cdots\rangle$). To simplify notations, let us re-denote $\langle\sigma\rangle$ by V with V =diag (a, a, b), where a and b in turn denote the vacuum expectation values of the strange and non-strange quarkonium, respectively, $a = \frac{1}{\sqrt{2}}\langle\sigma_{\rm ns}\rangle$, $b = \langle\sigma_{\rm s}\rangle$. We now shift, as usual, the old σ field to a new scalar field $S = \sigma - V$ such that $\langle S\rangle = 0$. In this way, new mass terms, three-meson interactions, and a linear term are generated. In particular, all these terms are affected – via the 't Hooft determinant – by the $U_A(1)$ anomaly which get coupled to the v.e.v' s of the scalar fields by the spontaneous breaking of chiral symmetry. The consequence of all these effects is the breakdown of the original symmetry down to $SU(2)_I$ isospin. The masses of the seven unmixed pseudoscalar and scalar mesons corresponding to the original Gell-Mann matrices λ_i $(i = 1, \ldots, 7)$, namely the *isovector* pseudoscalar (π) and scalar (a_0) mesons as well as the two *isodoublets* of pseudoscalar (K) and scalar (κ) mesons, are obtained as [4–6]

$$\begin{aligned} m_\pi^2 &= \xi + 2\beta b + \lambda a^2\,, & m_K^2 &= \xi + 2\beta a + \lambda(a^2 - ab + b^2)\,,\\ m_{a_0}^2 &= \xi - 2\beta b + 3\lambda a^2\,, & m_\kappa^2 &= \xi - 2\beta a + \lambda(a^2 + ab + b^2)\,, \end{aligned} \tag{6}$$

where we used the convenient short–hand notation $\xi \equiv \mu^2 + \lambda'(2a^2 + b^2)$. The elimination of the linear terms imposes the following constraints on the explicit-symmetry-breaking terms $c_{\rm ns}$, and $c_{\rm s}$:

$$c_{\rm ns} = \sqrt{2}am_\pi^2\,, \qquad c_{\rm s} = \frac{1}{\sqrt{2}}(a+b)m_K^2\,, \tag{7}$$

whereas PCAC relations yield

$$f_\pi = \sqrt{2}a\,, \qquad f_K = \frac{1}{\sqrt{2}}(a+b). \tag{8}$$

The mass term of the Lagrangian involving the *mixed isoscalar* pseudoscalar and scalar fields, which correspond to the $\lambda_{\rm ns}$ and $\lambda_{\rm s}$ matrices, reads

$$\begin{aligned} \mathcal{L}_{mass} = &-\frac{1}{2}(m_{P_{\rm ns}}^2 P_{\rm ns}^2 + m_{P_{\rm s}}^2 P_{\rm s}^2 + 2m_{P_{\rm s-ns}}^2 P_{\rm s} P_{\rm ns})\\ &-\frac{1}{2}(m_{S_{\rm ns}}^2 S_{\rm ns}^2 + m_{S_{\rm s}}^2 S_{\rm s}^2 + 2m_{S_{\rm s-ns}}^2 S_{\rm s} S_{\rm ns}) \end{aligned} \tag{9}$$

where

$$m^2_{P_{\rm ns}} = \xi - 2\beta b + \lambda a^2\,, \qquad m^2_{S_{\rm ns}} = \xi + 2\beta b + 3\lambda a^2 + 4\lambda' a^2\,, \tag{10}$$

$$m^2_{P_{\rm s}} = \xi + \lambda b^2\,, \qquad m^2_{S_{\rm s}} = \xi + 3\lambda b^2 + 2\lambda' b^2\,, \tag{11}$$

$$m^2_{P_{\rm s-ns}} = -2\sqrt{2}\beta a\,, \qquad m^2_{S_{\rm s-ns}} = 2\sqrt{2}(\beta + \lambda' b)a\,. \tag{12}$$

Here $m_{\chi_{\rm s}}$ and $m_{\chi_{\rm ns}}$ with $\chi \in \{P, S\}$ are the masses of the strange and non-strange (pseudo-)scalar quarkonia respectively, while $m_{\chi_{\rm s-ns}}$ denotes the transition mass-matrix elements of the strange–non-strange (pseudo-)scalar quarkonia.

Eqs. (12) show that *the mixing between strange and non-strange quarkonia is due to the instanton-induced interactions and the spontaneous breakdown of chiral symmetry.*

In the following we will first discuss the mixed pseudoscalar sector. The physical isoscalar pseudoscalar fields are linear combinations of $P_{\rm s}$, $P_{\rm ns}$ which diagonalize the pseudoscalar part of $\mathcal{L}_{mass}$:

$$\begin{aligned} \eta &= P_{\rm ns}\ \cos\phi_P - P_{\rm s}\ \sin\phi_P \\ \eta' &= P_{\rm ns}\ \sin\phi_P + P_{\rm s}\ \cos\phi_P\,. \end{aligned} \tag{13}$$

Here, ϕ_P stands for the isoscalar-pseudoscalar mixing angle in the flavor basis. This diagonalization of the mass matrix for the pseudoscalar mesons yields the relations

$$\sin\ 2\phi_P = \frac{2m^2_{P_{\rm s-ns}}}{m^2_{\eta'} - m^2_\eta}\,, \qquad \cos\ 2\phi_P = \frac{m^2_{P_{\rm s}} - m^2_{P_{\rm ns}}}{m^2_{\eta'} - m^2_\eta}\,. \tag{14}$$

In addition, one finds the following trace relation

$$m^2_{\eta'} + m^2_\eta = m^2_{P_{\rm s}} + m^2_{P_{\rm ns}} \tag{15}$$

to be valid. The parameters entering the *pseudoscalar* sector of the model (ξ, λ, β, a, b) can be fixed through the masses and the decay constants of the pseudoscalars ($m_{\eta'}$, m_η, m_π, m_K, f_π) and can be used to predict all the other properties of the pseudoscalar mesons such as the mixing of the strange and non-strange fields. Alternatively, the kaon decay constant f_K can be used as input, replacing the combination $(m^2_{\eta'} - m^2_\eta)$ of the above given quantities [5]. The latter procedure creates slightly different results for the pseudoscalar mixing angle. Finally, one could also have used the pseudoscalar mixing angle as input [6]. This leads to a different identification of the scalar nonet. The pertinent masses turn out to be highly sensitive to the choice for the input parameters. In particular, the latter version yield heavy scalars [6].

Now, after some algebraic manipulations of Eqs. (6, 10–15), the β parameter is expressed in terms of the pseudoscalar meson masses according to

$$-4\beta(a+b) = \frac{(m^2_{\eta'} - m^2_\pi)(m^2_\eta - m^2_\pi)}{(m^2_K - m^2_\pi)}. \tag{16}$$

The physical solution found here coincides with the one associated with the quadratic equation for β reported in an earlier work [4], namely $\beta \approx -1.55$

GeV. The parameter a can be directly fixed through the left Eq. (8), whereas b, parametrized as $b = (1 + 2x)\ a$, can be fixed either through the kaon mass as $x = x_N = 0.39$ [4], or through f_K in the right Eq. (8) as $x = x_T = 0.22$ [5]. Using Eq. (14) and $x = 0.39$ we obtain

$$\sin\ 2\phi_P = 0.9202\,, \qquad \cos\ 2\phi_P = -0.3911\,. \tag{17}$$

A careful analysis of the mass matrix shows that the actual mixing angle in the flavor basis is the one arising from the cosine relation in Eqs. (17). This angle is complementary to the one arising from the sine relation, which does not distinguish between $\pi/2 - \delta$ and δ. The mixing angle in the flavor basis thus turns out to lie within the value predicted for $x = 0.22$, $\phi_P = 49.7^\circ$ and that for $x = 0.39$, $\phi_P = 56.7^\circ$. The corresponding angle in the singlet-octet basis is $\theta_P = \phi_P - \alpha$, where $\alpha = 54.7^\circ$ is the relative angle between the singlet-octet and the strange-non-strange basis as arise from the following relations

$$\begin{aligned} P_8 &= P_{\rm ns}\ \cos\alpha - P_{\rm s}\ \sin\alpha \\ P_1 &= P_{\rm ns}\ \sin\alpha + P_{\rm s}\ \cos\alpha\,, \end{aligned} \tag{18}$$

with $cos\alpha = \sqrt{1/3}$ and $sin\alpha = \sqrt{2/3}$. The so-determined values of θ_P are in the range $\theta_P \in [-5^\circ,\ +2^\circ]$ and therefore close to zero. This finding, in combination with the fact that the flavor fields are mixed up solely by the instanton-induced interaction (see the left Eq. (12)), establishes the main result of this work: *'t Hooft's instanton-induced interaction mixes strange and non-strange pseudoscalar fields in such a way that one of the physical fields becomes a member of the octet, while the other one becomes an SU(3) singlet.*

If the coefficient λ' in Eq. (12) is ignored, the mixing between the *scalar* strange and non-strange quarkonia due to 't Hooft's instanton-induced interaction is predicted to be of the same size as the corresponding mixing in the pseudoscalar sector but with opposite sign. This is consistent with the results in [7,8]. In the scalar sector, however, one has to account for the additional effect brought about by one of the chiral invariants in Eq. (2) whose strength is measured by the above-mentioned λ' coupling as dictated by the right of Eqs.(12). As discussed in [5], this chiral invariant corresponds to OZI-rule violating disconnected hairpin diagrams. They represent one out of various examples of sub-leading OZI-rule violating mechanisms, the most important among them being probably Lipkin's non-planar hadronic loops [9]. In the scalar sector, hadronic loops cancellation is totally spoiled by parity conservation [10]. Thus, although sub-leading and suppressed with respect to instanton contributions, this effect acquires importance as *interferes destructively with the instanton-induced contribution to the mixing of the scalar mesons.* This renders the scalars less strongly mixed than pseudoscalars and thus closer to the flavor basis. Estimates based on meson spectrum and on recent data on radiative ϕ decays yield for the isoscalar-scalar mixing angle $\phi_S \in [-9^\circ, -14^\circ]$ [4,11,12]. Therefore one of the scalar isoscalar mesons can be nearly identified with the non-strange scalar and is strongly moved down relative to the scalar isovector (a_0) as can be seen from Eqs. (10, 6). This is also consistent with results in [7].

It is worth noting that the physical properties of mesons, belonging to sectors which are not affected by the instanton-induced interactions, such as the spin-1^{--} and tensor 2^{++} mesons, are well described in terms of almost pure flavor states. The small departure from the flavor basis in these sectors ($\phi_V \approx 4^\circ$) can be attributed to strong and yet incomplete cancellation of meson loops in this sector [10]. The same argument can be used for the actual deviation of the η from being a pure octet state. The effect is slightly larger here due to incomplete cancellations among hadronic loops [9,10] and it should lower the pseudoscalar mixing angle ϕ_P from the value dictated by the instanton dynamics ($\phi_P \approx 54.7^\circ$) to its physical value $\phi_P \approx 39.4$ that has been concluded recently from fitting $\eta(\eta')$ decays [13].

Summarizing we have shown that the SU(3) symmetry exhibited by the pseudoscalar mesons is an accidental symmetry having its root in the QCD instanton dynamics. The application of these ideas to the ηN coupling, will be considered in a forthcoming work.

ACKNOWLEDGEMENTS

We thank Andreas Wirzba for useful comments. One of us (M.N.) appreciates illuminating correspondence with G. 't Hooft about the predictions of the model used in this work, which lead him to reconsider the extraction of the pseudoscalar mixing angle. Work supported by CONACyT, Mexico under contract I27604E.

REFERENCES

1. G. 't Hooft, *Phys. Rev. Lett.* **37**,8 (1976); *Phys. Rev.* **D14**,3432 (1976); *Phys. Rept.* **142**,357 (1986).
2. M. Kirchbach, *Phys. Rev.* **D58**, 117901 (1998).
3. S. Okubo, *Phys. Lett.* **5**,165 (1963); G. Zweig, CERN Report No 8419 TH 412 (1964); J. Iizuka, *Prog. Theor. Phys. Suppl.* **37-38**,21 (1966).
4. M. Napsuciale, hep-ph/9803396.
5. N. A. Tornqvist, *Eur. Phys. J.* **C11**359 (1999) [hep-ph/9905282].
6. G. 't Hooft, hep-th/9903189.
7. T. Schäfer and E. Shuryak, hep-lat/0005025.
8. N. Isgur and H. B. Thacker, hep-lat/0005006.
9. H. J. Lipkin, *Nucl. Phys.* **B244**,147 (1984); *ibid.* **B291**,720 (1987).
10. H. J. Lipkin and B. Zou, *Phys. Rev.* **D53**,6693 (1996); B. Zou, *Phys. Atom. Nucl.* **59**, 1427 (1996) [hep-ph/9611238].
11. J. L. Lucio and M. Napsuciale, *Phys. Lett.* **B454**,365 (1999) [hep-ph/9903234].
12. J. L. Lucio and M. Napsuciale, hep-ph/0001136.
13. T. Feldmann, P. Kroll and B. Stech, *Phys. Rev.* **D58**,114006 (1998) [hep-ph/9802409]; T. Feldmann, *Int. J. Mod. Phys.* **A15**,159 (2000) [hep-ph/9907491].

$s - \bar{s}$ asymmetry in nucleons and the strange quark distribution in kaons and hyperons

E. Cuautle[†1], J. Magnin[†‡2]

[†]*Centro Brasileiro de Pesquisas Físicas, CBPF*
Rua Dr. Xavier Sigaud 150, 22290-180 Rio de Janeiro Brazil
[‡]*Depto de Física, Universidad de los Andes,*
AA 4976, Santafé de Bogotá, Colombia

Abstract. We study the s and $\bar{s}$ low Q^2 non-perturbative distributions in protons in the framework of the Meson Cloud Model. From a comparison of model predictions to a recent global fit from Deep Inelastic Scattering data on the s and $\bar{s}$ quark distribution functions in protons, we extract qualitative information on the strange quark distribution functions in hyperons and kaons.

Since 1987 there exist theoretical speculations on a $s - \bar{s}$ asymmetry in the nucleon sea [1]. However, the first experimental evidence came out in 1995 when the CCFR Collaboration [2] showed that such an asymmetry in nucleons cannot be excluded. Most recently, in a global reanalysis of Deep Inelastic Scattering (DIS) data, Barone *et al.* [3] showed the first clear evidence of such an asymmetry.

On the theoretical side, a number of models [1,4–6] attempted to predict the $s - \bar{s}$ asymmetry with different degrees of success. Among them, the most significant approach seems to be the Meson Cloud Model (MCM). In this model, fluctuations of the proton to kaon-hyperon virtual states are responsible for the $s - \bar{s}$ asymmetry. Since the s quark belong to the hyperon and the $\bar{s}$ quark is in the kaon, a the asymmetry arises naturally in the MCM due to the different momentum carried by kaons and hyperons in the fluctuation. Up to our knowledge, two different approaches exist within the hypothesis of the MCM. The first based in a form factor description of the extended proton-kaon-hyperon vertex [1,5], and a second in terms of parton degrees of freedom [6]. In the first, the knowledge of the exact form of the (unknown) form factors is crucial to get a reasonable description of the $s - \bar{s}$ asymmetry (see e.g. Ref. [5]). In the second, fluctuations are generated through gluon emission from the constituent valence quarks and its subsequent

1) ecuautle@lafex.cbpf.br
2) jmagnin@uniandes.edu.co

CP562, *Particles and Fields: Ninth Mexican School,* edited by G. Herrera Corral and L. Nellen

splitting to a $q-\bar{q}$ pair. This $q-\bar{q}$ pair then recombines with constituent quarks to form a kaon-hyperon bound state. Although the exact details of the recombination process, which are in the realm of non-perturbative QCD, are unknown, this models give a definite prediction for the sign of the $s-\bar{s}$ asymmetry [6]. This model is also able to give a good description, both qualitative and quantitative, of the $\bar{d}/\bar{u}$ asymmetry [7] in the proton sea recently measured by the E866 Collaboration [8], and predicts the existence of *valence-like* gluons inside the proton [9], an ingredient which is needed, as input at low Q^2, in fits to DIS data (see e.g. [3]).

Most recently, it has been shown that, in the framework of the MCM there exist, a close relationship between the low Q^2 structure of pions and kaons [10]. The same relationship can be found among the structure of nucleons, hyperons, pions and kaons, since these appear in the Fock state expansion of the proton wave function. Then, in principle, a precise study of the proton structure can give some insight on the structure of other hadrons.

In this work we shall try to extract some preliminary information on the low Q^2 structure of kaons and hyperons out of experimental data on the strange quark distributions in protons, within the MCM hypothesis.

To this end, let us start with a low Q^2 expansion of the proton wave-function as

$$|p\rangle = a_0|p_0\rangle + a_1|\hat{p}_0 g\rangle + a_2|\pi^+ n\rangle + a_3\left[c_{(1,-1/2)}|\pi^+\Delta^0\rangle + c_{(-1,3/2)}|\pi^-\Delta^{++}\rangle\right] + a_4|K^+\Lambda^0\rangle + a_5\left[c_{(1/2,0)}|K^+\Sigma^0\rangle + c_{(-1/2,1)}|K^0\Sigma^+\rangle\right] + \ldots \quad (1)$$

The s and $\bar{s}$ distributions are then given by

$$\begin{aligned} s(x) &= |a_4|^2 P_{|K\Lambda\rangle} \otimes v_{s/H} + \\ &\quad |a_5 c_{(1/2,0)}|^2 P_{|K\Sigma\rangle} \otimes v_{s/H} + |a_5 c_{(-1/2,1)}|^2 P_{|K\Sigma\rangle} \otimes v_{s/H} \\ \bar{s}(x) &= |a_4|^2 P_{|\Lambda K\rangle} \otimes v_{s/K} + \\ &\quad |a_5 c_{(1/2,0)}|^2 P_{|\Sigma K\rangle} \otimes v_{s/K} + |a_5 c_{(-1/2,1)}|^2 P_{|\Sigma K\rangle} \otimes v_{s/K} \,, \end{aligned} \quad (2)$$

where

$$P_{|MB\rangle} \otimes v_{s/B} = \int_x^1 \frac{dy}{y} P_{|MB\rangle}(y)\, v_{s/B}\left(\frac{x}{y}\right). \quad (3)$$

$v_{s/H}$ and $v_{s/K}$ are the strange constituent quark -*valon*- [11] distributions in the hyperon and kaon. Coefficients $|a_i|^2$ are the probability of the correspondings fluctuation. These coefficients are partly given by Clebsch-Gordan (C.G.) coefficients and partly by the internal QCD dynamics of the fluctuation. $c_{(m_1,m_2)}$ are the corresponding C.G. coefficients.

The $P_{|HK\rangle}$ probability density is given by [7]

$$P_{|HK\rangle}(x) = \frac{(1-x)^a}{x} \int_0^x dy\, y\bar{q}(y)\,(y-x) v_p(y-x)\,, \quad (4)$$

where $v_p = \frac{105}{16}\sqrt{x}(1-x)^2$, is the light ($u$ or d) valon distribution in the nucleon [11], and

$$\bar{q}(x) = N\frac{\alpha_{st}^2(Q_v^2)}{(2\pi)^2}\int_x^1 \frac{dy}{y} P_{qg}\left(\frac{x}{y}\right)\int_y^1 \frac{dz}{z} P_{gq}\left(\frac{y}{z}\right) v_p(z), \tag{5}$$

is the probability density of an antiquark obtained from the decay chain $v_p \to v_p + g$ and $g \to q + \bar{q}$. $P_{gq}(z) = \frac{4}{3}(1 + (1-z)^2)/z$ and $P_{qg}(z) = \frac{1}{2}(z^2 + (1-z)^2)$ are the Altarelli-Parisi [12] splitting functions (see Refs. [6,7] for details).

The probability densities of having a hyperon or a kaon inside the proton are related by $P_{|KH\rangle}(x) = P_{|HK\rangle}(1-x)$ due to momentum conservation, with an additional constraint in velocity,

$$\frac{\left\langle P_{|KH\rangle}\right\rangle}{m_H} = \frac{\left\langle P_{|HK\rangle}\right\rangle}{m_K}, \tag{6}$$

which fixes the exponent a in Eq.4. Thus, for the $|K\Lambda\rangle$ state we obtain $a = 2.4$ and for $|K\Sigma\rangle$ we have $a = 2.74$.

Collecting all the contributions, and after evolution to $Q^2 = 20$ GeV^2, from comparison to experimental data we obtain

$$\bar{s}_K = v_{\bar{s}/K} = 11.43x^6(1-x)^{(1/5)}; \quad s_B = v_{s/H} = 5.92x^3(1-x)^{(1/5)} \tag{7}$$

for the low Q^2 s and $\bar{s}$ valon distributions in hyperons and kaons. Our curves for the $s - \bar{s}$ and $s/\bar{s}$ distributions in protons are displayed in Fig. 1a and 1b respectively and compared to fits by Barone *et al.* [3]. In Fig. 1c, the s and $\bar{s}$ low Q^2 distributions in hyperons and kaons are shown. We have taken $Q_0^2 = 0.64$ GeV^2 as the starting point for the evolution, according with the valon model [11].

Note that fluctuations containing no strange quarks in the expansion of Eq.1 contribute to the $s/\bar{s}$ distribution at $Q^2 = 20$ GeV^2 due to strange and antistrange quarks generated in the evolution.

Once the s and $\bar{s}$ valon distributions in hyperons and kaons are determined, the light di-valon distribution in hyperons and the light valon distribution in kaons are fixed by momentum conservation. They are given by

$$q_K = 11.43x^{(1/5)}(1-x)^6; \quad dq_H = 5.92x^{(1/5)}(1-x)^3\,, \tag{8}$$

where q_K is for light valons in kaons and dq_H is the di-valon distribution in hyperons.

Summarizing, we have used the MCM to extract qualitative information on the constituent -*valon*- distributions in kaons and hyperons out of a measurement of the s and $\bar{s}$ distributions in protons.

Although the model is not able yet to give a precise description of the $s - \bar{s}$ asymmetry in the proton sea, this scheme seems to be significant in order to have a qualitative knowledge of hadron structure in general, even for unstable mesons and baryons, which are difficult to be measured.

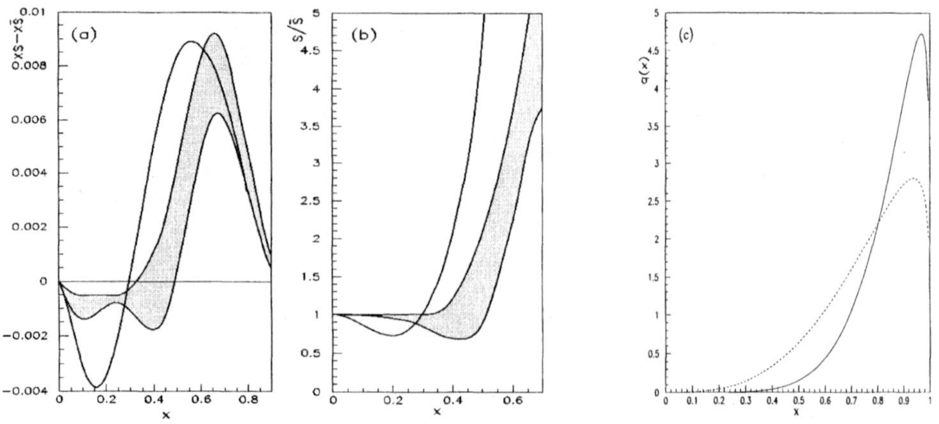

FIGURE 1. (a) $xs - x\bar{s}$ asymmetry, model presented here (full line) compared to results from the global fit (shaded area) from Ref. [3]. (b) $s/\bar{s}$ ratio. (c) strange valon distributions in kaons (full line) and in hyperons (dashed line).

ACKNOWLEDGEMENTS

E.C. would like to thank the conference organizers for the kind hospitality during the event. E.C. is supported by CONACyT (México) and CLAF/CNPq (Brazil), J.M. is partially supported by COLCIENCIAS (Colombia) under contract No. 242-99.

REFERENCES

1. A.I. Signal and A.W. Thomas, Phys. Lett. **B191**, 205 (1987).
2. A. O. Bazarko et al., (CCFR Collaboration), Z.Phys. **C65**, 189 (1995).
3. V. Barone, C. Pascaud, F. Zomer, Eur. Phys. J. **C12**, 243 (2000).
4. S.J. Brodsky and B.Q. Ma, Phys. Lett. **B381**, 317 (1996), M. Burkard and B.J. Warr, Phys. Rev. **D45**, 958 (1992).
5. W. Melnitchouk and M. Malheiro, Phys. Lett. **B451**, 224 (1999).
6. H. R. Christiansen, J. Magnin, Phys. Lett. **B445**, 8 (1998).
7. J. Magnin, H. R. Christiansen, Phys. Rev. **D61**, 054006 (2000).
8. E.A. Hawker *et al.* (E866 Collaboration), Phys. Rev. Lett. **80**, 3715 (1998).
9. H.R. Christiansen and J. Magnin, hep-ph/0003088.
10. C. Avila, J. Magnin and J.C. Sanabria, hep-ph/0005287.
11. R.C. Hwa, Phys. Rev. **D22**, 759 (1980); *ibid.* 1593.
12. G. Altarelli and G. Parisi, Nucl. Phys. **B126**, 298 (1977).
13. K.P. Das and R.C. Hwa, Phys. Lett. **B68**, 459 (1977).

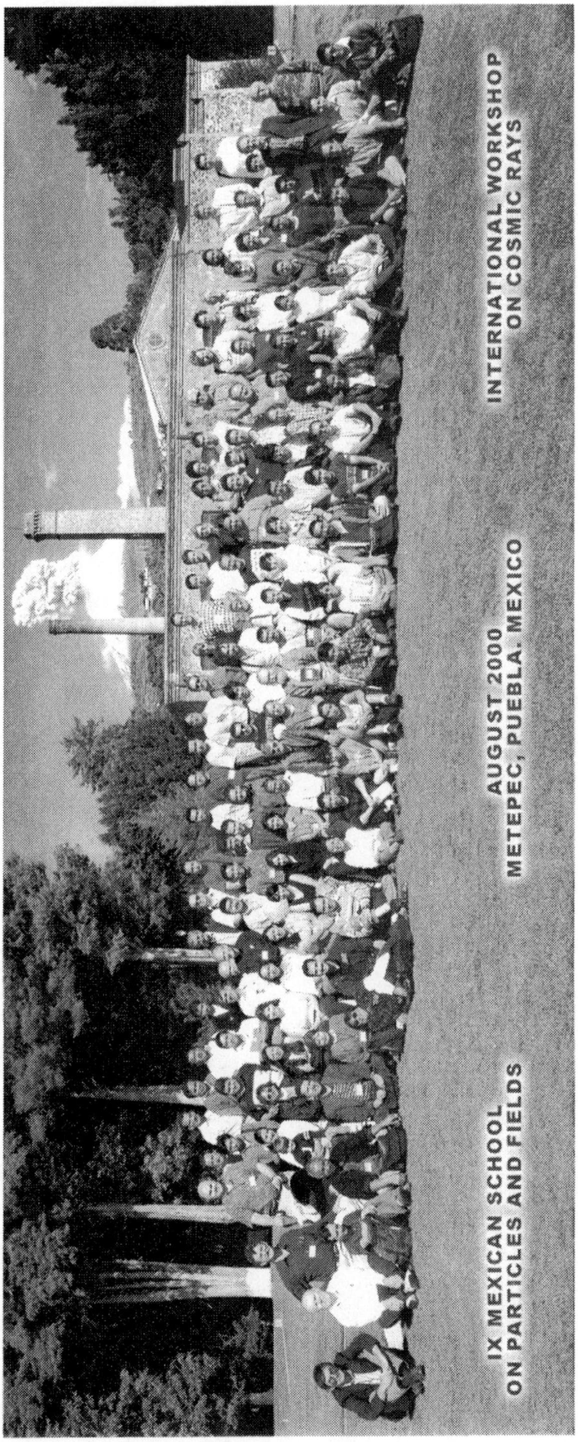
IX MEXICAN SCHOOL
ON PARTICLES AND FIELDS
AUGUST 2000
METEPEC, PUEBLA, MEXICO
INTERNATIONAL WORKSHOP
ON COSMIC RAYS

LIST OF PARTICIPANTS

Acosta Meza, Raul Martin	IF-UASLP, Mexico
Aldana Segura, Waleska	U San Carlos, Guatemala
Appel, Jeffrey	Fermilab, USA
Araiza García, Moisés Elías	IF-BUAP, Mexico
Aranda, Jorge Isidro	ECFM-UMSNH, Mexico
Avila, Manuel	FC-UAEM, Mexico
Calderon Polania, German A.	CINVESTAV, Mexico
Carena, Marcela	Fermilab, USA
Carreño Alvarado, Alexandra Stephanie	ICN-UNAM, Mexico
Contreras, Guillermo	CINVESTAV-Merida, Mexico
Cuautle F, Eleazar	CBPF, Brasil
D'Olivo Saez, Juan Carlos	ICN-UNAM, Mexico
Diaz Cruz, Lorenzo	IF-BUAP, Mexico
Diaz Sanchez, Rodolfo Alexander	UNAL, Colombia
Ellis, John	CERN, Switzerland
Escobar Acosta, Jose Henry	Universidad del Magdalena, Colombia
Félix Beltrán, Olga Guadalupe	IF-BUAP, Mexico
Flores Urbína, Julio César	CINVESTAV, Mexico
Gaitán Lozano, Ricardo	FESC-UNAM, Mexico
Gallegos Infante, Luis Armando	IFUG, Mexico
García, Augusto	CINVESTAV, Mexico
García Compean, Hector Hugo	CINVESTAV, Mexico
García Hidalgo, Rosa Magdalena	IF-BUAP, Mexico
García Luna, José Luis	CINVESTAV, Mexico
Godina Nava, Juan José	CINVESTAV, Mexico
Goiz Hernández, Klara	CINVESTAV, Mexico
Hernandez Galeana, Albino	ESFM-IPN, Mexico
Hernández López, Javier Miguel	FCFM-BUAP, Mexico
Hernández Moreno, Francisco Javier	FCFM-BUAP, Mexico
Hernández Sánchez, Jaime	IF-BUAP, Mexico
Herrera Corral, Gerardo	CINVESTAV, Mexico
Herrera Herrera, Simón Fernando	ESFM-IPN, Mexico
Huet Soto, Adolfo	ECFM-UMSNH, Mexico
Juarez Wysozka, S. Rebeca	ESFM-IPN, Mexico
Kirchbach, Mariana	EF-UAZ, Mexico
Larios Forte, Francisco C.	CINVESTAV-Merida, Mexico
León Monzán, Ildefonso	CINVESTAV, Mexico
Loaiza Brito, Oscar Gerardo	CINVESTAV, Mexico
López Castro, Gabriel	CINVESTAV, Mexico
Manrique Ascencio, Elisa	ECFM-UMSNH, Mexico
Mariano, Alejandro	CINVESTAV, Mexico

MARTÍNEZ, Jesús	ICN-UNAM, Mexico
MARTÍNEZ HERNÁNDEZ, Mario Iván	CINVESTAV, Mexico
MASPERI, Luis	CLAF, Brasil
MEDELLIN ZAPATA, Juan	IF-UASLP, Mexico
MIRANDA ROMAGNOLI, Omar	CINVESTAV, Mexico
MONDRAGÓN CEBALLOS, Myriam	IF-UNAM, Mexico
MORALES RUIZ, Benjamin	IF-UNAM, Mexico
MORENO SOTO, Jorge	CINVESTAV, Mexico
MUÑOZ SALAZAR, Laura	IFUG, Mexico
MURGUÍA ROMERO, Gabriela	IF-UNAM, Mexico
NAVARRO ESTRADA, Jorge Luis	Universidad del Atlantico, Colombia
NELLEN, Lukas	ICN-UNAM, Mexico
PALLARES PRIETO, Gabriel	ICN-UNAM, Mexico
PEREZ LORENZANA, Abdel	U Maryland, USA
RAMIREZ, Carlos	Universidad Industrial Santander, Colombia
RAMÍREZ, Cupatitzio	FCFM-BUAP, Mexico
RAMÍREZ ZAVALATA, Fernando I	FC-UAEM, Mexico
RANDRUP, Jorgen	LBL, USA
RAYA MONTAÑO, Alfredo	ECFM-UMSNH, Mexico
ROMÁN LÓPEZ, Sergio	IF-BUAP, Mexico
ROSADO SANCHEZ, Alfonso	IF-BUAP, Mexico
SAHU, Sarira	ICN-UNAM, Mexico
SÁNCHEZ CECILIO, Angel	FC-UNAM, Mexico
SÁNCHEZ COLON, Gabriel	CINVESTAV-Merida, Mexico
SÁNCHEZ HERNÁNDEZ, Alberto	CINVESTAV, Mexico
SANTORO, Alberto	CBPF, Brasil
SOLIS RODRIGUEZ, Hugo Gabriel	ICN-UNAM, Mexico
SWAIN, John	NEU, USA
TORRES AGUILAR, Ibrahim	IF-UASLP, Mexico
URRUTIA, Luis	ICN-UNAM, Mexico
VEGA CABRERA, José	ECFM-UMSNH, Mexico
VEGA CANO, Rubén	ECFM-UMSNH, Mexico
VOGT, Ramona	LBL, USA
WAGNER, Carlos	ANL, USA
WOLF, Günter	DESY, Germany
ZELENY VÁZQUEZ, Enrique	IF-BUAP, Mexico

Author Index

A

C

D

E

G

H

J

K

L

M

N

P

R

S

V